Quantum Theory, Groups, Fields and Particles

MATHEMATICAL PHYSICS STUDIES

A SUPPLEMENTARY SERIES TO
LETTERS IN MATHEMATICAL PHYSICS

VOLUME 4

Quantum Theory, Groups, Fields and Particles

Edited by

A. O. BARUT

Department of Physics, University of Colorado at Boulder, U.S.A.

D. Reidel Publishing Company

Dordrecht : Holland / Boston : U.S.A. / London : England

Library of Congress Cataloging in Publication Data

CIP

Main entry under title:

Quantum theory, groups, fields, and particles.

(Mathematical physics studies ; v. 4)
Includes index.
1. Mathematical physics–Addresses, essays, lectures.
I. Barut, A. O. (Asim Orhan), 1926- II. Title.
QC20.5.Q36 1983 530.1'5 82-25054
ISBN 90-277-1552-1

Published by D. Reidel Publishing Company
P.O. Box 17, 3300 AA Dordrecht, Holland

Sold and distributed in the U.S.A. and Canada
by Kluwer Boston Inc.,
190 Old Derby Street, Hingham, MA 02043, U.S.A.

In all other countries, sold and distributed
by Kluwer Academic Publishers Group,
P.O. Box 322, 3300 AH Dordrecht, Holland

D. Reidel Publishing Company is a member of the Kluwer Group

Table of Contents

Preface

Symmetry and Dynamics have played, sometimes dualistic, sometimes complimentary, but always a very essential role in the physicist's description and conception of Nature. These are again the basic underlying themes of the present volume. It collects self-contained introductory contributions on some of the recent developments both in mathematical concepts and in physical applications which are becoming very important in current research. So we see in this volume, on the one hand, differential geometry, group representations, topology and algebras and on the other hand, particle equations, particle dynamics and particle interactions.

Specifically, this book contains a complete exposition of the theory of deformations of symplectic algebras and quantization, expository material on topology and geometry in physics, and group representations. On the more physical side, we have studies on the concept of particles, on conformal spinors of Cartan, on gauge and supersymmetric field theories, and on relativistic theory of particle interactions and the theory of magnetic resonances.

The contributions collected here were originally delivered at two Meetings in Turkey, at Blacksea University in Trabzon and at the University of Bosphorus in Istanbul. But they have been thoroughly revised, updated and extended for this volume.

It is a pleasure for me to acknowledge the support of UNESCO, the support and hospitality of Blacksea and Bosphorus Universities for these two memorable Meetings in Mathematical Physics, and to thank the Contributors for their effort and care in preparing this work.

A. O. BARUT

Boulder, October 1982

A. O. Barut (ed.), Quantum Theory, Groups, Fields and Particles, vii.

Part I

Quantum Theory and Deformations

QUANTUM MECHANICS AND DEFORMATIONS OF GEOMETRICAL DYNAMICS

André LICHNEROWICZ

Collège de France - Paris - FRANCE

Abstract - For a dynamical system, it is possible to give an alternative form of the conventional Quantum Mechanics, in terms of deformations of the associative algebra of functions defined on the phase space. A quantization is determined here by a suitable deformation of the considered associative algebra, corresponding with a deformation of the Poisson Lie algebra, that is called a twisted product. This invariant approach appears as a generalization of the Weyl-Wigner quantization associated to the flat case and introduces cohomological problems and treatments. In particular, the role played by the De Rham cohomology of the phase space appears clearly. The questions concerning the invariance of Quantim Mechanics are solved by a study of the group of automorphisms of a twisted product.

INTRODUCTION

It is well-known that it is possible to give a complete description of Classical Mechanics in terms of symplectic geometry and Poisson brackets. It is the essence of the Hamiltonian formalism. In a common program with Bayen, Flato, Fronsdal, Sternheimer, J. Vey and others, we have study properties and applications *of the*

A. O. Barut (ed.), Quantum Theory, Groups, Fields and Particles, 3–82.

deformations of the Poisson Lie algebra and of deformations of the associative algebra defined by the usual product of functions. Such deformations give a new approach of Quantum Mechanics which corresponds to an autonomous generalization of the Weyl-Wigner quantization. The quantization of a system described by a symplectic manifold (phase space), is given by the choice of a so-called *twisted product*. I consider here only dynamical systems with a finite number of degrees of freedom, but the approach and a significative part of the results may be extended to physical fields.

In the part I, we recall shortly the basic tools, define the Lie algebras associated to a symplectic manifold and introduce the notion of *Poisson manifold*. This concept gives a good geometrical framework for the so-called Kirillov-Kostant-Souriau theorem.

The part II introduces *the two cohomologies* (Hochschild cohomology and Chevalley cohomology) which appear in the study of the derivations and of the deformations of the trivial associative algebra on the Poisson Lie algebra. We recall and adapt the main results of the theory of Gerstenhaber concerning the deformations of algebraic structures, in particular the results relative to the equivalence between deformations.

In the part III, we introduce the notion of *twisted product* on a symplectic manifold and study a main example which coïncides with the so-called Moyal product; this product appears in the Weyl-Wigner quantization. We introduce elements of a spectral theory and show how each twisted product determines a quantization. Natural extensions of Moyal product or bracket are given by *the Vey algebras* which are defined and studied.

The part IV is relative to the existence and the equivalence of twisted products and formal Lie algebras. We establish for the first time an important existence theorem obtained by Neroslavsky and Vlasov. We show that each twisted product is equivalent to a Vey twisted product and, proceeding by equivalence, we determine a condition under which a formal Lie algebra is generated by a twisted product.

The part V studies the group of the automorphisms of a twisted product. The knowledge of this group gives precious informations for *invariance problems* concerning Quantum Mechanics.

A list of references gives only the most important papers connected with the deformation theory.

I - SYMPLECTIC MANIFOLDS AND GEOMETRICAL DYNAMICS

1 - Preliminaries and notations.

All the basic tools needed for the geometrical study of the Hamiltonian mechanics and for the theory of the deformations of Classical mechanics which leads to Quantum mechanics are not developped here. We suppose known the foundations of differential geometry and of the tensor calculus. These tools are completely developped for example in [25]. In the first paragraphs, we recall only some necessary motions and we fix the notations.

a) We consider here *paracompact,* connected, differential manifolds of finite dimension and class C^∞. Such a manifold is sayed to be *smooth.* For brevity, we set $N = C^\infty(W;R)$. All the introduced elements are supposed to be of class C^∞.

Let W be a smooth manifold of dimension m. The C^∞-differentiable structure of W is defined by a set of charts $\{U,f\},\{V,g\}$... where U, V, ... are domains of W, the union of which coïncides with W, where f (resp. g) is a bijection of U (resp. V) on domains of $\mathbb{R}^m$, and where the following condition is satisfied.

In the intersection $U \cap V$ *of the domains of two charts, the coordinates* $x^i = f^i(x)$ $(i = 1,\ldots,m)$ *of a point* x *in the first chart are* C^∞*-functions, with a non-vanishing Jacobian, of the coordinates* $x^{j'} = g^{j'}(x)$ *of the same point* x *in the second chart.*

We obtain thus a so-called *atlas* (set of charts) of W. A chart is said to be *admissible* if its union with an atlas defining the differentiable manifold W is an atlas again. We denote by $\{x^i\}$ $(i = 1,\ldots,m)$ an admissible chart of W of domain U.

Let W, W' be two smooth manifolds. A *mapping* $\sigma : W \to W'$ is said to be C^∞*-differential* (or of class C^∞) if for each x in W and admissible chart $\{U',f'\}$ of W' with $\sigma(x) \subset U'$, there is a chart $\{U,f\}$ of W, with $x \in U$ and $\sigma(U) \subset U'$, such that the *local representative* of σ, let $\sigma_{ff'} = f' \circ \sigma \circ f^{-1}$ is of class C^∞ in R^m.
$\sigma_{ff'}$ can be represented by :

$$x^{j'} = x^{j'}(x^i)$$

where $\{x^i\} \in \mathbb{R}^m$ and $\{x^{j'}\} \in \mathbb{R}^{m'}$.

Let $v \in T_x W$ be a *vector* tangent to W at x; it has components $\{v^i\}$ in $\{U,f\}$. The quantities

$$v^{j'} = A^{j'}_i v^i \qquad (A^{j'}_i = \partial x^{j'}/\partial x^i)$$

are the components in $\{U',f'\}$ of a vector $v' \in T_{x'} W'$ tangent to W' at $x' = \sigma(x)$. We define thus a linear map, denoted by $\sigma^{(x)}_*$, of the tangent space $T_x W$ into the tangent space $T_{x'} W'$; $\sigma^{(x)}_*$ gives the *push-forward* of v by σ. The mapping σ_* which maps the tangent bundle TW into the tangent bundle T W' is said to be *the differential of the mapping* σ.

Let $p' \in T^*_{x'} W'$ be a covector (see §3) of W' at $x' = \sigma(x)$; it has components $\{p_{j'}\}$ in $\{U',f'\}$. The quantities :

$$p_i = A^{j'}_i \; p_{j'}$$

are the components in $\{U,f\}$ of a covector $p \in T^*_x W$ of the manifold W at x. We define thus a linear map, denoted by σ^*, of the cotangent bundle T^* W' into the cotangent bundle T^* W; $\sigma^*_{(x')} p'$ gives the pullback of p' by σ.

The maps σ_* (resp. σ^*) can be extended in a natural way to contravariant tensors (resp. covariant tensors).

A map $\sigma : W \to W'$ of class C^∞ is called a *diffeomorphism* if it is a bijection such that $\sigma^{-1} : W' \to W$ is also of class C^∞. If such is the case, σ_* is also bijective. The push-forward and the pull-back of any tensor are well-defined and we are led to write $\sigma^* = (\sigma^{-1})_*$.

b) A vector field X of the manifold W is a C^∞-section of TW, that is a C^∞-mapping from W to TW that assigns to each point x of W a vector in $T_x W$. Similar definitions for the tensor fields of various types.

A *curve* c at a point x of W is a differential map $c : I \to W$, where I is an open interval of $\mathbb{R}$, such that $0 \in I$ and $c(0) = x$. For such a curve, we may assign at each point c(t) of the curve $(t \in I)$ the push-forward of the unit vector of I, let $\dot{c}(t) = c^{(t)}_*(1)$; $\dot{c}(t)$ is the *tangent vector* at c(t) to the curve.

An integral curve of the vector field X at $x \in W$ is a curve c at x such $\dot{c}(t) = X(c(t))$ for each $t \in I$. The curve c is an integral curve of X iff the local representatives of c satisfy a system of first order ordinary differential equations that can be written in the chart $\{x^i\}$, with evident notations :

$$\frac{dc^i(t)}{dt} = X^i(c^1(t),\ldots,c^m(t)) \qquad (i = 1,\ldots,m) \qquad (1\text{-}1)$$

For such a system, there are well-known existence and uniqueness theorems. If t is sufficiently small, we have a unique solution c(t;x) of (1-1) taking the value x at t = 0 if $X(x) \neq 0$, that is an integral curve of X at $x \in W$. It follows from the uniqueness that for t,s sufficiently small :

$$c(t;c(s;x)) = c(t+s;x) \tag{1-2}$$

we say that $x \to c(t;x)$ gives *the flow* F_t defined by X.

c) Let X be a vector field and $u : W \to \mathbb{R}$ a scalar element of N; u admits the differential $du = u_*$. For t sufficiently small, we have :

$$(u \circ F_t)(x) = u(c(t;x)) \tag{1-3}$$

and the derivative of (1-3) with respect to t is equal, for t = 0, to :

$$\left\{ \frac{d}{dt} u \circ F_t \right\}_{t=o} (x) = i(\dot{c}(o;x)\ du(x) = (i(X)\ du)(x)$$

where i(.) is the inner product by a vector (or tensor). We set

$$\mathcal{L}(X)\ u = i(X)\ du \tag{1-4}$$

and we call $\mathcal{L}(X)u$ *the Lie derivative of* u *with respect to* X. It is easy to see that, for t sufficiently small :

$$\frac{d}{dt}(u \circ F_t) = \mathcal{L}(X)\ u \circ F_t$$

that is :

$$\frac{d}{dt}(F_t^* u) = F_t^* \mathcal{L}(X)\ u \tag{1-5}$$

The space N admits a structure of associative and commutative algebra defined by the usual product of functions ($u,v \in N \times N \to u.v \in N$); $\mathcal{L}(X)$ is a *derivation* of the algebra (N,.), that is an endomorphism of the space N satisfying :

$$\mathcal{L}(X)(u.v) = \mathcal{L}(X)\ u.v + u\ .\ \mathcal{L}(X)\ v$$

It is known that, conversely, for each derivation D of (N,.), there is a unique vector field X such that $D = \mathcal{L}(X)$ (see §10,a).

If $\sigma : W' \to W$ is *a diffeomorphism*, then $\mathcal{L}(X)$ is *natural with respect to pull-back by* σ, that is :

$$\mathcal{L}(\sigma^* X)(\sigma^* u) = \sigma^* \mathcal{L}(X)\ u$$

$\mathcal{L}(X)$ is also *natural with respect to the restrictions* (or local):

if U is a domain of W we have :

$$\mathcal{L}(X_{|U})(u_{|U}) = \mathcal{L}(X)u_{|U}$$

If X and Y are vector fields on W, then $[\mathcal{L}(X), \mathcal{L}(Y)] = \mathcal{L}(X).\mathcal{L}(Y) - \mathcal{L}(Y)\mathcal{L}(X)$ is a derivation on (N,.) and there is a unique vector field, denoted by $[X,Y]$, such that $\mathcal{L}([X,Y]) = [\mathcal{L}(X), \mathcal{L}(Y)]$.

We call $[X,Y] = \mathcal{L}(X)Y$ the *Lie derivative of* Y *with respect to* X.

The composition law $X,Y \to [X,Y]$ determines on the space of the vector fields a structure of *Lie algebra,* that is $[\ ,\ ]$ is linear on this space, is skewsymmetric and satisfies *Jacobi's identity* : if X, Y, Z are three vectors fields, we have :

$$S\,[[S,Y],\,Z] = 0 \tag{1-6}$$

where S is the summation after cyclic permutation on X, Y, Z; $[X,Y]$ is said to be *the bracket* of X, Y. $\mathcal{L}(X)$ acting on the vector fields is natural with respect to the diffeomorphisms and natural with respect to the restrictions.

d) Let $\mathcal{T}(W)$ be the full *tensor algebra* of the manifold W. A tensor derivation is a linear map D of $\mathcal{T}(W)$ in itself, respecting the types of the tensors, satisfying for each $T_1, T_2 \in \mathcal{T}(W)$

$$D(T_1 \otimes T_2) = DT_1 \otimes T_2 + T_1 \otimes DT_2 \tag{1-7}$$

and such that

$$D\,\delta = 0$$

where δ is the Kronecker tensor determined by the differentiable structure.

A *differential operator* on $\mathcal{T}(W)$ is a *tensor derivation* which is *natural with respect to the restrictions* (or local) in an evident sense.

It is possible to prove :

Proposition - If X *is a vector field, there is a unique differential operator* $\mathcal{L}(X)$ *on* $\mathcal{T}(W)$ *, called the Lie derivative with respect to* X*, such that* $\mathcal{L}(X)$ *coïncides with (1-4) on the scalars and with* $\mathcal{L}(X)Y = [X,Y]$ *on the vectors.*

If the vector field X defines the flow F_t, we can prove that, for an arbitrary tensor field T, we have :

$$\frac{d}{dt}(F_t^{*} T) = F_t^{*} \mathcal{L}(X)\, T \tag{1-8}$$

This identity is the key of the connection between flows and Lie derivatives. The operator $\mathcal{L}(X)$ is called also *the operator of infinitesimal transformation* by X. This operator can be extended in a unique way to various geometrical objects, in particular to the linear connections. The Lie derivative of a connection is a tensor of type (1,2).

2 - Exterior differential algebra. De Rham's cohomology.

a) We call p-*form* ($0 \leqslant p \leqslant m$) an antisymmetric covariant tensor field of order p on the manifold W. For $p = 0$, we obtain the scalar functions, elements of N; for $p = 1$, the 1-forms are the covector fields; for $p > m$ the p-forms are null. We denote by $\Lambda^p(W)$ the space of the p-forms, which can be also considered, with respect to N, as a N-module.

Then we may define the notion of *exterior product* as follows : if $\alpha \in \Lambda^p(W)$ and $\beta \in \Lambda^q(W)$ consider the tensor product $\alpha \otimes \beta$ which is a (p+q)-tensor and antisymmetrize this tensor. We obtain the exterior product, denoted by $\alpha \wedge \beta$, which is a (p+q)-form. Introduce a chart $\{x^i\}$ of W of domain U; in this chart $\alpha_{|U}$ admits the components $\{\alpha_{i_1 \ldots i_p}\}$ and $\beta_{|U}$ the components $\{\beta_{j_1 \ldots j_q}\}$. The (p+q)-form $\alpha \wedge \beta$ admits in U the components :

$$(\alpha \wedge \beta)_{k_1 .. k_{p+q}} = \frac{1}{p!q!} \varepsilon^{i_1 .. i_p\, j_1 .. j_q}_{k_1 .. k_{p+q}} \alpha_{i_1 .. i_p} \beta_{j_1 .. j_q} \tag{2-1}$$

where ε is the antisymmetric tensor indicator of Kronecker : $\varepsilon = 1$ (resp. -1) if $(i_1 .. i_p\, j_1 .. j_q)$ is an even (resp. odd) permutation of the sequence $(k_1 .. k_{p+q})$ of distinct indices and $\varepsilon = 0$ in all the other cases.

It is easy to see that the composition $\wedge$ is bilinear and that we have :

$$\alpha \wedge \beta = (-1)^{pq}\, \beta \wedge \alpha \tag{2-2}$$

Moreover we have the *associativity property :*

$$(\alpha \wedge \beta) \wedge \gamma = \alpha \wedge (\beta \wedge \gamma) = (\alpha \wedge \beta \wedge \gamma) \tag{2-3}$$

The direct sum $\Lambda(W)$ of the spaces $\Lambda^p(W)$ ($p = 0,1,\ldots,m$) is a vector space on which the product induced by $\wedge$ defines *the exterior differential algebra* of the manifold W.

b) We define now on $\Lambda(W)$ *the exterior differentiation operator* denoted by d.

The operator d assigns to each p-form α a (p+1)-form $d\alpha$ with the following conditions :

C_1) d *is linear and natural with respect to restrictions : if* U *is a domain of* W, *we have :*

$$d\alpha_{|U} = d(\alpha_{|U})$$

C_2) *If* $u \in N$, *we have :*

$$d(u\alpha) = du \wedge \alpha + u\, d\alpha \tag{2-4}$$

C_3) *If* u_λ $(\lambda = 1,\ldots,p)$ *are* p *elements of* N :

$$d(du_1 \wedge \ldots \wedge du_p) = 0 \tag{2-5}$$

These conditions determine one and only one operator d. Suppose that such a d exists. On the doamin U of chart $\{x^i\}$, we have for a p-form α :

$$\alpha_{|U} = \frac{1}{p!}\alpha_{i_1..i_p}\, dx^{i_1} \wedge \ldots \wedge dx^{i_p}$$

We deduce from the conditions C_1 and C_2 :

$$d(\alpha_{|U}) = \frac{1}{p!}\, d\alpha_{i_1\ldots i_p} \wedge (dx^{i_1} \wedge \ldots dx^{i_p}) + \frac{1}{p!}\alpha_{i_1\ldots i_p}\, d(dx^{i_1} \wedge \ldots \wedge dx^{i_p})$$

where the last term of the right side is null according to C_3. We obtain :

$$d\alpha_{|U} = \frac{1}{p!}\, d\alpha_{i_1\ldots i_p} \wedge dx^{i_1} \wedge \ldots \wedge dx^{i_p} = \frac{1}{p!}\,\partial_{i_o}\alpha_{i_1\ldots i_p}\, dx^{i_o} \wedge dx^{i_1} \wedge \ldots \wedge dx^{i_p} \tag{2-6}$$

where $\partial_i = \partial/\partial x^i$. Hence $d\alpha$ is uniquely determined. Conversely it is easy to verify that (2-6) defines an operator on the forms satisfying C_1, C_2, C_3. It follows from (2-6) that we have :

$$d \circ d = d^2 = 0 \tag{2-7}$$

(2-7) expresses that d is a so-called *cohomology operator.*

If $\alpha \in \Lambda^p(W)$ and $\beta \in \Lambda^p(W)$, we prove easily :

$$d(\alpha \wedge \beta) = d\alpha \wedge \beta + (-1)^p \alpha \wedge d\beta \tag{2-8}$$

(2-8) expresses that d is *an antiderivation of the exterior algebra.*

If $\sigma : W' \to W$ is a C^∞-mapping, then $\sigma^* : \Lambda(W) \to \Lambda(W')$ *is a homomorphism of algebra,* that is :

$$\sigma^*(\alpha \wedge \beta) = \sigma^*\alpha \wedge \sigma^*\beta \tag{2-9}$$

and d *is natural with respect to mappings :*

$$\sigma^* d\alpha = d(\sigma^*\alpha) \tag{2-10}$$

c) The Lie derivative $\mathcal{L}(X)$ by a vector field X is *a derivation of the exterior algebra,* that is :

$$\mathcal{L}(X)(\alpha \wedge \beta) = \mathcal{L}(X)\,\alpha \wedge \beta + \alpha \wedge \mathcal{L}(X)\beta \tag{2-11}$$

and this derivation commutes with d : $d\,\mathcal{L}(X) = \mathcal{L}(X)\,d$. In fact consider the flow F_t defined by X. We have :

$$d(F_t^* \alpha)/dt = F_t^* \mathcal{L}(X)\,\alpha$$

d commutes with F_t^* according to (2-10) and with d/dt since d is linear. We obtain :

$$d(F_t^* d\alpha)/dt = F_t^* d\,\mathcal{L}(X)\alpha$$

But we have :

$$d(F_t^* d\alpha)/dt = F_t^* \mathcal{L}(X)\,d\alpha$$

It follows that $d\,\mathcal{L}(X) = \mathcal{L}(X)\,d$.

Consider the inner product $i(X) : \alpha \in \Lambda^p(W) \to i(X)\,\alpha \in \Lambda^{p-1}(W)$ by the vector field X. It is easy to see that i(X) is also an *antiderivation* on the exterior algebra and we may deduce from this fact that the operator $D(X) = d\,i(X) + i(X)\,d$ is a *derivation* of this algebra which commutes with d. We will prove that for an arbitrary p-form :

$$\mathcal{L}(X)\,\alpha = d\,i(X)\,\alpha + i(X)\,d\alpha \tag{2-12}$$

that is $D(X) = \mathcal{L}(X)$. We proceed by induction on p. First we note that for p = 0, (2-12) reduces to :

$$\mathcal{L}(X)\, u = i(X)\, du$$

Now assume that (2-12) holds for $p \geqslant 0$, that is $D(X) = \mathcal{L}(X)$ on the p-forms. We have for $u \in N$ and $\alpha \in \Lambda^p(W)$:

$$D(X)(du \wedge \alpha) = D(X)(du) \wedge \alpha + du \wedge D(X)\alpha$$

But :

$$D(X)\, du = d\, D(X)\, u = d\, \mathcal{L}(X)\, u = \mathcal{L}(X)\, du$$

It follows :

$$D(X)(du \wedge \alpha) = \mathcal{L}(X)(du) \wedge \alpha + du \wedge \mathcal{L}(X)\, \alpha = \mathcal{L}(X)(du \wedge \alpha)$$

Note that on the domain U of a chart $\{x^i\}$ a (p+1)-form β may be written :

$$\beta_{|U} = \sum_i dx^i \wedge \alpha_i \qquad (\alpha_i \in \Lambda^p(U))$$

It follows that $D(X) = \mathcal{L}(X)$ on $\Lambda^{p+1}(W)$.

d) *A manifold* M *with boundary* is defined exactly as in §1 by atlases with the following difference : if (U,f) is a chart, we require that $f(U) \subset R^m_-$. Let $\mathcal{A} = \{(U,f)\}$ be an atlas of our manifold with boundary; define Int. $M = \bigcup_U f^{-1}(\text{Int } f(U))$ and $\partial M = \bigcup_U f^{-1}(\partial f(U))$ where $\partial f(U)$ is the boundary of f(U). *The interior of* M Int. M is open in M and so is an m-dimensional smooth manifold; the *boundary* ∂M of M is an (m-1)-dimensional smooth manifold (possibly empty) without boundary.

An *orientation* of a manifold (with or without boundary) is a smooth pseudoscalar ε of M such that $\varepsilon^2 = 1$. A manifold is *orientable* iff there is an orientation ε or iff there is a m-form $\eta \neq 0$ everywhere; such a form is called a volume element of M. If M is orientable, the dimension of $\Lambda^m(M)$, considered as a N-module, is equal to 1. A positive chart $\{x^i\}$ of domain U of the oriented manifold (M,ε) is a chart for which ε has the components 1. For such charts, the jacobians corresponding to the intersections $U \cap V$ of domains of the charts are positive.

Conversely an atlas of a manifold M such that the jacobians corresponding to the intersection of domains of the charts are positive, defines on M *an orientation.*

Let ∂M be the boundary of an oriented manifold (M,ε). Consider only charts $\{x^i\}$ of M of domains U for which $\partial M \cap U \neq \emptyset$ is given by $x^1 = 0$ and Int $M \cap U$ by $x^1 < 0$. We say that $\{x^2,\ldots,x^m\}$ is a positive chart of ∂M if $\{x^1,x^2,\ldots x^m\}$ is a positive chart of (M,ε) ; the set of these charts of ∂M defines on this manifold an orientation which is said to be *the induced orientation on* ∂M.

Let M be an oriented m-manifold and α a m-form of M with a *compact support* $S(\alpha)$. We will now define the integral of α on M.

Suppose that $S(\alpha) \subset U$, where (U,f) is a positive chart $f : x \in U \to \{x^i\}$. If α admits in this chart the components $\alpha_{1\ldots m}$ we define

$$\int_M \alpha = \int_{f(U)} \alpha_{1\ldots m}(x^i)\, dx^1 \ldots dx^m$$

where the right side denotes a Riemann integral. The change of variables rule in a multiple integral shows that if (U',f') is another positive chart such that $S(\alpha) \subset U'$, we obtain for the right side the same result.

Now, let $\mathcal{A}$ be an atlas of positive charts and $P = \{(U_\lambda, \varphi_\lambda)\}$ a differentiable partition of unity subordinate to $\mathcal{A}$. Define $\omega_\lambda = \varphi_\lambda\, \omega$ so ω_λ has compact support in U_λ. Then, we define

$$\int_M \alpha = \sum_\lambda \int_M \alpha_\lambda$$

where the sum contains only a finite number of non-zero terms. For any other atlas of positive charts and subordinate partition of the unity Q, we obtain the same result.

It is possible to prove :

Stokes' Theorem - Consider an oriented smooth m-*manifold* M *with boundary and a* (m-1)-*form* α *which has compact support. Let* $i : \partial M \to M$ *be the inclusion map so that* $i^*\alpha \in \Lambda^{m-1}(\partial M)$. *Then :*

$$\int_{\partial M} i^*\alpha = \int_M d\alpha$$

for the induced orientation of ∂M *or, for short,*

$$\int_{\partial M} \alpha = \int_M d\alpha \qquad (2\text{-}13)$$

This theorem implies the usual formulas of Gauss, Riemann, Stockes ... We see that if we consider the integration as defining a bilinear form on the differential forms and the fields of integration, the boundary operator ∂ and the operator d on the forms are dual; d is often sayed to be *a coboundary operator*.

e) *Definition - We call* $\alpha \in \Lambda^p(M)$ *closed if* $d\alpha = 0$ *and exact if there is a* $\beta \in \Lambda^{p-1}(M)$ *such that* $\alpha = d\beta$.

It follows from $d^2 = 0$ that *every exact form is closed.*

Conversely, we can establish :

Poincaré lemma - Let U *be a contractible domain of* W. *If* $\alpha_U \in \Lambda^p(U)$ *is closed, it is exact.*

We say that each closed p-form $\alpha \in \Lambda^p(M)$ is *locally exact.*

Consider the quotient space of the (infinite dimensional) space of the closed p-forms by the space of the exact p-forms. This quotient space, called *the* $p^{\underline{th}}$ *de Rham space,* is denoted by $H^p(M;R)$. We can prove that the spaces $H^p(M;\mathbb{R})$ ($p = 0,1,\ldots,m$) are *finite dimensional;* we put :

$$\dim H^p(M;\mathbb{R}) = b_p(M)$$

where $b_p(M)$ is called *the* $p^{\underline{th}}$ *Betti number of the manifold,* for the De Rham cohomology with unrestricted supports. The De Rham cohomology corresponds by duality to a so-called homology theory (sayed here to be with compact supports). If we consider only forms with compact supports, we obtain another De Rham cohomology which does not appear here.

De Rham cohomology has been historically the prototype of the cohomology theories. We shall see here many other examples. In the general terminology of these theories, De Rham cohomology is defined by the De Rham complex for which

1) *the cochains are the differential forms*
2) *the coboundary operator is the exterior differentiation* d.

The closed cochains are called *cocycles;* a cocycle is *exact* if it is the coboundary of a cochain.

Denote by $H(W;\mathbb{R})$ the direct sum of the spaces $H^p(W;\mathbb{R})$. It follows from the formula (2-8) that the exterior product can be defined on $H(W;\mathbb{R})$, that is on the cohomology classes defined by the closed forms; $H(W;\mathbb{R})$ admits a structure of *exterior algebra* or is *a cohomology algebra.*

3 - The Liouville forms.

a) Let M be a smooth differentiable manifold of dimension n; TM (resp. T^*M) is the tangent (resp. cotangent) bundle of M. We denote by $\pi : T^*M \to M$ the canonical projection corresponding to the cotangent bundle. The map π defines naturally a projection $\pi_* : T\,T^*M \to TM$; if X is a vector tangent to T^*M at $p \in T^*M$, consider the linear map :

$$\omega : X \in T\,T^*M \to p(\pi_* X) \in \mathbb{R}$$

where p is considered as a 1-form (or covector) at the point $x = \pi(p)$ of M. We define thus on T^*M a canonical global 1-form ω, the *Liouville 1-form* of T^*M.

b) It is useful to consider also ω in terms of local charts. Denote by $\{q^\alpha\}$ (α, each greek index $= 1,\dots,n$) a local chart of M of domain U. Let us consider the covectors $p \in T^*M$ such that $\pi(p) = x \in U$. If $\{p_\alpha\}$ are the components of p at x in this chart, we introduce on $\pi^{-1}(U)$ the local 1-form

$$\omega_U = p_\alpha \, dq^\alpha \qquad \text{(with summation)}$$

Similarly, for the domain V of another chart $\{q^{\beta'}\}$, we have on $\pi^{-1}(V)$ the 1-form :

$$\omega_V = p_{\beta'} \, dq^{\beta'}$$

If $U \cap V \neq \emptyset$, we set $A^{\beta'}_\alpha = \partial q^{\beta'}/\partial q^\alpha$, $A^\gamma_{\beta'} = \partial q^\gamma/\partial q^{\beta'}$, so that $A = (A^{\beta'}_\alpha)$ and $A^{-1} = (A^\gamma_{\beta'})$ are the corresponding jacobian matrices. We have :

$$dq^{\beta'} = A^{\beta'}_\alpha \, dq^\alpha \qquad\qquad p_{\beta'} = A^\gamma_{\beta'} \, p_\gamma$$

It follows :

$$\omega_V|_{\pi^{-1}(U\cap V)} = A^\gamma_{\beta'} A^{\beta'}_\alpha \, p_\gamma \, dq^\alpha = \delta^\gamma_\alpha \, p_\gamma \, dq^\alpha = p_\alpha \, dq^\alpha$$

so that :

$$\omega_V|_{\pi^{-1}(U\cap V)} = \omega_U|_{\pi^{-1}(U\cap V)}$$

If we introduce a locally finite covering of M by domains of charts, we see that the collection of the forms ω_U defines onto T^*M a global 1-form which is nothing other than the Liouville 1-form ω of T^*M.

c) Consider the exterior differential of the Liouville 1-form :

$$F = d\omega \tag{3-1}$$

This 2-form can be written on $\pi^{-1}(U)$:

$$F|_{\pi^{-1}(U)} = dp_\alpha \wedge dq^\alpha \tag{3-2}$$

and it is clear that F has the rank 2n everywhere : the exterior power F^n, which is a 2n-form onto T^*M of dimension 2n, is $\neq 0$

everywhere.
We say that the exact Liouville 2-form F defines *the canonical symplectic structure* of the cotangent bundle T^*M. The fact that each cotangent bundle admits a canonical symplectic structure plays a major role in analytic dynamics.

4 - Notion of symplectic manifold.

a) Let W be a connected, paracompact, differentiable manifold of class C^∞. All the considered elements are supposed of class C^∞ and we put, for simplicity $N = C^\infty(W;\mathbb{R})$. A *p-tensor* is, by definition, a skewsymmetrical covariant tensor of order p. A *symplectic structure* is defined in general on a manifold W of even dimension 2n by a *closed 2-form* F ($dF = 0$) *such that* F^n *is* $\neq 0$ *everywhere* (rank 2n). On the symplectic manifold (W,F), the 2n-form $\eta = F^n/n!$ defines a volume element, *the symplectic volume element*, and the manifold admits an orientation.

The symplectic structure is sayed to be *exact* if the 2-form F is exact. *It is never the case if* W *is compact*. In fact, we deduce from the Stokes formula that, for a compact manifold, the symplectic volume element can not be an exact form (we have $\int_W \eta > 0$); it follows that the closed form F (and all its powers) cannot be exact.

We denote by $\{x^i\}$ (i,j,... each latin index = 1,...,2n) a local chart for W of domain U. According to classical considerations of Darboux (see for instance Godbillon [6]), it is well known that there are on W atlases of so-called *canonical coordinates* $\{x^\alpha, x^{\bar\alpha}\}$ ($\alpha = 1,\ldots,n$; $\bar\alpha = \alpha + n$) such that F can be written on the domain U of any of these charts :

$$F|_U = \sum_\alpha dx^\alpha \wedge dx^{\bar\alpha} \tag{4-1}$$

In terms of G-structures, a symplectic structure is *an integrable* $Sp(n;\mathbb{R})$ *structure*, where $Sp(n;\mathbb{R})$ is the real symplectic group for 2n variables.

b) Let $\mu : TW \to T^*W$ be the isomorphism of vector bundles defined by $X \to -i(X)F$, where i(.) is the inner product. This isomorphism can be extended to the tensors in a natural way; for the skewsymmetrical tensors, it is compatible with the exterior product. In particular, we denote by Λ the 2-tensor $\mu^{-1}(F)$ of rank 2n.

On the domain U of an arbitrary chart $\{x^i\}$, let (F_{ij}) be the regular matrix given by the components of F and (Λ^{ij}) the inverse matrix (so that $\Lambda^{ik} F_{ij} = \delta^k_j$). For a contravariant p-tensor t, we have :

$$\mu(t)_{i_1 \dots i_p} = F_{i_1 r_1} \dots F_{i_p r_p} t^{r_1 \dots r_p} \tag{4-2}$$

and for a p-form α :

$$\mu^{-1}(\alpha)^{i_1 \dots i_p} = \Lambda^{r_1 i_1} \dots \Lambda^{r_p i_p} \alpha_{r_1 \dots r_p} \tag{4-3}$$

In particular, the Λ^{ij}'s are the components of the 2-tensor $\Lambda = \mu^{-1}(F)$.

5 - Lie algebras of vectors associated to a symplectic manifold.

a) Let (W,F) be a symplectic manifold. Study the action of an infinitesimal transformation (i.t.) given by a vector field X on the main 2-form. If $\mathcal{L}(X)$ is the corresponding Lie derivative, we have :

$$\mathcal{L}(X)\ F = d\ i(X)\ F + i(X)\ dF$$

that is, F being closed,

$$\mathcal{L}(X)\ F = d\ i(X)\ F \tag{5-1}$$

A *symplectic infinitesimal transformation* is defined by a vector field X such that F is invariant under X, that is $\mathcal{L}(X)\ F = 0$. Such is the case if and only if the 1-form $\mu(X) = -\ i(X)\ F$ is *closed* ; X is also called *a locally hamiltonian vector field*. We denote by L the (infinite dimensional) Lie algebra of the symplectic i.t.

b) Let X, Y be two *arbitrary* vector fields of W. We search for the image by μ of the bracket $[X,Y]$, let :

$$\mu([X,Y]) = -\ i([X,Y])\ F = -(\mathcal{L}(X)\ i(Y) - i(Y)\mathcal{L}(X))\ F$$

But :

$$\mathcal{L}(X)\ i(Y)\ F = d\ i(X)\ i(Y)\ F + i(X)\ d\ i(Y)\ F = d\ i(X)\ i(Y)\ F + i(X)\mathcal{L}(Y) F$$

We obtain :

$$\mu([X,Y]) = d\ i(X \wedge Y) F - i(X)\mathcal{L}(Y)\ F + i(Y)\mathcal{L}(X)\ F$$

It is easy to verify that :

$$i(X \wedge Y)\ F = i(\mu^{-1}(F))\ \mu(X \wedge Y) = i(\Lambda)(\mu(X) \wedge \mu(Y))$$

Therefore we have for two arbitrary vector fields X, Y of W :

$$\mu([X,Y]) = d\ i(\Lambda)(\mu(X)\wedge\mu(Y)) - i(X)\mathcal{L}(Y)F + i(Y)\mathcal{L}(X)F \quad (5\text{-}2)$$

c) If $X,Y \in L$, we have $\mathcal{L}(X)F = \mathcal{L}(Y)F = 0$ and it follows that :

$$\mu([X,Y]) = d\ i(\Lambda)(\mu(X)\wedge\mu(Y)) \quad (5\text{-}3)$$

Let L^* be the subspace of L defined by the converse images of the exact 1-forms ($X_u = \mu^{-1}(du)$, $u \in N$). An element of L^* is *a (globally) hamiltonian vector field.* Consider the commutator ideal $[L,L]$ of L : each element of $[L,L]$ is, by definition, a finite sum of brackets of elements of L. It follows trivially from (5-3) that $[L,L] \subset L^*$. It has been proved (Arnold, Calabi, myself) that we have exactly $[L,L] = L^*$. ([1]).

We can interpret L^* in terms of the De Rham theorem. Let ${}_0H_1(W;\mathbb{R})$ the first space of homology of W with compact supports. If X is an element of L and if γ is a compact differentiable 1-cycle defining the elements $[\gamma]$ of ${}_0H_1(W;\mathbb{R})$, we set :

$$J(X)\ [\gamma] = \int_\gamma \mu(X) \in \mathbb{R}$$

where the right member depends only upon X and $[\gamma]$. We define thus a linear map $J : X \to J(X)$ of L into $\hom({}_0H_1(W;\mathbb{R});\mathbb{R})$, that is into the first space $H^1(W;\mathbb{R})$ of cohomology with non restricted supports; $\mu(X)$ being an arbitrary closed 1-form, J is surjective and the kernel of J is L^*, according to the definition of this ideal. We see that the space L/L^* is isomorphic to the cohomology space $H^1(W;\mathbb{R})$. We have :

Proposition - The Lie algebra L^* *of the globally hamiltonian vector fields coincides with the commutator ideal of the Lie algebra* L *and* $\dim. L/L^* = b_1(W)$, *where* $b_1(W)$ *is the first Betti number of* W *for the cohomology with unrestricted supports.*

d) A vector field X defines a *conformal symplectic* i.t. if there is $a \in N$ such that $\mathcal{L}(X)\ F = a\ F$. It follows by exterior differentiation :

$$F \wedge da = 0 \quad (5\text{-}4)$$

F being everywhere of rank 2n, we deduce from a classical theorem of Lepage that, for $n > 1$, (5-4) implies $da = 0$, that is $a = \text{const}$.

For any n, we denote by L^c, and call *Lie algebra of the conformal symplectic infinitesimal transformations,* the Lie algebra of the vector fields X such that there exists a constant k(X) for which

$$\mathcal{L}(X)\ F + k(X)\ f = 0 \quad \text{or} \quad \mathcal{L}(X)\ \Lambda = k(X)\ \Lambda \tag{5-5}$$

For $X \in L^c$ and $Y \in L$, the formula (5-2) can be written :

$$\mu([X,Y]) = d\ i(\Lambda)(\mu(X) \wedge \mu(Y) + k(X)\ \mu(Y) \tag{5-6}$$

$\mu(Y)$ being closed, $\mu([X,Y])$ is closed and $[X,Y] \in L$. Therefore L *is an ideal of* L^c. If $Y \in L^*$, the right member of (5-6) is an exact 1-form and $[X,Y] \in L^*$. We see that L^* *is also an ideal of* L^c.

e) If $X \in L^c$, (5-5) can be written $k(X)F = d\mu(X)$.
If F *is not exact* (in particular if W is compact), we have necessarily $k(X) = 0$ for each $X \in L^c$ and L^c *coïncides with* L.
Suppose now that F *is exact*. We have $F = d\omega$, where ω is a 1-form of W. If $Z = \mu^{-1}(\omega)$, we have $F = d\mu(Z)$ and Z belongs to L^c with $k(Z) = 1$. If $X \in L^c$, we have :

$$d\,\mu\,(X - k(X)\ Z) = 0$$

and $(X - k(X)Z) \in L$. Each element X of L^c admits a unique decomposition in a sum of an element of L and an element proportional to Z

$$X = Y + k(X)\ Z \qquad (Y \in L)$$

We denote by C_Z the subspace of L^c of dimension 1 defined by the vectors kZ ($k \in \mathbb{R}$). It follows from the d and from this decomposition that $[L^c,L^c] \subset L$. Conversely, let Y be an arbitrary element of L ; if we write (5-6) for the vectors $Z \in L^c$ and $Y \in L$, we obtain :

$$Y = [Z,Y] - \mu^{-1}\{d\ i(\Lambda)\ (\mu(Z) \wedge \mu(Y))\} \tag{5-7}$$

The first term of the right member of (5-7) belongs to $[L^c,L^c]$, the second term belongs to L^* and so to $[L,L]$. It follows that $L \subset [L^c,L^c]$ and so that $L = [L^c,L^c]$. We have :

Proposition - If (W,F) *is an exact symplectic manifold, the Lie algebra* L^c *is, as vector space, the direct sum of* L *and of a subspace* C_Z *of dimension* 1; *in particular dim.*$L^c/L = 1$. *The Lie algebra* L^c *admits* L *as its commutator ideal.*
If (W,F) *is non exact,* L^c *coïncides with* L.

6 - Poisson brackets and Poisson Lie algebra.

a) Let $\bar{N}$ be the space of the classes of functions, elements of N, defined up to the additive real constants; we denote by

$\pi : u \in N \to \pi(u) = \bar{u} \in \bar{N}$ the canonical projection of N onto $\bar{N}$. If $u \in N$, its differential du depends only upon the class $\bar{u}$ of u; we denote eventually by $d\bar{u}$ this differential. To each hamiltonian vector field X_u corresponds an exact 1-form $\mu(X_u) = du = d\bar{u}$ and therefore a unique class $\bar{u}$, and conversely. We define thus a natural isomorphism of the vector space L^* onto the vector space $\bar{N}$.

It is easy to determine the Lie algebra structure induced on $\bar{N}$: if $\bar{u}, \bar{v} \in \bar{N}$, it follows from (5-3) that the function

$$w = i(\Lambda)(d\bar{u} \wedge d\bar{v}) \qquad (6\text{-}1)$$

defines a class $\bar{w}$ which is, for the Lie algebra on $\bar{N}$, the bracket of $\bar{u}$ and $\bar{v}$. The function (6-1) is, by definition, *the Poisson bracket* relatively to the symplectic structure of $\bar{u}$ and $\bar{v}$, or of two representants u and v in N. We denote by $\{u,v\}$ this Poisson bracket. Thus we have :

$$\{u,v\} = i(\Lambda)(du \wedge dv) = \mathcal{L}(X_u)v = P(u,v) \qquad (6\text{-}2)$$

where P (Poisson operator) is a bidifferential operator of order 1 on each argument, null on the constants.

b) Let $\{x^\alpha, x^{\bar{\alpha}}\}$ be a canonical chart of (W,F) of domain U. The components of Λ are on U

$$\Lambda^{\alpha\bar{\alpha}} = 1 \quad , \quad \Lambda^{\alpha\bar{\beta}} = 0 \text{ for } \beta \neq \alpha \quad \Lambda^{\alpha\beta} = \Lambda^{\bar{\alpha}\bar{\beta}} = 0 \qquad (6\text{-}3)$$

We obtain for the Poisson bracket the classical local expression :

$$\{u,v\}_U = \sum_\alpha (\partial_\alpha u\, \partial_{\bar{\alpha}} v - \partial_{\bar{\alpha}} u\, \partial_\alpha v) \qquad (\partial_i = \partial/\partial x^i) \qquad (6\text{-}4)$$

c) The Poisson bracket P defines on the space N itself a structure of Lie algebra. It is sufficient to prove that the Jacobi identity is satisfied. If $u,v,w \in N$, it follows from the Jacobi identity corresponding to the Lie algebra defined on $\bar{N}$

$$\{\{u,v\},w\} + \{\{v,w\},u\} + \{\{w,u\},v\} = \text{const.} = C$$

Let U be an arbitrary domain of W. Substitute to u a function $u' \in N$ such that $u|_U = u'|_U$ and which vanishs on another open set U' of W. We have thus :

$$\{\{u,v\},w\} + \{\{v,w\},u\} + \{\{w,u\},v\}|_U = 0$$

for any domain U. The Jacobi identity is proved. We have :

Proposition - The Poisson bracket P *defines on the space* N *a structure of Lie algebra and we have a natural homomorphism of* *the Lie algebra* (N,P), *called the Poisson Lie algebra of* (W,F), *on the Lie algebra* L^* *of the hamiltonian vector fields.*

$$X_{\{u,v\}} = [X_u, X_v] \qquad (u,v \in N) \tag{6-5}$$

7 - Classical Dynamics and Symplectic Manifolds.

a) Consider a dynamical system with time independent constraints and n degrees of freedom. The corresponding configuration space is an arbitrary differentiable manifold M of dimension n. The Lagrangian formalism corresponds to a description of the motions of the system in terms of points of the tangent bundle TM.

We consider here *the hamiltonian formalism*. We have seen that the cotangent bundle T^*M admits a natural symplectic structure defined by the Liouville 2-form F which may be written locally in terms of the classical variables (q^α, p_α) :

$$F_{|U} = \Sigma\, dp_\alpha \wedge dq^\alpha \tag{7-1}$$

For the hamiltonian formalism, a dynamical state of the system is nothing but a point of the manifold $W = T^*M$, which is the usual *phase space* of the mechanicians and physicists. The analysis of the equations of Mechanics has showed, for a long time, that it is essential to may introduce changes of the variables (q^α, p_α) which does not respect the cotangent structure of W.

b) We are thus led to introduce as *phase space* a symplectic manifold (W,F) of dimension 2n. If $\{x^\alpha, x^{\bar\alpha}\}$ is a canonical chart of W of domain U, we can put $x^\alpha = p_\alpha$, $x^{\bar\alpha} = q^\alpha$ and we obtain with these notations the local form (4-1) of F.

Dynamics is determined, on the phase space (W,F), by a function $H \in N$, *the hamiltonian* of the dynamical system, which defines the hamiltonian vector field X_H.

A motion of the dynamical system is given, by definition, by an integral curve c(t) *of the hamiltonian vector field* X_H , *the parameter* t *being the time.*

Such is the geometrical meaning of the classical equations of Hamilton. In fact, if $\{x^\alpha, x^{\bar\alpha}\}$ is a canonical chart of (W,F) of domain U, X_H admits in this chart the components :

$$X^{\bar\alpha} = \Lambda^{\alpha\bar\alpha}\, \partial_\alpha H = \partial_\alpha H \qquad X^\alpha = \Lambda^{\bar\alpha\alpha}\, \partial_{\bar\alpha} H = -\, \partial_{\bar\alpha} H$$

A curve c(t) of W is an integral curve of X_H iff we have, for any domain U of canonical coordinates :

$$\frac{dx^{\bar{\alpha}}}{dt}(c(t)) = (\partial_\alpha H)(c(t)) \qquad \frac{dx^\alpha}{dt}(c(t)) = -(\partial_{\bar{\alpha}} H)(c(t))$$

If we introduce the notations $p_\alpha = x^\alpha$, $q^\alpha = x^{\bar{\alpha}}$, we have on U

$$\frac{dq^\alpha}{dt}(c(t)) = \frac{\partial H}{\partial p_\alpha}(c(t)) \qquad \frac{dp_\alpha}{dt}(c(t)) = -\frac{\partial H}{\partial q^\alpha}(c(t)) \qquad (7\text{-}2)$$

or

$$\frac{dq^\alpha}{dt}(c(t)) = \{H,q^\alpha\}(c(t)) \qquad \frac{dp_\alpha}{dt}(c(t)) = \{H,p_\alpha\}(c(t)) \qquad (7\text{-}3)$$

that are the usual canonical *Hamilton equations*.

c) We can adopt another viewpoint. The space N admits the following two algebraic structures :

1) a structure of *associative algebra* defined by the usual product of functions (which is here commutative)

2) a structure of *Lie algebra* defined by the Poisson bracket. Since $\{u,v\} = \mathcal{L}(X_u)v$, the Poisson bracket defines derivations of the product : if $u,v,w \in N$ we have :

$$\{w,uv\} = \{w,u\}.v + u.\{w,v\} \qquad (7\text{-}4)$$

Consider a differentiable family u_t of elements of N satisfying the differential equation :

$$du_t/dt = \{H,u_t\} \qquad (7\text{-}5)$$

and taking the initial value $u_o \in N$ at $t = 0$. If F_t is the flow defined by X_H, we have

$$u_t = u_o \circ F_t \qquad (7\text{-}6)$$

In fact, it follows from (1-5) that :

$$\frac{d}{dt}(u_o \circ F) = \frac{d}{dt}(F_t^* u_o) = F_t^* \mathcal{L}(X_H)u_o = \mathcal{L}(X_H)(u_o \circ F_t)$$

We deduce from the uniqueness theorem for (7-5) that $u_t = u_o \circ F_t$. Therefore the evolution in the time of u_t comes from the flow of the integral curves of X_H which appears in the first point of view. The equation (7-5) can be considered as *the coordinate free, intrinsic equation of Classical Dynamics.*

d) We have completely described Classical Mechanics in terms of the two laws of composition introduced on N : associative product and bracket connected by (7-4). It is natural to study if it is possible to deform in a suitable way these two algebraic laws so that we obtain a model isomorphic to the conventional Quantum Mechanics.

The answer is positive ([2]).

8 - Schouten bracket and Poisson Manifolds.

a) Let W be an arbitrary manifold of dimension m. We denote by $\{x^A\}$ (A,B,... = 1, ..., m) a local chart of W of domain U. For the antisymmetric contravariant tensors, Schouten and Nijenhuis have introduced an useful tool, the so-called *Schouten bracket*. (see [16] and [18]).

If A (resp. B) is a p-tensor (resp. a q-tensor), $[A,B]$ is a (p+q-1)-tensor defined in the following way : for each *closed* (p+q-1)-form β, we have :

$$i([A,B])\beta = (-1)^{pq+q}\, i(A)\, d\, i(B)\beta + (-1)^{p}\, i(B)\, d\, i(A)\beta \qquad (8\text{-}1)$$

For p = 1, $[A,B] = \mathcal{L}(A)B$, where $\mathcal{L}(A)$ is the operator of Lie derivative corresponding to the vector A. We have :

$$[A,B] = (-1)^{pq}\,[B,A] \qquad (8\text{-}2)$$

Moreover, if C is a r-tensor, we have the pseudo Jacobi identity :

$$S(-1)^{pq}\left[[B,C],A\right] = 0 \qquad (8\text{-}3)$$

where S is the summation after cyclic permutation. We have also, with respect to the exterior product, the following identity :

$$[A,B\wedge C] = [A,B]\wedge C + (-1)^{pq+q}\, B\wedge[A,C] \qquad (8\text{-}4)$$

An elementary calculus gives the components of $[A,B]$ for an arbitrary chart :

$$[A,B]^{K_2\ldots K_{p+q}} = \frac{1}{(p-1)!q!}\,\varepsilon^{K_2\ldots\ K_{p+q}}_{I_2\ldots I_p\ J_1\ldots J_q}\, A^{RI_2\ldots I_p}\,\partial_R B^{J_1\ldots J_q}$$

$$+\ \frac{(-1)^p}{p!(q-1)!}\,\varepsilon^{K_2\ldots\ K_{p+q}}_{I_1\ldots I_p\ J_2\ldots J_q}\, B^{RJ_2\ldots J_q}\,\partial_R A^{I_1\ldots I_p} \qquad (8\text{-}5)$$

where $\partial_R = \partial/\partial x^R$ and where ε is the antisymmetric Kronecker indicator.

b) Introduce on W a 2-tensor Λ and consider on the space N the bracket $\{\ ,\ \}_\Lambda$ (generalized Poisson bracket) defined by :

$$\{u,v\}_\Lambda = i(\Lambda)(du \wedge dv) \qquad (u,v \in N) \tag{8-6}$$

If u,v,w $\in$ N, study the function :

$$t = S\ \{\{u,v\}_\Lambda\ ,w\}_\Lambda$$

A direct calculus gives :

$$t_{|U} = (\Lambda^{RA}\,\partial_R\,\Lambda^{BC} + \Lambda^{RB}\,\partial_R\,\Lambda^{CA} + \Lambda^{RC}\,\partial_R\,\Lambda^{AB})\,\partial_A u\,\partial_B v\,\partial_C w$$

It follows from (8-5) that the components of $[\Lambda;\Lambda]$ on U are given by :

$$\frac{1}{2}\,[\Lambda,\Lambda]^{ABC} = \Lambda^{RA}\,\partial_R\,\Lambda^{BC} + \Lambda^{RB}\,\partial_R\,\Lambda^{CA} + \Lambda^{RC}\,\partial_R\,\Lambda^{AB}$$

We obtain :

$$t = \frac{1}{2}\,i([\Lambda,\Lambda])(du \wedge dv \wedge dw)$$

The Jacobi identity is satisfied by (8-6) iff $[\Lambda,\Lambda] = 0$. We are led to the following definition :

Definition - A structure of Poisson manifold is defined on a manifold W *of dimension* m *by a 2-tensor* Λ *such that* $[\Lambda,\Lambda] = 0$ ([24]).

For a Poisson manifold (W,Λ), $\{\ ,\ \}_\Lambda = \{\ ,\ \}$ defines on N a Lie algebra structure; (N, $\{\ ,\ \}$) is called again *the Poisson Lie algebra* defined by Λ.

A Poisson manifold is called *regular* if Λ has a constant rank 2n (that is $\Lambda^n \neq 0$, $\Lambda^{n+1} = 0$ everywhere); if such is the case , h = m-2n is sayed to be *the codimension* of the manifold.

c) Let (W,F) be a symplectic manifold of dimension 2n ; $\Lambda = \mu^{-1}(F)$ is the corresponding 2-tensor and the usual Poisson bracket of (W,F) is defined by (8-6). We have $[\Lambda,\Lambda] = 0$. Therefore a symplectic structure is nothing but a regular Poisson structure of codimension 0. Moreover, if A is a p-tensor, it is easy to verify in canonical coordinates that :

$$\mu([\Lambda,A]) = d\,\mu(A) \tag{8-7}$$

From the existence of the atlases of canonical charts and the argument of Darboux, we deduce the following lemma :

Lemma - Let V *be a domain of* $\mathbb{R}^{2n}$ *and* Λ *a 2-tensor of rank* 2n *on* V *satisfying* $[\Lambda,\Lambda] = 0$. *If* x $\in$ V, *there exists a chart* $\{x^i\}=\{x^\alpha, x^{\bar\alpha}\}$

of domain U *satisfying* x $\in$ U $\subset$ V, *such that* Λ *admits only as non-vanishing components :*

$$\Lambda^{\alpha\bar{\alpha}} = -\Lambda^{\bar{\alpha}\alpha} = 1 \qquad (8\text{-}8)$$

d) Let (W,Λ) be a regular Poisson manifold of codimension h $\neq$ 0. If U is a contractible domain of W, it follows from the condition concerning the rank of Λ that there exists h linearly independant 1-forms $\omega^{(a)}$ (a = 1,...,h) on U such that :

$$\Lambda^{AB}\,\omega^{(a)}_A = 0$$

It follows from this property and from $[\Lambda,\Lambda] = 0$ that :

$$\Lambda^{AC}\,\Lambda^{BD}\,(d\omega^{(a)})_{CD} = 0 \qquad (8\text{-}9)$$

Adopt on U coframes $\{\omega^{(A)}\}$ of the type $\{\omega^{(a)},\omega^{(i)}\}$ (i = 1,...,2n). For these coframes $\Lambda^{aB} = 0$ and $\det(\Lambda^{ij}) \neq 0$. The invariant relation (8-9) can be written :

$$(d\omega^{(a)})_{jk} = 0 \qquad (8\text{-}10)$$

This relation expresses that the Pfaffian system $\omega^{(a)} = 0$ (a = 1,...,h) defined by Λ is *integrable*. There are on U local charts $\{x^a,x^i\}$ such that, for these charts, $\Lambda^{aB} = 0$.
The tensor Λ define thus an integrable field on 2n-plans onto W and so a foliation of W of codimension h. The restriction of Λ to an arbitrary leaf Σ determines on Σ a 2-tensor Λ_Σ of rank 2n everywhere with satisfies $[\Lambda_\Sigma,\Lambda_\Sigma]_\Sigma = 0$. Therefore Σ is a symplectic manifold. We have proved :

Proposition - A regular Poisson manifold (W,Λ) *of codimension* h *admits a natural foliation of codimension* h *by symplectic manifolds.*

e) In the chart $\{x^a,x^i\}$ of domain U for which we have $\Lambda^{aB} = 0$, $[\Lambda,\Lambda] = 0$ can be written :

$$\frac{1}{2}[\Lambda,\Lambda]^{ijk} = \Lambda^{\ell i}\,\partial_\ell\,\Lambda^{jk} + \Lambda^{\ell j}\,\partial_\ell\,\Lambda^{hi} + \Lambda^{\ell k}\,\partial_\ell\,\Lambda^{ij} = 0$$

It follows from the above lemma that, for x^a = const. (a=1,...,h), we can introduce changes of the coordinates x^i such that the components of Λ are canonical in the sense of (8-8). Therefore there are changes of charts $x^{a'} = x^a$, $x^{j'} = x^{j'}(x^a,x^i)$ such that, with respect to $\{x^{a'},x^{j'}\} = \{x^a,x^\alpha,x^{\bar{\alpha}}\}$, the 2-tensor Λ of W admits only as non-vanishing components $\Lambda^{\alpha\bar{\alpha}} = -\Lambda^{\bar{\alpha}\alpha} = 1$.

Proposition - There exist on the regular Poisson manifold (W,Λ) *atlases of so-called canonical charts* $\{x^a,x^i\} = \{x^a,x^\alpha,x^{\bar{\alpha}}\}$ (α = 1, ...,n ; $\bar{\alpha} = \alpha + n$; a = 2n+1,...,m) *such that* Λ *has only as non-vanishing components :*

$$\Lambda^{\alpha\bar{\alpha}} = - \Lambda^{\bar{\alpha}\alpha} = 1 \qquad (8\text{-}11)$$

In particular $\Lambda^{aB} = 0$ *and* x^a = *const. along a leaf.*

More generally, they are on the Poisson manifold (W,Λ) atlases of charts $\{x^A\}$ such that the components of Λ in the chart are all *constant*. Such a chart is called *a natural chart* for a regular Poisson manifold (in particular for *a symplectic manifold*).

9 - An example of Poisson manifold connected with the coadjoint representation of a Lie algebra.

a) Let A be a finite dimensional Lie algebra of dimension m on the real field; A^* is the dual vector space of A and < , > the bilinear duality form. Introduce the 2-tensor Λ of A^* defined by :

$$i(\Lambda)(X \wedge Y) = <[X,Y],\xi> \qquad (\forall X,Y \in A,\ \xi \in A^*) \qquad (9\text{-}1)$$

We have :

Proposition - The 2-tensor Λ *defines on* A^* *a Poisson structure.*

In fact, let $\{e^A\}$ $(A,B,\ldots = 1,\ldots,m)$ a basis of A, $\{\lambda_A\}$ the dual basis of A^*. We have

$$[e^A, e^B] = C^{AB}_C\, e^C$$

where $C = \{C^{AB}_C\}$ is the structure tensor of the Lie algebra A. The Jacobi identity can be translated by :

$$S\, C^{RA}_D\, C^{BC}_R = 0 \qquad (9\text{-}2)$$

where S is the summation after cyclic permutation on A,B,C. If $\xi \in A^*$ admits the components $\{\xi^C\}$, the 2-tensor Λ defined by (9-1) admits the components

$$\Lambda^{AB} = C^{AB}_C\, \xi^C \qquad (9\text{-}3)$$

and we verify immediately that the Jacobi identity (9-2) implies $[\Lambda,\Lambda] = 0$.

Conversely let V be a vector space of dimension m and suppose that there is on V a 2-tensor Λ depending linearly upon the vectors $\xi \in V$ and satisfying $[\Lambda,\Lambda] = 0$.
According to (9-3) we deduce from Λ a tensor C which may be considered as the structure tensor of a Lie algebra defined on the dual space V^* of V.

For the Poisson manifold (A^*,Λ), we denote by $2r(\xi)$ the rank of the 2-tensor Λ at the point $\xi \in A^*$. We put :

$$r = \underset{\xi \in A^*}{\mathrm{Max}}\ r(\xi)$$

The set of the points ξ such that $r(\xi) = r$ is an open submanifold W of A^*, the complementary set of W in A^* being a cone. The restriction of Λ to W defines on W a structure of regular Poisson manifold.

The identity map $\xi \to \xi$ defines on A^* a vector field - the vector field of the infinitesimal homotheties of A^* - which is denoted again by ξ. We have immediately :

$$\mathcal{L}(\xi)C = -C \qquad \mathcal{L}(\xi)\Lambda = [\Lambda,\xi] = -\Lambda \tag{9-4}$$

b) Let G be a connected Lie group admitting A as Lie algebra. The group G acts naturally on A^* by its coadjoint representation Ad^*G defined in the following way : if $g \in G$, Ad^*g is determined by :

$$\langle Y, Ad^*g.\xi\rangle = \langle Ad(g^{-1})\,Y,\xi\rangle \qquad (\forall\ Y \in A,\ \xi \in A^*) \tag{9-5}$$

Study the Lie algebra Ad^*A of the group Ad^*G. Let X be an element of A, $\exp(uX)$ the one-parameter subgroup of G defined by X. The action of this group on $\xi \in A^*$ gives :

$$\langle Y, (Ad^*\exp(uX)).\xi\rangle = \langle (Ad\exp(-uX))Y,\xi\rangle$$

After differentiation with respect to u, we obtain at $u = 0$:

$$\langle Y, \frac{d}{du}(Ad^*\exp(uX)\ .\ \xi\rangle = \langle [X,Y],\xi\rangle \qquad (\ Y \in A) \tag{9-6}$$

We deduce from (9-6) that to each element $X = \{X_A\} \in A$ corresponds an element $X^A = \{X^{*B}\} \in Ad^*A$ given by :

$$X^{*B} = \Lambda^{AB}\, X_A \tag{9-7}$$

The map $X \in A \to X^* \in Ad^*A$ is a homomorhism of Lie algebras admitting as kernel the centre of A. It is easy to deduce from the Jacobi identity :

Proposition - The vector field ξ *and the tensor field* Λ *of* A^* *are invariant under the Lie algebra* Ad^*A.

c) Choose a point ξ of A^* and consider the linear map $\nu_\xi : X \in A \to X^*(\xi) \in A^*$. We obtain at ξ a vector subspace $\mathfrak{W}_\xi = \nu_\xi(A)$ and we have :

$$\dim. \mathfrak{O}_\xi = \text{rank of } \Lambda(\xi) = 2r(\xi) \qquad \dim. \operatorname{Ker} \nu_\xi = m - 2r(\xi)$$

An element ζ of A^* belongs to $\mathfrak{O}_\xi$ if and only if $\langle Z,\zeta\rangle = 0$ for every $Z \in \operatorname{Ker} \nu_\xi$. It follows that $\Lambda(\xi)$ may be interpreted as a 2-tensor on $\mathfrak{O}_\xi$ of rank $2r(\xi)$.

Let $M(\xi_o)$ be the orbit of $\xi_o \in A^*$ for the coadjoint action of the group G ; $M(\xi_o)$ is a connected manifold embedded in A^*. It follows from the above proposition that the rank of Λ is constant along $M(\xi_o)$. The tangent space $T_\xi M(\xi_o)$ at each point $\xi \in M(\xi_o)$ coïncides with $\mathfrak{O}_\xi$. Therefore $M(\xi_o)$ has the even dimension $2r(\xi_o)$ and the system of the orbits depend only upon the Lie algebra A and is independant of the choice of the group G.

The restriction of Λ to $M(\xi_o)$ determines a 2-tensor $\Lambda_{M(\xi_o)}$ of rank $2r(\xi_o)$ of $M(\xi_o)$ satisfying :

$$\left[\Lambda_{M(\xi_o)}, \Lambda_{M(\xi_o)}\right] = 0$$

and defining thus on $M(\xi_o)$ a *symplectic structure*. We obtain :

Theorem - Let A *be a finite dimensional Lie algebra and* A^* *the dual space. The structure tensor of* A *determines on* A^* *a 2-tensor* Λ *defining* A^* *as a Poisson manifold. If* G *is a connected Lie group admitting* A *as Lie algebra, the orbits of* G *in* A^* *for the coadjoint representation depend only upon* A *and admit natural invariant symplectic structures defined by* Λ.

We have thus obtained for (A^*,Λ) a singular foliation of A^* in symplectic manifolds.

d) A *coadjoint cocycle* σ of the group G is, by definition, a map $\sigma : G \to A^*$ such that the identity :

$$\sigma(g.g') = \sigma(g) + Ad^*(g^{-1}) \,.\, \sigma(g') \tag{9-8}$$

holds for all $g,g' \in G$. In particular such a cocycle is said to be a coboundary if there is $\eta \in A^*$ such that :

$$\sigma(g) = \eta - Ad^*(g^{-1}) \,.\, \eta$$

The group G acts also on A^* by the representation Φ^σ defined by :

$$\Phi^\sigma(g) = \xi \to Ad^*(g^{-1}) \,.\, \xi + \sigma(g) \tag{9-9}$$

Define a skewsymmetric bilinear form Σ on A by :

$$\Sigma(X,Y) = \langle X, d\hat{\sigma}_Y(e) \rangle$$

where $\hat{\sigma}_Y : g \in G \to \langle Y,\sigma(g)\rangle$. It follows from (9-8) that Σ satisfies Jacobi's identity :

$$S\,\Sigma(X, [Y,Z]) = 0 \qquad (\forall X,Y,Z \in A) \qquad (9\text{-}10)$$

Σ is a constant 2-tensor on A^* and (9-10) may be translated by :

$$[\Lambda,\Sigma] = 0 \qquad (9\text{-}11)$$

Σ is said to be a coadjoint cocycle of the Lie algebra A; it is a coboundary if there is $\eta \in A^*$ such that :

$$\Sigma = \Lambda(\eta)$$

It follows from the equation (9-11) that the 2-tensor of A^*

$$\hat{\Lambda} = \Lambda + \Sigma \qquad (9\text{-}12)$$

defines also on A^* a *Poisson structure* ($[\hat{\Lambda},\hat{\Lambda}] = 0$).

The Φ^σ-representation of the group G defines a Φ^σ-representation of its Lie algebra A. This representation may be defined in the following way : to each element $X = \{X_A\}$ of A corresponds the element $\hat{X} = \{\hat{X}^B\}$ of $\Phi^\sigma(A)$ given by :

$$\hat{X}^B = \hat{\Lambda}^{AB} X_A$$

It is easy to see that $\hat{\Lambda}$ is invariant under the Lie algebra $\Phi^\sigma(A)$.

The argument of §c shows that *the orbits of* G *in* A^* *for the* Φ^σ*-representation admit invariant symplectic structures defined by* $\hat{\Lambda}$. If σ is a coboundary, the orbits are deduced from the coadjoint orbits by a translation in A^* defined by η.

These results are another form of classical results known as *the Kirillov-Kostant-Souriau theorem*. (See [17],[20],[25]).

In the following, I limit myself to the symplectic manifolds, but many notions and results concerning these manifolds can be extended to the *Poisson manifolds*.

II - COHOMOLOGIES AND DEFORMATIONS.

10 - Hochschild and Chevalley cohomologies - Derivations.

a) Derivations and deformations of an *associative algebra* arise from a same cohomology, the so-called *Hochschild cohomology* of algebra. Let W be an arbitrary smooth differentiable manifold and

$(N = C^\infty(W;R),.)$, the associative algebra defined on N by the usual product of functions. The Hochschild cohomology of $(N,.)$ is defined by the following complex : a p-cochain C of $(N,.)$ is a p-linear map of N^p into N, the 0-cochains being identified with the elements of N. The coboundary of the p-cochain C is the (p+1)-cochain $\tilde{\partial}C$ defined by :

$$\tilde{\partial}C(u_o,..,u_p) = u_o.C(u_1,..,u_p) - C(u_o.u_1,u_2,..,u_p) +$$
$$+ C(u_o,u_1.u_2,..,u_p) + (-1)^p C(u_o,u_1,..,u_{p-1}.u_p) \quad (10\text{-}1)$$
$$+ (-1)^{p+1} C(u_o,..,u_{p-1}).u_p$$

where $u_\lambda \in N$. The associativity implies that we have $\tilde{\partial}^2 = 0$ (cohomology operator). A p-cocycle C is a p-cochain such that $\tilde{\partial}C = 0$; an exact p-cocycle C is a p-cocycle such that there is a (p-1)-cochain $\bar{C}$ for which $C = \tilde{\partial}\bar{C}$. The general $p^{\underline{th}}$ cohomology space is the quotient space of the space of the p-cocycles by the space of the exact p-cocycles.

If T is a 1-cochain of $(N,.)$, that is an endomorphism of N, we have :

$$\tilde{\partial}T(u,v) = u.T(v) + T(u).v - T(uv) \quad (10\text{-}2)$$

It follows that a one-cocycle of $(N,.)$ is a *derivation* of this algebra.

A p-cochain C is called *local* if, for each $u_\lambda \in N$ such that $u_\lambda|_U = 0$ for a domain U, we have $C(u_1,..,u',..,u_p)|_U = 0$. If C is local, $\tilde{\partial}C$ is local. A p-cochain C is called *d-differential* $(d \geq 0)$ if it is defined by a multi-differential operator of maximum order d in each argument. If T is a (d+1)-differential one-cochain, it is easy to see that $\tilde{\partial}T$ is d-differential. Conversely study the endomorphisms T of N such that $\tilde{\partial}T$ is d-differential $(d \geq 0)$.

It follows from (10-2) that if T is an endomorphism of N such that $\tilde{\partial}T$ is null on the constants, it is the same for T. If T is an endomorphism such that $\tilde{\partial}T$ is *local*, (10-2) implies that T is local itself. From this locality, it follows by a nontrivial argument [11].

Proposition - If T *is an endomorphism of* N *such that* $\tilde{\partial}T$ *is a* d-*differential* $(d \geq 0)$ *two-cochain,* T *is* (d+1)-*differential itself. If* T *is an endomorphism such that* $\tilde{\partial}T$ *is null on the constants, it is the same for* T.

In particular, for a derivation T of $(N,.)$, we have $d = 0$ and T is defined by a differential operator of order 1, null on the constants. Such an operator is given by the Lie derivative by an arbitrary vector field.

Corollary - A derivation of (N,.) *is given by the Lie derivative by an arbitrary vector field.*

b) We consider only in the following part *differential Hochschild cochains which are null on the constants*. For this Hochschild complex, we denote by $\tilde{H}^p(N;N)$ the corresponding $p^{\underline{th}}$ cohomology space ; J. Vey [22] has deduced from results of Gelfand and I have given a relatively elementary proof of the following :

Theorem (Vey) - The space $H^p(N;N)$ *is isomorphic to the space of the antisymmetric contravariant* p*-tensors of* W.

This result has been extended to the local cohomology in dimension 2 and 3 by M. Cahen, S. Gutt and De Wilde (to appear).

c) In a symmetric way, derivations and deformations of a Lie algebra arise from a same cohomology, the so-called *Chevalley cohomology* of the Lie algebra. It is the cohomology with values in the algebra, corresponding to the adjoint representation.

Let (W,F) be a symplectic manifold and (N,P) the corresponding Poisson Lie algebra. A p-cochain C of (N,P) is an *alternating* p-linear map of N^p in N, the 0-cochains being identified with the elements of N. The coboundary of the p-cochain C is the (p+1)-cochain ∂C defined by :

$$\partial C(u_o,\ldots,u_p) = \varepsilon^{\lambda_o\ldots\lambda_p}_{o\ldots p}\Big(\frac{1}{p!}\,\{u_{\lambda_o},C(u_{\lambda_1},\ldots,u_{\lambda_p})\} - \frac{1}{2(p-1)!}\,C(\{u_{\lambda_o},u_{\lambda_1}\},u_{\lambda_2},\ldots,u_{\lambda_p})\Big) \qquad (10\text{-}3)$$

where $u_\lambda \in N$ and where ε is the skewsymmetric Kronecker indicator; Jacobi's identity implies that we have $\partial^2 = 0$. Same definitions for the notions of cocycle and exact cocycle.

For the locality or for the d-differential character of a cochain C, we have definitions similar to the definitions concerning the Hochschild complex. But *we suppose here* $d \geqslant 1$; if C is a d-differential ($d \geqslant 1$) p-cochain, ∂C is also d-differential. Conversely, in dimension 1, I have proved [11] :

Proposition - If T *is a local endomorphism of* N *such that* ∂T *is a* d*-differential* ($d \geqslant 1$) *two-cochain,* T *is* d*-differential itself.*

If W is *noncompact,* it is easy to prove that if T is an endomorphism such that ∂T is a local two-cochain, then T is local itself. We obtain :

Lemma 1 - Let (W,F) *be a noncompact symplectic manifold. If* T *is an endomorphism of* N *such that* $C = \partial T$ *is* d-*differential* ($d \geq 1$), T *is* d-*differential itself.*

The study of the compact case gives [11] :

Lemma 2 - Let (W,F) *be a compact symplectic manifold. If* T *is an endomorphism of* N *such that* $C = \partial T$ *is* d-*differential* ($d \geq 1$), T *has the form :*

$$Tu = Qu + k \int_W u F^n$$

where $k \in \mathbb{R}$ *and where* Q *is a differential operator of order* d.

We deduce from these two lemmas the general proposition :

Proposition 2 - If C *is an exact* d-*differential two-cocycle* ($d \geq 1$) *of the Poisson Lie algebra* (N,P) *there is, on the symplectic manifold* (W,F), *a differential operator* T *of order* d *such that* $C = \partial T$.

(Compare with the proposition of a).

d) If $w \in N$ defines a zero-cochain, its coboundary C is given by :

$$C(u) = \{u,w\} = \{-w,u\} \qquad (10\text{-}4)$$

If T is a one-cochain, we have :

$$\partial T(u,v) = \{u,Tv\} + \{Tu,v\} - T\{u,v\} \qquad (10\text{-}5)$$

It follows from (10-5) that one-cocycle of (N,P) is a *derivation* of the Lie algebra. According to (10-4), an exact one-cocycle is an *inner derivation* of (N,P).

The results of C can be applied to the determination of the derivations D of (N,P). In the noncompact case, all the derivations of (N,P) are local and are given by first order differential operators. In the compact case, the corresponding operator Q (lemma 2) is a first order differential operator. It follows from the identity defining the derivations ($\partial D = 0$) that these first order differential operators are given by $\mathcal{L}(X) + k(X)$, where $X \in L^c$. We obtain [1] :

Theorem - If W *is noncompact, the derivations of the Poisson Lie algebra* (N,P) *are given by :*

$$Du = \mathcal{L}(X)\, u + k(X)\, u \qquad (10\text{-}6)$$

where $X \in L^c$.

If W *is compact, the derivations of* (N,P) *are given by :*

$$Du = \mathcal{L}(X)\,u + k \int_W u\,F^n \qquad (10\text{-}7)$$

where $k \in \mathbb{R}$ *and* $X \in L$.

This theorem implies :

Corollary - For an arbitrary symplectic manifold, the derivation D *of the Poisson Lie algebra* (N,P) *which are null on the constants are given by* $D = \mathcal{L}(X)$, *where* $X \in L$.

e) In the following, we mainly consider *differential cochains of* (N,P) *which are null on the constants*. For this Chevalley complex, we denote by $H^p(N;N)$ the corresponding p^{th} cohomology space.

The first cohomology space $H^1(N;N)$ is the quotient space of the derivations which are null on the constants by the space of the inner derivations; it is isomorphic to L/L^* and we have :

$$\dim H^1(N;N) = b_1(W)$$

The cohomologies spaces $H^2(N;N)$ and $H^3(N;N)$ play an important role in the theory of the formal deformations of the Poisson Lie algebra (N,P).

11 - Deformations of the associative algebra (N,.).

I will now recall and extend the main elements of the theory of Gerstenhaber [5] concerning the deformations of the algebraic structures, in particular of the associative algebras.
We consider here the associative algebra (N,.) defined on the vector space N by the usual product of the functions.

a) Let $E(N;\nu)$ be the space of the formal functions of $\nu \in \mathbb{C}$ with coefficients in N; ν is said to be the deformation parameter. An element u_ν of $E(N;\nu)$ has the form :

$$u_\nu = \sum_{r=0}^{\infty} \nu^r u_r \qquad (u_r \in N)$$

Consider a bilinear map $N \times N \to E(N;\nu)$ which gives a formal series in ν :

$$u *_\nu v = \sum_{r=0}^{\infty} \nu^r C_r(u,v) = uv + \sum_{r=1}^{\infty} \nu^r C_r(u,v) \qquad (11\text{-}1)$$

where the C_r ($r \geqslant 1$) are differential two-cochains of (N,.) which are *null on the constants*. These cochains can be extended to $E(N;\nu)$ in a natural way. We say that, (11-1) defines *a formal deformation* of the associative algebra (N,.) if the assocítivity identity holds formally. If such is the case, (11-1) defines on $E(N;\nu)$ a structure of *formal associative algebra.*

If the map (11-1) is arbitrary, we have for $u,v,w \in N$

$$(u \star_\nu v) \star_\nu w - u \star_\nu (v \star_\nu w) = \sum_{t=1}^{\infty} \nu^t \tilde{D}_t(u,v,w) \qquad (11\text{-}2)$$

where we have introduced the three-cochains :

$$\tilde{D}_t(u,v,w) = \sum_{r+s=t} (C_r(C_s(u,v),w) - C_r(u,C_s(v,w))) \quad (r,s \geqslant 0) \qquad (11\text{-}3)$$

We have a formal deformation if and only if $\tilde{D}_t = 0$ for $t = 1,2,\ldots$ If we will introduce a recursion process, we are led to put :

$$\tilde{E}_t(u,v,w) = \sum_{r+s=t} (C_r(C_s(u,v),w) - C_r(u,C_s(v,w)) \quad (r,s \geqslant 1) \qquad (11\text{-}4)$$

and we have the identity :

$$\tilde{D}_t \equiv \tilde{E}_t - \tilde{\partial} C_t$$

If (11-1) is limited to the order q with respect to ν, we say that we have a deformation of order q if the associativity identity is satisfied up to the order (q+1); we have $\tilde{D}_t = 0$ for $t = 1,2,\ldots,q$. It is easy to prove by recursion that, if such is the case, $\tilde{E}_{q+1}$ is automatically a three-cocycle of (N,.). We can find a two-cochain C_{q+1} satisfying $\tilde{D}_{q+1} \equiv \tilde{E}_{q+1} - \tilde{\partial} C_{q+1} = 0$ iff $\tilde{E}_{q+1}$ is *exact;* $\tilde{E}_{q+1}$ defines a cohomology class, element of $\tilde{H}^3(N;N)$, which is *the obstruction at the order* (q+1) to the construction of a formal deformation of (N,.).

A deformation of order 1 of (N,.) is called *an infinitesimal deformation.* We have $\tilde{E}_1 = 0$ and so only $\tilde{\partial} C_1 = 0$. The first term C_1 is a two-cocycle of (N,.).

b) Consider a formal series in ν :

$$T_\nu = \sum_{s=0}^{\infty} \nu^s T_s = Id_N + \sum_{s=1}^{\infty} \nu^s T_s \qquad (11\text{-}5)$$

where the T_s's ($s \geqslant 1$) are endomorphisms of N; T_ν acts naturally on $E(N;\nu)$. Consider also another bilinear map $N \times N \to E(N;\nu)$ corresponding to the formal series :

$$u \star'_\nu v = u.v + \sum_{r=1}^{\infty} \nu^r C'_r(u,v) \qquad (11\text{-}6)$$

where the C'_r's are differential two-cochains, null on the constants again.

Suppose that (11-5) and (11-6) are such that we have formally the identity :

$$T_\nu(u \star'_\nu v) = T_\nu u \star_\nu T_\nu v \qquad (11\text{-}7)$$

We shall prove that (11-6) is then a formal deformation of (N,.); this new deformation is said to be *equivalent to* (11-1).

I will now establish some universal formulas which are useful in different questions. Suppose that (11-5) and (11-6) are arbitrary. We have :

$$T_\nu(u \star'_\nu v) - T_\nu u \star_\nu T_\nu v = \sum_{t=1}^{\infty} \nu^t \tilde{F}_t \qquad (11\text{-}8)$$

where we have introduced the two-cochains :

$$\tilde{F}_t(u,v) = \sum_{r+s=t} T_s\, C'_r(u,v) - \sum_{r+s+s'=t} C_r(T_s u, T_{s'} v) \quad (r,s,s' \geqslant 0)$$

It is easy to see that we have :

$$\tilde{F}_t \equiv C'_t - C_t - \partial T_t + \tilde{G}_t \qquad (11\text{-}9)$$

where $\tilde{G}_1 = 0$ and where

$$\tilde{G}_t(u,v) = \sum_{r+s=t} T_s\, C'_r(u,v) - \sum_{s+s'=t} T_s\, u.T_{s'} v$$

$$- \sum_{r+s+s'=t} C_r(T_s u, T_{s'} v) - \sum_{r+s=t} (C_r(T_s u, v) \qquad (11\text{-}10)$$

$$+ C_r(u, T_s v)) \qquad (r,s,s' \geqslant 1)$$

c) Search for evaluate $T_\nu((u \star'_\nu v) \star'_\nu w) - T_\nu(u \star'_\nu (v \star'_\nu w))$ in two different ways. We have with notations similar to the notations of (11-2) :

$$(u \star'_\nu v) \star'_\nu w - u \star'_\nu (v \star'_\nu w) = \sum_{r=0} \nu^r \tilde{D}'_r(u,v,w)$$

We obtain by action of T_ν the relation :

$$T_\nu((u \star'_\nu v) \star'_\nu w) - T_\nu(u \star'_\nu (v \star'_\nu w)) = \sum_{t=0} \nu^t \sum_{r+s=t} T_s\, \tilde{D}'_r(u,v,w)$$

On the other hand :

$$T_\nu((u \star'_\nu v) \star'_\nu w) = T_\nu(u \star'_\nu v) \star_\nu T_\nu w + \sum_{t=0} \nu^t \tilde{F}_t(u \star'_\nu v, w)$$

let :

$$T_\nu((u \star'_\nu v)\star'_\nu w) = (T_\nu u \star_\nu T_\nu v)\star_\nu T_\nu w) + \sum_{t=0}^{\infty} \nu^t \tilde{F}_t(u,v)\star_\nu T_\nu w$$
$$+ \sum_{t=0}^{\infty} \nu^t \tilde{F}_t(u \star'_\nu v,w)$$

Similarly :

$$T_\nu(u \star'_\nu(v \star'_\nu w)) = T_\nu u \star_\nu(T_\nu v \star_\nu T_\nu w) + \sum_{t=0}^{\infty} \nu^t T_\nu u \star_\nu \tilde{F}_t(v,w)$$
$$+ \sum_{t=0} \nu^t \tilde{F}_t(u,v \star'_\nu w)$$

The map (11-1) being an associative deformation, it follows by difference that :

$$T_\nu((u \star'_\nu v) \star'_\nu w) - T_\nu(u \star'_\nu(v \star'_\nu w)) =$$
$$+ \sum_{t=0}^{\infty} \nu^t \sum_{r+s+s'=t} (C_r(\tilde{F}_s(u,v),T_{s'}w) - C_r(T_{s'}u,\tilde{F}_s(u,v)))$$
$$+ \sum_{t=0}^{\infty} \nu^t \sum_{r+s=t} (\tilde{F}_r(C'_s(u,v),w) - \tilde{F}_r(u,C'_s(v,w))) \quad (r,s,s' \geqslant 0)$$

We obtain the identities :

$$\sum_{r+s=t} T_s \tilde{D}'_r(u,v,w) = \sum_{r+s=t} (\tilde{F}_r(C'_s(u,v),w) - \tilde{F}_r(u,C'_s(v,w)))$$
$$+ \sum_{r+s+s'=t} (C_r(\tilde{F}_s(u,v),T_{s'}w)$$
$$- C_r(T_{s'}u,\tilde{F}_s(v,w))) \qquad (r,s,s' \geqslant 0)$$

We deduce from these identities the following :

Lemma - The formal deformation (11-1) *being given, for each formal series* (11-5) *and for each bilinear map* (11-6). *We have for* $t = 1,2,\ldots,$ *the identities :*

$$\tilde{D}'_t(u,v,w) + \sum_{r+s=t} T_s \tilde{D}'_r(u,v,w) = - \partial\tilde{\tilde{F}}_t(u,v,w) \qquad (11\text{-}11)$$
$$+ \sum_{r+s=t} (\tilde{F}_r(C'_s(u,v),w) - \tilde{F}_r(u,C'_s(v,w)))$$
$$+ \sum_{r+s=t} (\tilde{F}_r(u,v).T_s w - T_s u.F_r(v,w)) + \sum_{r+s=t} (C_r(\tilde{F}_s(u,v),w)$$
$$- C_r(u,\tilde{F}_s(v,w))) + \sum_{r+s+s'=t} (C_r(\tilde{F}_s(u,v),T_{s'}w)$$
$$- C_r(T_{s'}u, \tilde{F}_s(v,w))) \qquad (r,s,s' \geqslant 1)$$

Suppose that (11-5) and (11-6) are such that we have (11-7), that if $\tilde{F}_t = 0$ for $t = 1,2,\ldots$ According to (11-9), we have :

$$C'_t = C_t - \tilde{G}_t + \tilde{\partial} T_t \qquad (11\text{-}12)$$

C_t and C'_t being supposed differential two-cochains which are null on the constants, T_t is necessarily a differential operator null on the constants. Conversely, if such is the case, (11-12) shows that the series (11-5) determines the bilinear map (11-6), satisfying (11-7) in a unique way. For this map (11-6), we deduce from the identities (11-11) :

$$\tilde{D}'_t(u,v,w) + \sum_{r+s=t} \dot{T}_s \tilde{D}'_r(u,v,w) = 0 \qquad (r,s \geqslant 1;\ t = 1,2,..)$$

These relations imply by recursion that $\tilde{D}'_t = 0$ for each t. Therefore (11-6) is a formal deformation. We have :

Proposition - The associative deformation (11-1) *of* (N,.) *being given, each formal series* (11-5) *where* T_s $(s \geqslant 1)$ *are necessarily differential operators which are null on the constants, generates a unique bilinear map* (11-6) *satisfying* (11-7)*; this map is a new associative deformation which is sayed to be equivalent to* (11-1). *In particular a deformation is called trivial if it is equivalent to the identity deformation* ($C_r = 0$ for every $r \geqslant 1$).

According to (11-7), T_ν is sayed to transform $*'_\nu$ in $*_\nu$.

Consider a formal deformation $u *'_\nu v$ of (N,.) ($\tilde{D}'_t = 0$ for $t = 1,2, \ldots$) and suppose that this deformation is equivalent to the deformation $u *_{\nu} v$ up to order (q+1) : we have by assumption for suitable T'_ts, $\tilde{F}_t = 0$ for $t = 1,2,\ldots,q$, that is

$$C'_t - C_t + \tilde{G}_t = \tilde{\partial} T_t \qquad (t = 1,2,\ldots q)$$

For t = q+1, the identity (11-11) gives only :

$$\tilde{\partial} \tilde{F}_{q+1} = \tilde{\partial}(C'_{q+1} - C_{q+1} + \tilde{G}_{q+1}) = 0$$

We see that $(C'_{q+1} - C_{q+1} + \tilde{G}_{q+1})$ is a Hochschild two-cocycle. There is equivalence at the order (q+1) iff this two-cocycle is exact. The two-cocycle $(C'_{q+1} - C_{q+1} + \tilde{G}_{q+1})$ defines a cohomology class, element of $H^2(N;N)$, which is *the obstruction to the equivalence for order* (q+1).

For infinitesimal deformations we have $\tilde{G}_1 = 0$. Therefore an infinitesimal deformation is trivial if the corresponding two-cocycle is exact. Two infinitesimal deformations defined respectively by the two-cocycles C_1 and C'_1 are equivalent iff the two-cocycle $(C'_1 - C_1)$ is exact.

12 - Deformations of the Poisson Lie algebra.

The Chevalley cohomology plays exactly the same role for the deformations of a Lie algebra as the Hochschild cohomology for the deformations of an associative algebra.

We consider here the Poisson Lie algebra (N,P) of a symplectic manifold (W,F). ([4]).

a) Let $E(N;\lambda)$ be the space of the formal functions of $\lambda \in C$ with coefficients in N. Consider an alternating bilinear map $N \times N \to E(N;\lambda)$ which gives a formal series in λ

$$[u,v]_\lambda = \sum_{r=0}^{\infty} \lambda^r \Gamma_r(u,v) = \{u,v\} + \sum_{r=1}^{\infty} \lambda^r \Gamma_r(u,v) \qquad (12\text{-}1)$$

where the Γ_r $(r \geqslant 1)$ are differential two-cochains of (N,P) which are *null on the constants*. We say that (12-1) defines a *formal deformation* of the Poisson Lie algebra (N,P) if Jacobi's identity holds formally. If such is the case, (12-1) defines on $E(N;\lambda)$ a structure of *formal Lie algebra*.

If the map (12-1) is arbitrary, we have for $u,v,w \in N$:

$$S\left[[u,v]_\lambda, w\right]_\lambda = \sum_{t=1}^{\infty} \lambda^t D_t(u,v,w) \qquad (12\text{-}2)$$

where we have introduced the Chevalley three-cochain :

$$D_t(u,v,w) = \sum_{r+s=t} S\,\Gamma_r\,(\Gamma_s(u,v),w) \quad (r,s \geqslant 0) \qquad (12\text{-}3)$$

We have a formal deformation iff $D_t = 0$ for $t = 1,2,\dots$ We put :

$$E_t(u,v,w) = \sum_{r+s=t} S\,\Gamma_r\,(\Gamma_s(u,v),w) \quad (r,s \geqslant 1) \qquad (12\text{-}4)$$

and we have the identity :

$$D_t \equiv E_t - \partial\,\Gamma_t$$

If (12-1) is limited to the order q, we have a deformation of order q if Jacobi's identity is satisfied up to order (q+1). If such is the case, E_{q+1} is automatically a 3-cocycle of (N,P). We can find a two-cochain Γ_{q+1} satisfying $D_{q+1} \equiv E_{q+1} - \partial\Gamma_{q+1}$ iff E_{q+1} is *exact;* E_{q+1} defines a cohomology class, element of $H^3(N;N)$ which is the obstruction of a formal deformation of (N,P).

For an infinitesimal deformation, the first term Γ_1 is a two-cocycle of (N,P).

b) Consider a formal series in λ

$$T_\lambda = \sum_{s=0}^{\infty} \lambda^s T_s = Id_N + \sum_{s=1}^{\infty} \lambda^s T_s \qquad (12\text{-}5)$$

where the T_s 's ($s \geqslant 1$) are local endomorphisms of N which are null on the constants. Consider also another alternating bilinear map $N \times N \to E(N;\lambda)$ corresponding to the formal series :

$$[u,v]'_\lambda = \{u,v\} + \sum_{r=1}^{\infty} \lambda^r \Gamma'_r(u,v) \qquad (12\text{-}6)$$

where the Γ'_r 's are differential two-cochains, null on the constants again. Suppose that (12-5) and (12-6) are such that we have formally the identity :

$$T_\lambda [u,v]'_\lambda = [T_\lambda u, T_\lambda v]_\lambda \qquad (12\text{-}7)$$

This identity can be translated to :

$$\Gamma'_t - \Gamma_t + G_t = \partial T_t \qquad (t = 1,2,\ldots) \qquad (12\text{-}8)$$

where $G_1 = 0$ and :

$$G_t(u,v) = \sum_{r+s=t} T_s \Gamma'_r(u,v) - \sum_{s+s'=t} \{T_s t, T_{s'} v\} - \qquad (12\text{-}9)$$

$$\sum_{r+s+s'=t} \Gamma_r (T_s u, T_{s'} v) - \sum_{r+s=t} (\Gamma_r(T_s u,v) +$$

$$\Gamma_r(u,T_s v)) \qquad (r,s,s' \geqslant 1)$$

It follows from the proposition of §10,c that the T'_s are differential operators. Using universal formulas similar to those of the associative case, we prove by recursion the following :

Proposition - The deformation (12-1) *of the Lie algebra* (N,P) *being given, each formal series* (12-5) *where the* T_s ($s \geqslant 1$) *are differential operators which are null on the constants, generates a unique bilinear map* (12-6) *satisfying* (12-7);*this map is a new deformation of* (N,P) *which is sayed to be equivalent to* (12-1). *In particular a deformation is called trivial if it is equivalent to the identity deformation* ($\Gamma_r = 0$ *for every* $r \geqslant 1$).

If two deformations are equivalent at order q, there appears the Chevalley two-cocycle $(\Gamma'_{q+1} - \Gamma_{q+1} + G_{q+1})$ which is *the obstruction to equivalence for order* (q+1). In particular an infinitesimal deformation is trivial iff the two-cocycle Γ_1 defining it is exact.

III - NOTION OF $*_\nu$ - PRODUCT AND APPLICATIONS.

13 - Notion of parity.

We consider here the Hochschild complex of the algebra (N,.) corresponding to the differential cochains which are null on the constants.

a) A Hochschild two-cochain C(u,v) is said to *even* if it is symmetric in u,v, *odd* if it is antisymmetric in u,v. Similarly a Hochschild three-cochain B(u,v,w) is called *even* if it is symmetric in u,w, *odd* if it is antisymmetric in u,w.

If A is a one-cochain, A is even. If C is a two-cochain, we have :

$$\tilde{\partial}C(u,v,w) = u\, C(v,w) - C(uv,w) + C(u,vw) - C(u,v)w$$

It follows :

Lemma - If the two-cochain C *is even (resp. odd),* $\tilde{\partial}$C *is odd (resp. even).*

In fact we have :

$$\tilde{\partial}C(w,v,u) = w\, C(v,u) - C(vw,u) + C(w,uv) - C(w,v)u$$

We see immediately that if C is even, we have :

$$\tilde{\partial}C(w,v,u) = - \tilde{\partial}C(u,v,w)$$

If C is odd, we have :

$$\tilde{\partial}C(w,v,u) = \tilde{\partial}C(u,v,w)$$

b) If $T^{(2)}$ is an antisymmetric contravariant 2-tensor, we denote by the same notation the bidifferential operator of order 1 defined by $T^{(2)}$:

$$T^{(2)}(u,v) = i(T^{(2)})(du \wedge dv)$$

Let C *be a Hochschild two-cocycle.* It follows from the theorem of Vey that C admits the decomposition :

$$C = T^{(2)} + \tilde{\partial}A \qquad (13\text{-}1)$$

where $T^{(2)}$ is given by an antisymmetric 2-tensor and where A is a differential operator. The two terms of the right member of (13-1) are respectively odd and even. It follows that *if* C *is an even (resp. odd) Hochschild two-cocycle, it is exact (resp. nonexact and* 1*-differential).*

Let B *be a Hochschild three-cocycle.* We have according to the theorem of Vey :

$$B = T^{(3)} + \tilde{\partial} C \qquad (13\text{-}2)$$

where $T^{(3)}$ is given by an antisymmetric 3-tensor and where C is a two-cochain. Decompose C in an even and an odd part :

$$C = C^{(e)} + C^{(o)}$$

If B *is an even three-cocycle,* we have $B = \tilde{\partial} C^{(o)}$ and B is *exact*
If B *is an odd three-cocycle,* we have $B = T^{(3)} + \tilde{\partial} C^{(e)}$

14 - Definition of a $*_\nu$-product.

a) On the symplectic manifold (W,Λ), the Poisson operator P corresponding to the tensor Λ is a *nonexact Hochschild two-cocycle* and

$$uv + \nu P(u,v)$$

is a nontrivial associative infinitesimal deformation of (N,.). We consider only, in the following, the associative deformations of the form :

$$u *_\nu v = uv + \nu P(u,v) + \sum_{r=2}^{\infty} \nu^r C_r(u,v) \qquad (14\text{-}1)$$

where the C_r's which are *null on the constants* satisfy the following assumption :

Parity assumption : C_r *is even if* r *is even, odd if* r *is odd (or* $u *_\nu v = v *_\nu u$).

If such is the case, *we say that* (14-1) *defines a* $*_\nu$*-product (or twisted product)* on (W,Λ) [11].
Consider the bracket defined by antisymmetrization :

$$[u,v]_{\nu^2} = (2\nu)^{-1}(u *_\nu v - v *_\nu u) \qquad (14\text{-}2)$$

which satisfies trivially Jacobi's identity. We obtain (with $\lambda = \nu^2$) :

$$[u,v]_\lambda = P(u,v) + \sum_{r=1} \lambda^r C_{2r+1}(u,v) \qquad (14\text{-}3)$$

which is *a deformation of the Poisson Lie algebra* (N,P) (with $\Gamma_r = C_{2r+1}$). We say that the twisted product (14-1) generates the Lie algebra deformation (14-3).

b) Introduce the space $N^c = C^\infty(W;C)$ and suppose that ν is *purely imaginary;* a $*_\nu$-product may be extended naturally to a map $N^c \times N^c \to E(N^c;\nu)$ and defines on $E(N^c;\nu)$ a formal associative algebra. The nullity of the cochains on the constants expresses that we have :

$$u *_\nu 1 = 1 *_\nu u = u \qquad (u \in N^c) \qquad (14\text{-}4)$$

The parity assumption is equivalent to the relation :

$$\overline{u *_\nu v} = \bar{v} *_\nu \bar{u} \qquad (u,v \in N^c) \qquad (14\text{-}5)$$

where $\bar{u}$ is the complex conjugate element of u.

c) Consider the formal Lie algebra (14-3) which is generated (with $\lambda = \nu^2$) by the twisted product (14-1). We will show that if (14-3) is given, such a twisted product is *unique*. Introduce another twisted product :

$$u *'_\nu v = \sum_{r=0}^{\infty} \nu^r C'_r(u,v) = uv + \nu P(u,v) + \sum_{r=2}^{\infty} \nu^r C'_r(u,v) \qquad (14\text{-}6)$$

which generates the same Lie algebra. We have $C'_{2r+1} = C_{2r+1}$ for each r. We will proceed by recursion; we assume that :

$$C'_{2r} = C_{2r}$$

for $0 \leqslant r \leqslant (t-1)$ and we will compare C'_{2t} and C_{2t}. We put :

$$M_{2t} = C'_{2t} - C_{2t}$$

where M_{2t} is an even bidifferential operator which is null on the constants. We have with ν clear notations :

$$\tilde{\partial} C_{2t+1} = \tilde{E}_{2t+1} \qquad \tilde{\partial} C_{2t+1} = \tilde{E}'_{2t+1}$$

and so :

$$\tilde{E}'_{2t+1} - \tilde{E}_{2t+1} = 0 \qquad (14\text{-}7)$$

We can write :

$$\tilde{E}_{2t+1}(u,v,w) = \sum_{r+s=t} (C_{2r+1}(C_{2s}(u,v),w) + C_{2s}(C_{2r+1}(u,v),w))$$

$$- \sum_{r+s=t} (C_{2r+1}(u,C_{2s}(v,w)) + C_{2s}(u,C_{2r+1}(v,w))) \quad (r \geqslant 0, s \geqslant 1)$$

It follows from the recursion assumption that (14-7) can be written :

$$P(M_{2t}(u,v) + M_{2t}(P(u,v),w) - P(u,M_{2t}(v,w)) \tag{14-8}$$
$$- M_{2t}(u,P(v,w)) = 0$$

d) We have the following lemma :

Lemma - Let M *be an even bidifferential operator which is null on the constants.*

The relation :

$$P(M(u,v),w) + M(P(u,v),w) - P(u,M(v,w)) \tag{14-9}$$
$$- M(u,P(v,w)) = 0$$

implies $M = 0$.

The assumption of nullity on the constants is essential (take $M(u,v) = u.v$). We suppose that the term of M which has the largest bidifferential type has the type (r,s) $(1 \leqslant s \leqslant r)$; the other terms of type (r,s') $(s' < s)$ or (r',s') $(r' < r,\ s' \leqslant r')$ are sayed to be of inferior type (t.i.t). We consider different cases :

1) *First case* $2 \leqslant s \leqslant r$.

Let $\{x^i\}$ be a *natural chart* (for which the Λ^{ij} are const.) of (W,Λ) of domain U. We have on U :

$$M(v,w)\Big|_U = A^{i_1..i_r,j_1..j_s}(\partial_{i_1..i_r} v\, \partial_{j_1..j_s} w +$$
$$\partial_{i_1..i_r} w\, \partial_{j_1..j_s} v) + t.i.t$$

where A is a tensor which is symmetric in the indices i and symmetric in the indices j. If $r = s$, we suppose that A is symmetric in $(i_1..i_r)$ and $(j_1..j_r)$. A point x of U being choosed, take u and w such the 1-jet $j^1u(x)$ and the r-jet $j^rw(x)$ are null at x. The relation (14-9) can be written at x :

$$M(u,P(v,w))(x) = 0$$

that is :

$$\Lambda^{k\ell} A^{i_1..i_r,j_1..j_s}\, \partial_{j_1..j_s} u\, \partial_k v\, \partial_{i_1..i_r\ell} w = 0$$

It follows from the arbitrary character of u,v that :

$$S\,\Lambda^{i_o k}\,A^{i_1\ldots i_r, j_1\ldots j_s} = 0$$

where S is the symmetrization on the indices $(i_o,\ldots,i_r)$. We obtain by contracted product by $F_{i_o k}$ that $A = 0$.

2) *Second case* $s = 1$, $r \geqslant 2$.

We have on U :

$$M(v,w)\Big|_U = A^{i_1..i_r,j}(\partial_{i_1..i_r} v\,\partial_j w + \partial_{i_1..i_r} w\,\partial_j v) + \text{t.i.t}$$

Take w again such that the r-jet $j^r w(x)$ is null at x. The relation (14-9) can be written at x :

$$(\Lambda^{k\ell} A^{i_1..i_r,j} + \Lambda^{j\ell} A^{i_1..i_r,k})\partial_j u\,\partial_k v\,\partial_{i_1..i_r\ell} w = 0$$

It follows that :

$$S\,\Lambda^{i_o k} A^{i_1..i_r,j} + S\,\Lambda^{i_o j} A^{i_1..i_r,k} = 0 \qquad (14\text{-}10)$$

where S is the symmetrization on the indices i. If we put $B^{i_2\ldots i_r} = A^{i_1\ldots i_r,k} F_{i_1 k}$, we obtain by product of (14-10) by $F_{i_o k}$:

$$(2n + r + 1)\,A^{i_1..i_r,j} + S\,\Lambda^{i,j} B^{i_2\ldots i_r} = 0 \qquad (14\text{-}11)$$

The product of (14-11) by $F_{i_1 j}$ gives $B = 0$ and so $A = 0$, according to (14-11).

3) *Third case* $r = s = 1$.

The same argument gives $A = 0$.
The lemma is proved.

e) We see now that $M_2 = C_2' - C_2$ satisfies the assumptions of the lemma and it follows that $C_2' = C_2$. We deduce by recursion from (14-8) and from the lemma that $C_{2t}' = C_{2t}$ for each t and $\star_\nu'$ coïncides necessarily with $\star_\nu$. We have :

Uniqueness theorem - If a formal Lie algebra (14-3), *where the* C_{2r+1} *are null on the constants, is generated by a* $\star_\nu$*-product, this* $\star_\nu$*-product is unique.*

Consider now two twisted products such that for all $u \in N$:

$$u \star_\nu' u = u \star_\nu u \qquad (14\text{-}12)$$

It follows that we have formally for all $u,v \in N$:

$$u \star'_\nu v + v \star'_\nu u = u \star_\nu v + v \star_\nu u$$

and we have $C'_{2r} = C_{2r}$ for each r. A similar argument shows that we have $C'_{2r+1} = C_{2r+1}$ for all r. We obtain :

Proposition - Two twisted products such that $u \star'_\nu u = u \star_\nu u$ *for all* $u \in N$ *coïncide.*

15 - Inverse - Automorphisms and equivalence.

a) Consider a $\star_\nu$-product on the symplectic manifold (W,F). Let $\hat{E}(N;\nu)$ be the subset of $E(N;\nu)$ defined by the elements $a_\nu = \Sigma\, \nu^s a_s$ *such that* $a_o > 0$ on W. Search for an element $b_\nu = \Sigma\, \nu^s b_s$ such that $a_\nu \star_\nu b_\nu = 1$. We have :

$$a_o b_o = 1 \qquad \sum_{r+s+s'=t} C_r(a_s, b_{s'}) = 0 \quad (t \geqslant 1)$$

This last relation can be written :

$$a_o b_t + a_t b_o + \sum_{s+s'=t} a_s b_{s'} + C_t(a_o, b_o) + \tag{15-1}$$

$$\sum_{r+s=t} (C_r(a_s, b_o) + C_r(a_o, b_s)) + \sum_{r+s+s'=t} C_r(a_s, b_{s'}) = 0$$

$$(r,s,s' > 1)$$

(15-1) determines b by recursion in a unique way. Similarly there is a unique $c_\nu \in E(N;\nu)$ such that $c_\nu \star_\nu a_\nu = 1$. It follows from the associativity that :

$$c_\nu = c_\nu \star_\nu a_\nu \star_\nu b_\nu = b_\nu$$

The element $b_\nu = c_\nu = a_\nu^{(\star)-1}$ is the *unique inverse* of $a_\nu \in \hat{E}(N;\nu)$ in the sense of the $\star_\nu$-product.

b) Consider an automorphism of the space $E(N;\nu)$:

$$A_\nu = A_o + \sum_{s=1} \nu^s A_s \tag{15-2}$$

where A_o is an automorphism and the A_s $(s \geqslant 1)$ are endomorphisms of N; A_ν *is an automorphism of the* $\star_\nu$*-product if we have :*

$$A_\nu(u \star_\nu v) = A_\nu u \star_\nu A_\nu v \tag{15-3}$$

If $A_o = Id_N$, we say that A_ν has *a trivial main part.* If $a_\nu \in \hat{E}(N;\nu)$, *an inner automorphism of the* $*_\nu$ is given by :

$$u \to a_\nu *_\nu u *_\nu a_\nu^{(*)-1}$$

It is easy to see that its main part is trivial.

Moreover we have the useful following :

Lemma - Each automorphism A_ν *of the space* $E(N;\nu)$*, which has a trivial main part, admits a unique square root. Each element* a_ν *of* $E(N;\nu)$ *admits in* $\hat{E}(N;\nu)$ *a unique square root.*

In fact, if $A_\nu = Id. + \Sigma\, \nu^s A_s$ is such an automorphism of $E(N;\nu)$, search for $B_\nu = Id + \Sigma\, \nu^s B_s$ such that :

$$B_\nu^2 = A_\nu \qquad (15\text{-}4)$$

(15-4) can be translated to :

$$\sum_{s+s'=t} B_s B_{s'} = A_t \qquad (s,s' \geqslant 0)$$

let :

$$2\, B_t = A_t - \sum_{s+s'=t} B_s B_{s'} \quad (s,s' \geqslant 1) \qquad (15\text{-}5)$$

The relations (15-5) determine B_ν by recursion in a unique way. We denote B_ν by $A_\nu^{1/2}$. The same argument establishs the second part of the lemma.

c) Let $*_\nu$, $*'_\nu$ be two equivalent twisted products and denote by T_ν an equivalence operator transforming $*'_\nu$ in $*_\nu$. We have :

$$T_\nu(u *'_\nu v) = T_\nu u *_\nu T_\nu v \qquad (u,v \in N)$$

and so changing ν in $-\nu$:

$$T_{-\nu}(v *'_\nu u) = T_{-\nu} v *_\nu T_{-\nu} u \qquad (u,v \in N)$$

It follows that $T_{-\nu}$ is also an equivalence operator transforming $*'_\nu$ in $*_\nu$. If we put :

$$A_\nu = T_{-\nu}\, T_\nu^{-1}$$

A_ν is an automorphism of $*_\nu$ with a trivial main part satisfying trivially :

$$A_{-\nu} = A_\nu^{-1} \qquad (15\text{-}6)$$

Introduce the automorphisms $A_\nu^{1/2}$ of the space $E(N;\nu)$. It follows from the lemma that we have with evident notations :

$$A_\nu^{1/2} = A_\nu^{-1/2}$$

If $A_\nu^{1/2}$ transforms $*_\nu$ in a twisted product $\hat{*}_\nu$, $A_\nu^{1/2}$ transforms $\hat{*}_\nu$ in $*_\nu$ and we deduce from (11-12) that $A_\nu^{1/2}$ is *an automorphism of* $*_\nu$.
Consider the equivalence operator $T_\nu^{(\ell)} : *'_\nu \to *_\nu$ defined by :

$$T_\nu^{(\ell)} = A_\nu^{1/2} \; T_\nu$$

we have :

$$T_{-\nu}^{(\ell)} = A_{-\nu}^{1/2} \; T_{-\nu} = A_\nu^{1/2} \; A_\nu^{-1} \; T_{-\nu} \; T_\nu^{-1} \; T_\nu = A_\nu^{1/2} \; A_\nu^{-1} \; A_\nu \; T_\nu = A_\nu^{1/2} \; T_\nu = T_\nu^{(\ell)}$$

and $T^{(\ell)}$ is *even in* ν. If we set $B_\nu = A_{-\nu}^{1/2}$, we have :

Proposition - Each equivalence operator $T_\nu : *'_\nu \to *_\nu$ *admits a unique decomposition of the form :*

$$T_\nu = B_\nu \; T_\nu^{(\ell)} \qquad (15\text{-}7)$$

where $T_\nu^{(\ell)} : *'_\nu \to *_\nu$ *is even in* ν *and where* B_ν *is an automorphism of* $*_\nu$ *satisfying* $B_{-\nu} = B_\nu^{-1}$.

The existence being proved, it is sufficient to establish the uniqueness. For two decompositions of T_ν corresponding to B_ν and B'_ν, $B_\nu^{-1} \; B'_\nu$ is an even automorphism of $*_\nu$.

Therefore :

$$B_\nu^{-1} \; B'_\nu = B_\nu \; B'^{-1}_\nu \qquad \text{or} \qquad B'^{2}_\nu = B_\nu^2$$

It follows from the lemma that $B'_\nu = B_\nu$.
We see that *two equivalent twisted products are even-equivalent with respect to* ν. It follows that *two formal Lie algebras generated by two equivalent twisted products are effectively equivalent* with respect to $\lambda = \nu^2$.

d) Let $G = \mathrm{Aut}_t(*_\nu)$ be the group of the automorphisms of the twisted product $*_\nu$ which have a trivial main part ; the group G admits the involution $S : A_\nu \in G \to A_{-\nu} \in G$. We denote by $H = \mathrm{Aut}_{te}(*\ )$ the subgroup of the elements of G invariant under S, that is even in ν. It follows from the decomposition (15-7) (where $*'_\nu = *_\nu$) that the space of the automorphisms B_ν satisfying

$S(B_\nu) = B_\nu^{-1}$ admits a structure G/H of symmetric homogeneous space, which it is possible to study.

The $\star_\nu$-product defines on $\hat{E}(N;\nu)$ a structure of group $\hat{G}$ for which 1 is the unit element and $a_\nu^{(\star)-1}$ the inverse element of a_ν. Each element $a_\nu \in \hat{E}(N;\nu)$ admits a unique decomposition of the form :

$$a_\nu = b_\nu \star_\nu c_\nu \tag{15-8}$$

where b_ν *is even in* ν and where c_ν *satisfies* $c_{-\nu} = c_\nu^{(\star)-1}$; it is sufficient to note that we have necessarily $b_\nu^{(\star)2} = a_\nu \star_\nu a_{-\nu}$ where the right member is even in ν ; this relation determines b_ν.

The group $\hat{G}$ admits the involution $\hat{S} : a_\nu \to (a_{-\nu})^{(\star)-1}$. We denote by $\hat{H}$ the subgroup of the element c_ν of $\hat{G}$ invariant under $\hat{S}$, that is such that $c_{-\nu} = c_\nu^{(\star)-1}$. It follows from the decomposition (15-8) that the set of the *even* elements b_ν of $\hat{E}(N;\nu)$ admits a structure $\hat{G}/\hat{H}$ of *symmetric homogeneous space*.

Let $Aut_i(\star_\nu)$ be *the group of the inner automorphisms of* $\star_\nu$ and ρ the natural homomorphism of $\hat{G}$ onto $Aut_i(\star_\nu)$:

$$\rho : a_\nu \to a_\nu \star_\nu u \star_\nu a_\nu^{(\star)-1}$$

If $b_\nu \in \hat{G}$ is even in ν, $\rho(b_\nu) = B_\nu$ satisfies $B_{-\nu} = B_\nu^{-1}$; if $c_\nu \in \hat{G}$ is such that $c_{-\nu} = c_\nu^{(\star)-1}$, $\rho(c_\nu) = C_\nu$ is even in ν.

16 - The main example.

a) Let (W,Λ) be a symplectic manifold. A *symplectic connection* Γ on (W,Λ) is a linear connection *without torsion such that* $\nabla\Lambda = 0$ (or $\nabla F = 0$) where ∇ is the operator of covariant differentiation defined by Γ. If $\{\Gamma^i_{jk}\}$ are the usual coefficients of a linear connection Γ *in a natural chart* $\{x^i\}$, we introduce the coefficients $\Gamma_{ijk} = F_{i\ell}\Gamma^\ell_{jk}$. We have :

$$\nabla_k F_{ij} = - F_{\ell j}\Gamma^\ell_{ik} - F_{i\ell}\Gamma^\ell_{jk} = \Gamma_{jik} - \Gamma_{ijk} \tag{16-1}$$

It follows from (16-1) that such coefficients $\{\Gamma_{ijk}\}$ define a symplectic connection iff they are *completely symmetric* for every natural chart.

A standard argument shows the existence of symplectic connections on every (paracompact) symplectic manifold. It follows that a symplectic manifold admits infinitely many symplectic connections, the difference between two symplectic connections being given by means of a symmetric covariant 3-tensor.

b) Let Γ be a symplectic connection on (W,Λ). We put $P^o(u,v) = u.v$, $P^1 = P$ and we introduce the bidifferential operator P^r of maximum order r on each argument defined, for each domain U of a chart $\{x^i\}$ by the following expression :

$$P^r(u,v)\Big|_U = P^r_\Gamma(u,v)\Big|_U = \tag{16-2}$$

$$\Lambda^{i_1 j_1}..\Lambda^{i_r j_r}\, \nabla_{i_1..i_r} u\, \nabla_{j_1..j_r} v \qquad (u,v \in N)$$

Suppose that (W,Λ) admits a symplectic connection Γ *without curvature*; if such is the case, the manifold (W,Λ,Γ) is called *a flat symplectic manifold*. The simplest example is the cotangent bundle of $\mathbb{R}^n$, that is $\mathbb{R}^n \times \mathbb{R}^n$; the connection whose the coefficients in the canonical coordinates of $\mathbb{R}^{2n}$ are all null is a flat connection.

For such a flat connection $\nabla_{i_1..i_r}$ is symmetric with respect to the indices i.

Given a formal function f(z) of z with constant coefficients such that f(o) = 1, substitute P^r to z^r in the expansion of $f(\nu z)$; we obtain a bilinear map $(u,v) \to N \times N \to u \ast_\nu v = f(\nu P)(u,v) \in E(N;\nu)$. We wish to choose f so that we define thus a *twisted product* on (W,Λ). A direct calculus shows that the answer is given by the following [2] :

Proposition - If (W,Λ,Γ) *is a flat symplectic manifold, there is only one formal function of the Poisson bracket* P *(up to a linear change of the deformation parameter* ν*) that generates a twisted product on the manifold : it is the exponential function.*

We have :

$$u \ast_\nu v = \sum_{r=0}^{\infty} (\nu^r/r!)\, P^r(u,v) = \exp(\nu P)(u,v) \tag{16-3}$$

which generates *the deformation of the Poisson Lie algebra* $(\lambda = \nu^2)$:

$$[u,v]_\lambda = \sum_{r=0}^{\infty} (\lambda^r/(2r+1)!)\, P^{2r+1}(u,v) = \nu^{-1}\sinh(\nu P)(u,v) \tag{16-4}$$

It is remarkable that for $\nu = \hbar/2i$, we deduce from (16-4) a bracket $(2/\hbar)\sin(\hbar/2\, P)$ given in 1949 by Moyal [14] in the context of the Hermann Weyl-Wigner quantization [21],[23]. We say that (16-3) (resp. (16-4)) is the Moyal twisted product (resp. the Moyal bracket).

Consider the term P^3 of the Moyal bracket (16-4). If the Chevalley two-cocycle were exact, it would be the coboundary of a one-cochain, which can be supposed *3-differential*, according to Proposition 2 of §10,c. But it is easy to see that such a coboundary has no term of bidifferential type (3,3). It is possible to prove that, for $\mathbb{R}^{2n}$, the second Chevalley cohomology space $H^2(N;N)$ has the dimension 1 ; P^3 defines a cohomology 2-class which is a generator of $H^2(N;N)$. We see that *the deformations* (16-3) *and* (16-4) *are non trivial even for the order* 1.

c) Suppose now that Γ is an *arbitrary linear connection on* (W, Λ) *such that* $\nabla\Lambda = 0$, but which may have a torsion. Consider :

$$uv + \nu P(u,v) + (\nu^2/2)\, P^2_\Gamma(u,v) \qquad (16\text{-}5)$$

(16-5) defines an associative deformation of order 2 iff :

$$\frac{1}{2}\tilde{\partial} P^2_\Gamma = \tilde{E}_2$$

A direct calculus gives :

$$(\frac{1}{2}\tilde{\partial} P^2_\Gamma - \tilde{E}_2)(u,v,w) = -\frac{1}{2}\Lambda^{ij}\Lambda^{k\ell}((\nabla_{ik} - \nabla_{ki})u\, \nabla_j v\, \nabla_\ell w + (\nabla_{ik} - \nabla_{ki})w\, \nabla_\ell u\, \nabla_j v)$$

The right member vanishs for all u,v,w iff $\nabla_{ik}u$ is symmetric for all u, that is if Γ is without torsion; Γ is then a symplectic connection. We see that (16-5) *defines an associative deformation of order* 2 *for an arbitrary symplectic connection* Γ.

Now let Γ be a symplectic connection and consider :

$$uv + \nu P(u,v) + (\nu^2/2)\, P^2_\Gamma(u,v) + (\nu^3/6)\, P^3_\Gamma(u,v) \qquad (16\text{-}6)$$

(16-6) defines an associative deformation of order 3 iff :

$$\frac{1}{6}\tilde{\partial} P^3_\Gamma = \tilde{E}_3 \qquad (16\text{-}7)$$

A small calculus shows that (16-7) is satisfied iff ∇_{ijk} is completely symmetric for all u, that is if Γ is *without curvature*. We see that the formulas (16-2) doe not give direct generalizations to nonflat symplectic manifolds, for the problem of the construction of associative deformations.

17 - The invariant β.

a) Consider on the symplectic manifold (W,Λ) an arbitrary symplectic connection Γ. If $u \in N$, denote by $\mathcal{L}(X_u)\Gamma$ the symmetric covariant 3-tensor defined by means of the Lie derivative of the symplectic connection Γ by the hamiltonian vector field X_u.

We have locally for a *natural* chart :

$$(\mathcal{L}(X_u)T)_{i_1i_2i_3} = \partial_{i_1i_2i_3} u - S\,\Lambda^{k\ell}\,\Gamma_{ki_1i_2}\,\partial_{\ell i_3} u - \Lambda^{k\ell}\,\partial_k\,\Gamma_{i_1i_2i_3}\,\partial_\ell\, u \tag{17-1}$$

where S is the summation after cyclic permutation on i_1, i_2, i_3. If T is a symmetric covariant 3-tensor, we have :

$$(\mathcal{L}(X_u)T)_{i_1i_2i_3} = -\,S\,\Lambda^{k\ell}\,T_{ki_1i_2}\,\partial_{\ell i_3}\,u - \Lambda^{k\ell}\,\partial_k\,T_{i_1i_2i_3}\,\partial_\ell\,u \tag{17-2}$$

Consider the Chevalley two-cochain S^3_Γ defined on each domain U of an arbitrary chart $\{x^i\}$ by :

$$S^3(u,v)\Big|_U = \Lambda^{i_1j_1}\,\Lambda^{i_2j_2}\,\Lambda^{i_3j_3}\,(\mathcal{L}(X_u)\Gamma)_{i_1i_2i_3}\,(\mathcal{L}(X_v)\Gamma)_{j_1j_2j_3} \tag{17-3}$$

This two-cochain admits the same principal symbol as P^3_Γ and coïncides with P^3 if is flat; it is null on the constants. We deduce from $X_{P(u,v)} = [X_u, X_v]$ that :

$$S\,S^3_\Gamma(P(u,v),w) = S\,\Lambda^{i_1j_1}\,\Lambda^{i_2j_2}\,\Lambda^{i_3j_3}\,(\mathcal{L}([X_u,X_v])\Gamma)_{i_1i_2i_3}\,(\mathcal{L}(X_w)\Gamma)_{j_1j_2j_3}$$

On the other hand we have :

$$S\,P(w, S^3_\Gamma(u,v)) = S\,\mathcal{L}(X_w)\,S^3_\Gamma(u,v)$$

and we deduce from the properties of the Lie derivative that :

$$S\,P(w,S^3_\Gamma(u,v)) = S\,\Lambda^{i_1j_1}\,\Lambda^{i_2j_2}\,\Lambda^{i_3j_3}\,(\mathcal{L}(X_w)\mathcal{L}(X_u)\Gamma)_{i_1i_2i_3}\,(\mathcal{L}(X_v)\Gamma)_{j_1j_2j_3} + S\,\Lambda^{i_1j_1}\,\Lambda^{i_2j_2}\,\Lambda^{i_3j_3}\,(\mathcal{L}(X_u)\Gamma)_{i_1i_2i_3}\,(\mathcal{L}(X_w)\mathcal{L}(X_v)\Gamma)_{j_1j_2j_3}$$

that is :

$$S\, P(w,S^3_\Gamma(u,v)) = S\, \Lambda^{i_1j_1}\Lambda^{i_2j_2}\Lambda^{i_3j_3}$$

$$(\mathcal{L}([X_u,X_v])\Gamma)_{i_1i_2i_3}\,(\mathcal{L}(X_w)\Gamma)_{j_1j_2j_3}$$

We obtain :

$$S\, P(u,S^3_\Gamma(v,w)) - S\, S^3_\Gamma(P(u,v),w) = 0$$

that is :

$$\partial\, S^3_\Gamma = 0$$

and S^3_Γ is a *Chevalley two-cocycle* of (N,P). The same argument as for the flat case shows that the two-cocycle S^3_Γ is *non exact.*

b) Let T be a symmetric covariant 3-tensor. We can associate to the 3-tensor T and to the symplectic connection Γ the third order differential operator $A_{T,\Gamma}$ given by :

$$A_{T,\Gamma}(u)\Big|_U = \Lambda^{i_1j_1}\Lambda^{i_2j_2}\Lambda^{i_3j_3}\, T_{i_1i_2i_3}(\mathcal{L}(X_u)\Gamma)_{j_1j_2j_3} \qquad (17\text{-}4)$$

$$= T^{j_1j_2j_3}(\mathcal{L}(X_u)\Gamma)_{j_1j_2j_3}$$

and to the tensor T the second order differential operator B_T given by :

$$B_T(u)\Big|_U = \Lambda^{i_1j_1}\Lambda^{i_2j_2}\Lambda^{i_3j_3}\, T_{i_1i_2i_3}(\mathcal{L}(X_u)T)_{j_1j_2j_3} \qquad (17\text{-}5)$$

$$= T^{j_1j_2j_3}\,(\mathcal{L}(X_u)T)_{j_1j_2j_3}$$

We can prove with these notations :

Proposition - The cohomology 2-class β, element of $H^2(N;N)$ *defined by the two-cocycle* S^3_Γ *is independant of the choice of the symplectic connection Γ.*

In fact consider two symplectic connections Γ and Γ' and take for T the symmetric 3-tensor defining the difference between the two connections. We verify immediately that :

$$S^3_{\Gamma'} - S^3_\Gamma = \partial\,(A_{T,\Gamma} + \frac{1}{2}B_T)$$

We see that *the cohomology 2-class* $\beta \in H^2(N;N)$ *is an invariant of the symplectic structure of the manifold.*

For the differential cochains which are null on the constants, it is possible to establish (S. Gutt or myself) :

Theorem - The second space $H^2(N;N)$ *of Chevalley cohomology admits as generators the class* β *and the classes given by the converse images of the closed 2-forms of* W *by the map* μ.

We see that $H^2(N;N)$ is determined by β and by the De Rham cohomology of dimension 2; $H^3(N;N)$ admits similar generators. In particular each 1-differential Chevalley 3-cocycle is the image of a closed 3-form μ^{-1} .

c) It is easy to verify that, for an arbitrary symplectic connection Γ :

$$u\nu + \nu P(u,v) + (\nu^2/2!)\, P_\Gamma^2(u,v) + (\nu^3/3!)\, S_\Gamma^3(u,v)$$

determines a $*_\nu$-product of order 3.

18 - The Vey twisted products.

a) Introduce now the following notations : we denote by Q^r a bidifferential operator of maximum order r on each argument, null on the constants, satisfying the parity assumption and such that *its principal symbol coincides with the principal symbol of* P_Γ^r. We take in particular $Q^o(u,v) = u\nu$ and $Q^1 = P$. We are led to the following :

Definition - A Vey twisted product is a twisted product of the form :

$$u *_\nu v = \sum_{r=0}^{\infty} (\nu^r/r!)\, Q^r(u,v) \qquad (18\text{-}1)$$

A Vey Lie algebra is a formal Lie algebra given by a bracket of the form :

$$[u,v]_\lambda = \sum_{r=0}^{\infty} (\lambda^r/(2r+1)!)\, Q^{2r+1}(u,v)$$

A Vey twisted product (resp. a Vey Lie algebra) are said to be strong if $Q^3 \in \beta$.

We note that our viewpoint differs appreciably from the viewpoint of J. Vey. We will study in the part IV the problems concerning the existence and the equivalence of such twisted products.

c) Consider a twisted product of order 2 :

$$uv + \nu P(u,v) + \nu^2\, C_2(u,v) \qquad (18\text{-}2)$$

It follows from the study of §16,c, that we have for an arbitrary symplectic connection $\tilde{\partial}(C_2 - (1/2)\, P_\Gamma^2) = 0$. The even Hochschild two-cocycle $C_2 - 1/2\, P_\Gamma^2$ being exact, there exists a differential operator A such that :

$$C_2 = \frac{1}{2} P_\Gamma^2 + \tilde{\partial} A \qquad (18\text{-}3)$$

Suppose that (18-2) is a Vey twisted product of order 2 : we have $C_2 = Q^2/2$. It is easy to prove that there exists then a unique symplectic connection Γ such that :

$$Q^2 = P_\Gamma^2 + \tilde{\partial}\, H \qquad (18\text{-}4)$$

where H is a differential operator of maximum order 2, defined up to the addition of a first order differential operator.

Suppose now that Q^3 is a Chevalley two-cocycle ($\partial Q^3 = 0$); there exists similarly a unique symplectic connection Γ such that :

$$Q^3 = S_\Gamma^3 + M + 3\,\partial\, H \qquad (18\text{-}5)$$

where M is a 2-tensor image of a closed 2-form of W by μ^{-1}, and where H is a differential operator of maximum order 2, defined also up to the addition of a first order differential operator. Consider :

$$uv + \nu P(u,v) + (\nu^2/2!)\, Q^2(u,v) + (\nu^3/3!)\, Q^3(u,v) \qquad (18\text{-}6)$$

(18-6) gives a twisted product of order 3 *iff the symplectic connections and the operators* H *which appear in* (18-4) *and* (18-5) *coincide*.

c) The Vey $*_\nu$-products constitute natural generalizations of the notion of Moyal product (see §16). We shall prove the existence of such twisted products under a general assumption. Explicit expressions of the cochains which appear in the twisted products are in general unknown.

But it is useful to obtain effective construction processes of Vey $*_\nu$-products for classes of natural symplectic manifolds. I have given [10] such processes for large classes of cotangent bundles of classical groups or homogeneous spaces. I will limite myself to the simplest example.

Consider the flat symplectic manifold defined by the cotangent bundle of the space $\mathbb{R}^n - \{0\}$, let $E = (\mathbb{R}^n - \{0\}) \times \mathbb{R}^n$. The solvable group G_2 of dimension 2 acts on E in the following way :

$$(x,y) \in E = (\mathbb{R}^n - \{0\}) \times \mathbb{R}^n \to (x' = e^\rho x,\ y' = e^{-\rho}(y+\sigma x)) \quad (\rho,\sigma \in \mathbb{R})$$

The group G_2 leaves invariant the natural symplectic structure of E and the flat symplectic connection Γ. It follows that it preserves the P^r's defined by (16-2). The space of the orbits of E by the group G_2 is isomorphic to T^*S^{n-1}, where $S^{n-1}=SO(n)/SO(n-1)$ is the sphere of dimension (n-1). We deduce from the Moyal $*_\nu$-product, invariant under G_2, defined on E by the P^r's, a natural strong Vey $*_\nu$-product on T^*S^{n-1}, which is invariant under the action of SO(n). It is well-known that the regularized Kepler problem of dimension n admits as phase space the so-called *Moser manifold*, that is $T_o^*S^n$ (the cotangent bundle of the sphere without the null section). We have obtained for the Moser manifold a natural invariant Vey twisted product.

We can deduce from this quotient process the construction of natural $*_\nu$-products for example for the cotangent bundles of the Stiefel manifolds or of the Grassmann manifolds.

19 - Introduction to a spectral theory and quantization (see [2])

a) Come back to the flat symplectic manifold $\mathbb{R}^n \times \mathbb{R}^n$. Under suitable assumptions, Hermann Weyl has defined in this case, in terms of Fourier transform, a map Ω (the Weyl map) which associates with each element u of a large class of classical functions (or distributions) an operator $\hat{u}$ of a Hilbert space and conversely. The conventional quantization processes in terms of these operators. But the Moyal product defined by (16-3) corresponds by Ω to the product of operators (for $\nu = \hbar/2i$). If :

$$u * v = \exp\left(\frac{\hbar}{2i} P\right)(u,v) \quad \text{we have} \quad \Omega(u * v) = \Omega(u).\Omega(v) \quad .$$

The Moyal bracket is (up to the factor $2\nu = \hbar/i$) the image by Ω^{-1} of the natural commutator of the corresponding operators.

In the Moyal case, consider :

$$P^r(u,v) = \Lambda^{i_1 j_1} \ldots \Lambda^{i_r j_r} \nabla_{i_1 \ldots i_r} u \, \nabla_{j_1 \ldots j_r} v$$

We have :

$$\Lambda^{i_1 j_1} \ldots \Lambda^{i_r j_r} \nabla_{i_2 \ldots i_r} u \, \nabla_{i_1 j_1 \ldots j_r} v = 0$$

and we can write :

$$P^r(u,v) = \nabla_{i_1} (\Lambda^{i_1 j_1} \ldots \Lambda^{i_r j_r} \nabla_{i_2 \ldots i_r} u \, \nabla_{j_1 \ldots j_r} v)$$

and we see that, for $r \geqslant 1$, $P^r(u,v)$ is the divergence of a vector.

Suppose for example that the intersection $S(u) \cap S(v)$ of the supports of u and v is compact and set :

$$\int_W (u *_\nu v)\eta = \sum_{r=0}^{\infty} \nu^r \int_W P^r(u,v)r!)\eta$$

where η is the symplectic volume element. We obtain in the Moyal case :

$$\int_W (u *_\nu v)\eta = \int_W uv \, . \, \eta \qquad (19\text{-}1)$$

It appears as possible to develop directly Quantum Mechanics in terms of ordinary functions (or distributions) and twisted products, without reference to some Ω and to operators, in a complete and autonomous way.

b) Consider a symplectic manifold (W,Λ) which is the *phase space* of a dynamical system, and on this manifold *a twisted product*. Let H be the classical hamiltonian of our problem. If we consider the value $\nu = \hbar/2i$ of the parameter of deformation suggested by the Moyal product, we are led to translate the quantum dynamical equation to :

$$\frac{du_t}{dt} = \frac{i}{\hbar} 2\nu \left[H,u_t\right]_{\nu^2} \qquad (u_t \in E(N^c;\nu)\times\mathbb{R}) \; (19\text{-}2)$$

If we put $\tilde{H} = iH/\hbar$, we have :

$$du_t/dt = \tilde{H} *_\nu u_t - u_t *_\nu \tilde{H} \qquad (19\text{-}3)$$

Introduce the $*_\nu$-powers of $\tilde{H}$ by recursion $(\tilde{H}^{(*)p} = \tilde{H}^{(*)p-1} *_\nu \tilde{H})$. It is easy to see that it follows from the parity assumption that $\tilde{H}^{(*)p}$ depends only upon the even powers of ν. We can define the $*_\nu$-exponential of $\tilde{H}t$ in the following way :

$$\mathrm{Exp}_*(\tilde{H}t) = \sum_{p=0}^{\infty} (t^p/p!)\tilde{H}^{(*)p} \qquad (19\text{-}4)$$

where $\mathrm{Exp}_*(\tilde{H}t)$ is an even element of $\hat{E}(N;\nu)$; we define thus a one parameter group of the group $\hat{G} = (\hat{E}(N;\nu),*_\nu)$ introduced in §15. Consider the corresponding inner automorphisms of $*_\nu$ in $E(N^c;\nu)$:

$$u_0 \in E(N^c;\nu) \to u_t = \mathrm{Exp}_*(\tilde{H}t) *_\nu u_0 *_\nu \mathrm{Exp}_*(-\tilde{H}t) \qquad (19\text{-}5)$$

(19-5) gives the formal solution of (19-3) taking the value u_0 at t = 0.

c) Consider now *the viewpoint of the mathematical analysis* and give to ν the value $\hbar/2i$. We denote by $*$ the twisted product for this value. Suppose that H is such that, for t in a complex neighborhood of the origin, the right side of (19-4) converges to a distribution on W denote by $Exp_{*}(\tilde{H}t)$ again. Suppose for simplicity that, for t fixed in a neighborhood of the origin, $Exp_{*}(\tilde{H}t)$ has a unique Fourier-Dirichlet expansion :

$$Exp_{*}(\tilde{H}t) = \sum_{\lambda \in I} \pi_{\lambda} e^{i\lambda t/\hbar} \tag{19-6}$$

where I is a set of C and where $\pi_{\lambda} \in N^{c}$. This expansion is similar to the spectral expansion of an operator : for t = 0, we have :

$$\sum_{\lambda \in I} \pi_{\lambda} = 1 \tag{19-7}$$

We obtain by differentiation of (19-6) with respect to t :

$$\frac{\hbar}{i} \frac{d}{dt} Exp_{*}(\tilde{H}t) = \sum_{\lambda \in I} \lambda \pi_{\lambda} e^{i\lambda t/\hbar}$$

let :

$$\frac{\hbar}{i} \frac{d}{dt} Exp_{*}(\tilde{H}t) = \sum_{\lambda \in I} H * \pi_{\lambda} e^{i\lambda t/\hbar} = \sum_{\lambda \in I} \pi_{\lambda} * H e^{i\lambda t/\hbar}$$

It follows by uniqueness :

$$H * \pi_{\lambda} = \pi_{\lambda} * H = \lambda \pi_{\lambda} \tag{19-8}$$

and we obtain by summation :

$$H = \sum_{\lambda \in I} \lambda \pi_{\lambda} \tag{19-9}$$

If λ , $\lambda' \in I$, we have, according to (19-8) :

$$\pi_{\lambda} * H * \pi_{\lambda'} = \lambda \pi_{\lambda} * \pi_{\lambda'} = \lambda' \pi_{\lambda} * \pi_{\lambda'}$$

It follows that if $\lambda' \neq \lambda$, we have :

$$\pi_{\lambda} * \pi_{\lambda'} = 0$$

Moreover, according to (19-7), we have :

$$\pi_{\lambda} = \left(\sum_{\lambda' \in I} \pi_{\lambda'} \right) * \pi_{\lambda} = \pi_{\lambda} * \pi_{\lambda}$$

We have proved that :

$$\pi_{\lambda} * \pi_{\lambda'} = \delta_{\lambda\lambda'} \pi_{\lambda} \tag{19-10}$$

where $\delta_{\lambda\lambda'} = 0$ for $\lambda' = \lambda$, $\delta_{\lambda\lambda'} = 1$ for $\lambda' = \lambda$.
We see that I *can be interpreted as the spectrum of* H, $\pi_{\lambda} \neq 0$ *being*

an eigenprojector corresponding to $\lambda \in I$.
Show that the spectrum I and the π_λ's are characterized *in a unique way* by the relations :

$$H \star \pi_\lambda = \pi_\lambda \star H = \lambda\, \pi_\lambda \qquad \sum_{\lambda \in I} \pi_\lambda = 1 \tag{19-11}$$

In fact consider another set $(\lambda', \pi'_{\lambda'})$, where $\lambda' \in I'$, such that :

$$H \star \pi'_{\lambda'} = \pi'_{\lambda'} \star H = \lambda'\, \pi'_{\lambda'} \qquad \sum_{\lambda \in I'} \pi'_{\lambda'} = 1$$

We have :

$$\pi'_{\lambda'} \star H \star \pi_\lambda = \lambda'\, \pi'_{\lambda'} \star \pi_\lambda = \lambda\, \pi'_{\lambda'} \star \pi_\lambda$$

It follows that if $\lambda' \neq \lambda$:

$$\pi'_{\lambda'} \star \pi_\lambda = 0$$

If $\lambda \notin I'$, we have $\left(\sum_{\lambda' \in I'} \pi'_{\lambda'}\right) \star \pi_\lambda = \pi_\lambda = 0$. We see that $I \subset I'$ and $I' \subset I$. Therefore $I' = I$.
Moreover the relations :

$$\left(\sum_{\lambda' \in I} \pi'_{\lambda'}\right) \star \pi_\lambda = \pi_\lambda \qquad \pi'_\lambda \star \left(\sum_{\lambda' \in I} \pi_{\lambda'}\right) = \pi'_\lambda$$

imply $\pi'_\lambda \star \pi_\lambda = \pi_\lambda = \pi'_\lambda$ and the eigenprojectors are equal.

d) a twisted product is said to be *non degenerated* if the relation $\bar{u} \star u = 0$, for a function $u \in N^c$, implies $u = 0$. We have :

Proposition 1 - For a flat symplectic manifold, the Moyal product is non degenerated.

In fact, let $\varphi \in N$ an arbitrary function with a compact support. Consider the integral :

$$\mathfrak{J} = \int_W |u \star \varphi|^2\, \eta = \int_W (\overline{u \star \varphi}) . (u \star \varphi)\eta = \int_W (\varphi \star \bar{u}) . (u \star \varphi)\eta$$

It follows from (19-1) and from the associativity that :

$$\mathfrak{J} = \int_W (\varphi \star \bar{u} \star u \star \varphi)\, \eta$$

If $\bar{u} \star u = 0$, we have $\mathfrak{J} = 0$ and so $u \star \varphi = 0$ for an arbitrary φ. If we choose φ such that $\varphi_{|K} = 1$, where K is an arbitrary compact set, we obtain $u_{|K} = 0$. Therefore $u = 0$ on W.

Proposition 2 - A twisted product deduced from a Moyal product by the quotient process is non degenerated.

In fact, let E be a flat symplectic manifold admitting the Moyal product $\star$. If (W,Λ) is a symplectic manifold for which the projection $p : E \rightarrow W$ defines a twisted product $\star'$, we have :

$$p^{\star}(u \star' v) = p^{\star} u \star p^{\star} v \qquad (u,v \in N^{c}(W))$$

Suppose that u is such $\bar{u} \star' u = 0$. We have $p^{\star} \bar{u} \star p^{\star}u = 0$, which implies $p^{\star}u = 0$ and so $u = 0$. The proposition is proved.

Consider now a symplectic manifold (W,Λ) admitting a *nondegenerated* twisted product $\star$. For this product, we deduce from (19-11) by complex conjugation :

$$\bar{\pi}_{\lambda} \star H = H \star \bar{\pi}_{\lambda} = \bar{\lambda}\, \bar{\pi}_{\lambda} \qquad \Sigma\, \bar{\pi}_{\lambda} = 1 \qquad (19\text{-}12)$$

If $\bar{\lambda} \neq \lambda$, it follows from (19-11 and (19-12) that we have $\bar{\pi}_{\lambda} \star \pi_{\lambda} = \underline{0}$ which implies $\pi_{\lambda} = 0$, and we obtain a contradiction. Therefore $\bar{\lambda} = \lambda$ and the spectrum of H is real. We deduce from the characterization of the (λ, π_{λ}) by (19-11) that $\bar{\pi}_{\lambda} \star \pi_{\lambda} = \pi_{\lambda} = \bar{\pi}_{\lambda}$ and we see that the π_{λ} 's are real-valued.

Theorem - The spectrum, for a nondegenerated twisted product, of all real-valued function admitting a spectral expansion in the sense of (19-6) *is real and the corresponding eigenprojector* π_{λ}*'s are real-valued.*

e) Define N_{λ} by the formula :

$$N_{\lambda} = \int_{W} \pi_{\lambda}\, \tilde{\eta} \qquad (19\text{-}13)$$

where $\tilde{\eta} = \eta/(2\pi\hbar)^{n}$, if η is the symplectic volume element. If N_{λ} is finite, a normalized state ρ_{λ} of the dynamical system is defined by $\rho_{\lambda} = \pi_{\lambda}/N_{\lambda}$ and N_{λ} is the multiplicity of the state in the usual sense of Quantum Mechanics.

More generally, we may consider the Fourier transform in the sense of distributions :

$$\mathrm{Exp}_{\star}(\tilde{H}t) = \int e^{i\lambda t/\hbar}\, d\mu\,(\lambda, x)$$

and the support of $d\mu$ will be referred as the spectrum of H. It is the spectrum in the sense of Schwartz of $\mathrm{Exp}_{\star}(\tilde{H}t)$ considered as a distribution in t, up to the factor $\hbar/i$.

A *state* ρ is here a real-valued (pseudoprobability) distribution on the phase space (W,Λ) normalized by the condition :

$$\int \rho \tilde{\eta} = 1$$

and such that :

$$\rho \star \rho = (1/N)\, \rho \qquad (19\text{-}14)$$

where N is the multiplicity. The measured value $< u >_t$ of the observable u at time t for the state ρ is given by :

$$< u >_t = \int_W (u_t \star \rho)\tilde{\eta} \qquad (19\text{-}15)$$

f) The previous algorithm directly applied to the flat case gives, *for the* n-*dimensional harmonic oscillator*, the energy levels $E_m = \hbar\,(m + n/2)$ with the correct multiplicities. For *the Hydrogen atom*, we can consider the Moser-Kepler manifold $T_o^{\star} S^3$ as phase space and introduce the corresponding natural twisted product invariant under SO(4). We obtain then the complete spectrum, that is the negative discrete spectrum and the positive continuous spectrum [2].

IV - EXISTENCE AND EQUIVALENCE OF TWISTED PRODUCTS AND LIE ALGEBRAS.

20 - The theorem of Neroslavsky and Vlassov.

We will prove the existence of a twisted product on a symplectic manifold (W,Λ) satisfying a general assumption.

a) We have seen that if we have a twisted product (14-1), let :

$$u \star_\nu v = u\nu + \nu P(u,v) + \sum_{r=2} \nu^r C_r(u,v) \qquad (C_1 = P) \qquad (20\text{-}1)$$

we have :

$$\tilde{\partial} C_t = \tilde{E}_t \qquad (20\text{-}2)$$

where :

$$\tilde{E}_t(u,v,w) = \sum_{\substack{r+s=t \\ (r,s \geqslant 1)}} (C_r(C_s(u,v),w) - C_r(u,C_s(v,w))) \qquad (20\text{-}3)$$

We will prove the following :

Lemma - If t *is odd (resp. even),* $\tilde{E}_t$ *is an even (resp. odd) Hochschild three-cochain.*

In fact :

$$\tilde{E}_t(w,v,u) = \sum_{r+s=t} (C_r(C_s(w,v),u) - C_r(w,C_s(v,u))) \qquad (r,s \geq 1)$$

Suppose that t is odd; if r is odd, s is even and if i is even, s is odd. Therefore :

$$\tilde{E}_t(w,v,u) = \sum_{r+s=t} (-C_r(u,C_s(v,w)) + C_r(C_s(u,v),w)) = \tilde{E}_t(u,v,w)$$

Suppose now that t is even; if r is odd, s is odd; if r is even, s is even :

$$\tilde{E}_t(w,v,u) = \sum_{r+s=t} (C_r(u,C_s(v,w)) - C_r(C_s(u,v),w)) = -\tilde{E}_t(u,v,w)$$

Our lemma is proved.

b) Consider a twisted product of order 2. It can be written :

$$\nu u + \nu P(u,v) + \nu^2 C_2(u,v) \qquad (20\text{-}4)$$

where $C_2(u,v) = (1/2) P^2_\Gamma(u,v) + \tilde{\partial} A$ (with the notations of (18-3)).

We will process by recursion. Choose *an odd integer* t and suppose that there exists a twisted product of even order $t-1 = 2t'$. The Hochschild three-cocycle $\tilde{E}_t$ is even and so exact (see §13) : there is an odd differential two-cochain C_t, null on the constants, such that the relation (20-2) is satisfied.

Choose the odd two-cochain C_t *which is defined up to an odd bidifferential operator* $T^{(2)}$ of order 1. We obtain a twisted product of odd order $t = 2t'+1$. Study the odd Hochschild three-cocycle :

$$\tilde{E}_{t+1} = T^{(3)} + \partial A^{(\ell)} \qquad (20\text{-}5)$$

where $T^{(3)}$ is given by an antisymmetric 3-tensor and where $A^{(\ell)}$ is an even two-cochain. *We will show that the alternate three-cochain* $T^{(3)}$ *is a Chevalley three-cocycle.*

We set for an arbitrary Hochschild three-cochain B :

$$(\hat{\Sigma} B)(u,v,w) = B(u,v,w) - B(v,u,w) - B(u,w,v)$$

It is easy to see that $\hat{\Sigma}\, T^{(3)} = 3T^{(3)}$, $\hat{\Sigma}\, \tilde{\partial}\, A^{(\ell)} = 0$. We deduce from (20-5)

$$3\, T^{(3)} = \hat{\Sigma}\, \tilde{E}_{t+1} \qquad (20\text{-}6)$$

We wish to evaluate the right side ; (t+1) is even and, in the expression (20-3) of E_{t+1}, r and s are either even or odd. In $\Sigma \tilde{E}_{t+1}$, the contribution of the terms corresponding to the even r, s vanishs and we obtain with (r = 2r'+1, s = 2s'+1) :

$$(\widehat{\Sigma \tilde{E}}_{t+1})(u,v,w) = 2\ S \sum_{r'+s'=t'} C_{2r'+1}(C_{2s'+1}(u,v),w) \quad (r',s' \geqslant 0)$$

where S is the summation after cyclic permutation on u,v,w. If we set $\Gamma_{r'} = C_{2r'+1}$, we obtain with the notations of §12 :

$$\hat{\Sigma}\ \tilde{E}_{t+1} = 2\ D_{t'} \qquad (t = 2t'+1) \tag{20-7}$$

where :

$$D_{t'} = E_{t'} - \partial\ \Gamma_{t'} = E_{t'} - \partial C_t$$

We have similarly the identities :

$$\hat{\Sigma}\ \tilde{E}_{2r+2} = 2\ D_r \qquad (r \leqslant t' - 1)$$

or

$$\hat{\Sigma}\ \tilde{D}_{2r+2} = 2\ D_r \qquad (r \leqslant t' - 1) \tag{20-8}$$

It follows from (20-8) that the antisymmetrization of the considered twisted product of order t-1 = 2t' gives a Poisson Lie algebra deformation of order (t'-1); $E_{t'}$ is thus a Chevalley three-cocycle and $D_{t'}$ also. It follows from (20-6) and (20-7) that :

$$T^{(3)} = \frac{2}{3}\ D_{t'}$$

and $T^{(3)}$ is a Chevalley three-cocycle. Such a cocycle is the image of a closed 3-form of W by the map μ^{-1} (see §17,b) and it is exact if the 3-form is exact.

c) We search for the possibility of a modification of C_t such that $T^{(3)}$ vanishs. If we change C_t in $C_t + T^{(2)}$, where $T^{(2)}$ corresponds to an antisymmetric 2-tensor, we have :

$$D_{t'} \to D_{t'} - \partial T^{(2)} \qquad \text{and} \qquad T^{(3)} \to T^{(3)} - (2/3)\ \partial T^{(2)}$$

and it is possible to annul $T^{(3)}$ if this Chevalley three-cocycle is exact.

It is certainly the case *if we suppose that* $b_3(W) = 0$. If such is the case, $\tilde{E}_{t+1}$ is an exact Hochschild three-cocycle for the good choice of C_t, and we can choose an even cochain $C_{t+1} = A^{(\ell)}$ (§13) such that we obtain a twisted product of order t+1 = 2t'+2.

We have proved by induction the existence of a twisted product. We have :

Theorem - Each symplectic manifold (W,F) *such that* $b_3(W) = 0$ *admit twisted products.*

This existence theorem has been obtained in 1979 by O.M. Neroslavsky and A.T. Vlassov.

21 - Vey twisted products and equivalence.

a) We have the following on the domain U of an arbitrary chart.

Lemma - Let C *be an even (resp. odd) differential Hochschild two-cochain which has exactly the bidifferential type* (t,s) *(with* $s \leq t$, $t > 2$). *If* $s > 1$, *the three-cocycle* ∂C *has effectively maximum bidifferential type* (t,s-1) *in* u, w; *if* s = 1, ∂C *has effectively maximum bidifferential type* (t-1,1) *in* u,w.

We have on U :

$$C(u,v) = C^{i_1..i_t,j_1..j_s}(\partial_{i_1..i_t} u\, \partial_{j_1..j_s} v + \varepsilon\, \partial_{i_1..i_t} v\, \partial_{j_1..j_s} u)$$

where $\varepsilon = \pm 1$ according to the parity of C. We have :

$$\partial C(u,v,w) \simeq - C^{i_1..i_t,j_1..j_s}(\partial_{i_1..i_t}(uv)\, \partial_{j_1..j_s} w + \varepsilon\, \partial_{i_1..i_t} w\, \partial_{j_1..j_s}(uv) - \partial_{i_1..i_t}\, \partial_{j_1..j_s}(vw) - \varepsilon\, \partial_{i_1..i_t}(vw)\, \partial_{j_1..j_s} u)$$

modulo terms depending upon u,v or w themselves.

For $s > 1$ the term of maximum bidifferential type in u,w of ∂C has the type (t,s-1) and can be written :

$$s\, C^{i_1..i_t,j_1..j_s}(\partial_{i_1..i_t} u\, \partial_{j_1} v\, \partial_{j_2..j_s} w - \varepsilon\, \partial_{i_1..i_t} w\, \partial_{j_1} v\, \partial_{j_2..j_s} u)$$

This term vanishs iff $C^{i_1..i_t,j_1..j_s} = 0$

For s = 1 the term of maximum bidifferential type in u,w of $\tilde{\partial} C$ has the type (t-1,1) and can be written :

$$- t\, C^{i_1..i_t,j} (\partial_{i_2..i_t} u\, \partial_{i_1} v\, \partial_j w - \varepsilon\, \partial_{i_2..i_t} w\, \partial_{i_1} v\, \partial_j w)$$

The lemma is proved.

b) We can now prove the following

Theorem - Each twisted product on a symplectic manifold (W,F) *is equivalent to a Vey twisted product.*

Proof. We can suppose by equivalence that $C_2 = Q^2/2!$. If Q^3 corresponds to Q^2 by (18-4),(18-5) $C_3 - Q^3/3!$ is an Hochschild two-cocycle of type (1,1) and, changing the notation, we can suppose $C_3 = Q^3/3!$. Proceed by induction and suppose that, by equivalence, our $*_\nu$-product is now a Vey twisted product of order 2q-1 ($q \geqslant 2$). We have :

$$\tilde{\partial} C_{2q} = \tilde{E}_{2q}$$

and it is easy to deduce from the induction assumption that $\tilde{E}_{2q}$ has maximum bidifferential type (2q,2q-1) in u,w.
Suppose that C_{2q} has a maximum bidifferential type (t',s) ; it follows from the lemma that it is impossible that t' > 2q and s > 1. We see that either C_{2q} has maximum bidifferential type (2q,2q) or maximum bidifferential type (t',1).

We consider this last case ; $\tilde{\partial} C_{2q}$ has a maximum bidifferential type which is at most (t'-1,s) and we have necessarily 2q < t'. On the domain U of a natural chart we set :

$$C_{2q|U} = B_U + C_U$$

where B_U has exactly the type (t',1) (with $t' \geqslant 2q+1$) and C_U the maximum order (t'-1). We have :

$$\partial C_{2q+1} = E_q$$

where E_q has the maximum order (2q+1). If C_{2q+1} has an order > (2q+1) it follows from calculus of S. Gutt concerning the Chevalley two-cocycles that the main part of C_{2q+1} is the main part of a Chevalley two-cocycle and so of ∂A, where A is a differential operator. If such is the case, transform the $*_\nu$-product by means of $T_\nu = Id - \nu^{2q} A$, which preserves C_r for $r \leqslant 2q-1$ and changes C_{2q} and C_{2q+1} so that we can suppose that C_{2q+1} has an order $\leqslant 2q+1$ and so that $\tilde{\partial} C_{2q+1}$ has an order $\leqslant 2q+1 \leqslant t'$ in u,w. The relation :

$$\tilde{\partial} C_{2q+1} = \tilde{E}_{2q+1} \qquad (21\text{-}1)$$

can be written :

$$\tilde{C}_{2q+1} = \tilde{F}_{2q+1} + M' + M$$

where we have set :

$$\tilde{F}_{2q+1}(u,v,w) = \sum_{r+s=2q+1} (1/r!s!)(Q^r(Q^s(u,v),w)$$

$$- Q^r(u,Q^s(v,w)) \qquad (r,s \geqslant 2)$$

$$M'(u,v,w) = P(C_U(u,v),w) + C_U(P(u,v),w) - P(u,C_U(v,w))$$

$$- C_U(u,P(v,w))$$

$$M(u,v,w) = P(B_U(u,v),w) + B_U(P(u,v),w) - P(u,B_U(v,w)$$

$$- B_U(u,P(v,w))$$

$\tilde{F}_{2q+1} + M'$ has an order $\leqslant t'$ in u,w and it is easy to verify that if B_U is $\neq 0$, M has effectively order $(t' + 1)$ in these arguments. It follows from (21-1) that $B_U = 0$. We see that C_{2q} has necessarily maximum bidifferential type $(2q,2q)$. Similarly C_{2q+1} has maximum bidifferential type $(2q+1,2q+1)$.

Introduce on U a flat symplectic connection Γ which defines the $P^r = P^r_\Gamma$ according to (16-2). We have on U :

$$\tilde{\partial}(P^{2q}/(2q)!) = \tilde{E}_{2q}(P^r/r!) \qquad (r \leqslant 2q-1)$$

We have also :

$$\tilde{\partial} C_{2q} = \tilde{E}_{2q}(Q^r/r!)$$

It follows by difference that $\tilde{\partial}(C_{2q} - P^{2q}/(2q)!)$ has a maximum type in u,w inferior to $(2q,2q-1)$, so that $C_{2q} - P^{2q}/(2q)!$ has a maximum bidifferential type $(2q,r)$ with $r < 2q$. Therefore $C_{2q} = Q^{2q}/(2q)!$ and similarly $C_{2q+1} = Q^{2q+1}/(2q+1)$; our twisted product defines now a Vey twisted product of order $(2q+1)$. Our theorem is proved.

c) We can deduce from the theorem the following

Proposition - Each symplectic manifold (W,F) *such that* $b_3(W) = 0$ *admits strong Vey twisted products.*

Consider a $\star_\nu$-product of order 3 of the form :

$$uv + \nu P(u,v) + (\nu^2/2!)\, Q^2(u,v) + (\nu^3/3!) Q^3(u,v) \qquad (21\text{-}2)$$

where $Q^3 \in \beta$. By means of the induction argument of §20, it is possible to deduce from (21-2) a twisted product and so a Vey product. This Vey product is a strong Vey twisted product, according to the proposition of b.

We deduce from this theorem by antisymmetrization of this Vey twisted product.

Corollary - Each symplectic manifold (W,F) *such that* $b_3(W) = 0$ *admits strong Vey Lie algebras.*

This statement has been proved by J. Vey in a direct and completely different way.

22 - The case where $b_2(W) = 0$.

For a symplectic manifold (W,F) such that $b_2(W) = 0$, F is exact and $H^2(N;N)$ admits only β as generator.

a) We shall prove :

Theorem - For a symplectic manifold (W,F) *such that* $b_2(W) = 0$, *all existing twisted products are mutually equivalent to a strong Vey twisted product.*

Consider on W two twisted products :

$$u \star_\nu v = u.v + \nu P(u,v) + \sum_{r=2}^{\infty} \nu^r C_r(u,v) \qquad (22\text{-}1)$$

and

$$u \star'_\nu v = u.v + \nu P(u,v) + \sum_{r=2}^{\infty} \nu^r C'_r(u,v)$$

We use for the transformations the formulas (11-10) and (11-12). We proceed by induction and suppose that, by transformation of (22-1), we have $C'_r = C_r$ for $r \leqslant 2q-1$ (with $q \geqslant 1$); the even Hochschild two-cocycle $C'_{2q} - C_{2q}$ is exact, so that :

$$C'_{2q} = C_{2q} + \tilde{\partial} A$$

where A is a differential operator. Transforming (22-1) by $T_\nu = Id + \nu^{2q} A$ we may suppose now :

$$C'_r = C_r \qquad (r = 1, \ldots, 2q)$$

If such is the case :

$$C'_{2q+1} - C_{2q+1} = T$$

where T is an antisymmetric 2-tensor image of a closed 2-form of W by the map μ^{-1}. All closed 2-form of W being exact, T is an exact Chevalley two-cocycle and there is a vector Z such that :

$$T = \partial \mathcal{L}(Z)$$

Consider the new twisted product, denoted by $*''_\nu$, deduced from (22-1) by means of

$$T_\nu = Id + \nu^{2q} \mathcal{L}(Z)$$

We have, with evident notations, $C''_r = C_r = C'_r$ for $r = 1,..,2q$. Moreover :

$$C''_{2q+1} - C_{2q+1} = - \tilde{G}_{2q+1}$$

with :

$$\tilde{G}_{2q+1}(u,v) = \mathcal{L}(Z)\ P(u,v) - P(\mathcal{L}(Z)u,v) - P(u,\mathcal{L}(Z)v)$$

that is :

$$\tilde{G}_{2q+1} = - \partial \mathcal{L}(Z)$$

We obtain :

$$C''_{2q+1} = C_{2q+1} + \partial\ \mathcal{L}(Z)$$

and we see that $C''_{2q+1} = C'_{2q+1}$. We have proved by induction that all $*_\nu$-products are equivalent.
It follows from Theorem 1 of §21 that all these products are equivalent to a Vey twisted product such that, according to the formula (18-5) :

$$Q^3 = S^3_\Gamma + 3\ \partial\ H$$

for a suitable symplectic connection. We have thus $Q^3 \in \beta$ and the Vey twisted product is a strong Vey twisted product.

b) Up to now, we have considered only equivalence operators which are defined by differential operators *which are null on the constants*. For the study of the formal Lie algebras, we are led to introduce equivalence operators *which doe not satisfy this condition*.

Consider formal Lie algebras (12-1), let :

$$[u,v]_\lambda = P(u,v) + \sum_{r=1}^{\infty} \lambda^r \Gamma_r(u,v) \tag{22-2}$$

where the Γ_r $(r \geqslant 1)$ are differential two-cochains of (N,P) which are *null on the constants.* If such two Lie algebras are deduced one out of the other by an equivalence operator *which is not null on the constants,* we say that these Lie algebras are *weakly equivalent.*

The two-form F being exact, the manifold W is non compact. The following lemma is useful in this case :

Lemma - If $D = \mathcal{L}(Z) + k(Z)$, *where* $Z \in L^c$, *is a derivation of the Poisson Lie algebra* (N,P), *we have for an arbitrary symplectic connection* Γ :

$$D\, S^3_\Gamma(u,v) - S^3_\Gamma(Du,v) - S^3_\Gamma(u,Dv) = 2\,k(Z)\, S^3_\Gamma(u,v) \tag{22-3}$$

$$+ (\partial\, A_{T,\Gamma})(u,v)$$

where $A_{T,\Gamma}$ *is defined by* (17-4) *with* $T = \mathcal{L}(Z)\,\Gamma$.

In fact, we see easily that :

$$\mathcal{L}(Z)\, X_u = k(Z)\, X_u + X_{\mathcal{L}(Z)u} \tag{22-4}$$

The left side of (22-3) is equal to :

$$\mathcal{L}(Z)\, S^3_\Gamma(u,v) - S^3_\Gamma(\mathcal{L}(Z)u,v) - S^3_\Gamma(u,\mathcal{L}(Z)v) - k(Z)\, S^3_\Gamma(u,v)$$

let, according to (22-4) and the definition of $A_{T,\Gamma}$:

$$2\,k(Z)\, S^3_\Gamma(u,v) + (\partial A_{T,\Gamma})(u,v)$$

We shall prove [9] :

Proposotion (S. Gutt) - If $b_2(W) = 0$, *all the formal Lie algebras* (22-2) *such that* $\Gamma_1 = Q^3/3!$ *are weakly equivalent. In particular, all the Vey Lie algebras are weakly equivalent and strong.*

For the transformations, we use the formulas (12-8) and (12-9). Consider two formal Lie algebras (22-2) and :

$$[u,v]'_\lambda = P(u,v) + \sum_{r=1}^{\infty} \lambda^r \Gamma'_r(u,v) \tag{22-5}$$

satisfying the assumptions of the proposition; we have $\Gamma_1 = Q^3/3!$ and $\Gamma'_1 = Q'^3/3!$ where :

$$Q^3 = S^3_\Gamma + 3\,\partial H \qquad\qquad Q'^3 = S^3_{\Gamma'} + 3\,\partial H'$$

Γ and Γ' being suitable symplectic connections. The Chevalley two-cocycle :

$$Q'^3 - Q^3 = S^3_{\Gamma'} - S^3_{\Gamma} + 3\,\partial(H' - H)$$

is exact. We may suppose by transformation using operators which are null on the constants :

$$Q'^3 = Q^3 = S^3_{\Gamma}$$

for the symplectic connection Γ.

Proceed by induction. By transformation of (22-2), we may suppose that $\Gamma' = \Gamma_r$ for $r \leqslant q$ $(q \geqslant 1)$. For the Chevalley two-cocycle $(\Gamma'_{q+1} - \Gamma_{q+1})$ we have :

$$\Gamma'_{q+1} - \Gamma_{q+1} = h\, S^3_{\Gamma} + \partial B \qquad (h \in \mathbb{R}) \qquad (22\text{-}6)$$

where B is a differential operator null on the constants.

Consider the new Lie algebra corresponding to a bracket denoted by $[\ ,\ ]''_{\lambda}$, deduced from (22-2) by means of the transformation :

$$T_{\lambda} = \mathrm{Id} + \lambda^q D \qquad (22\text{-}7)$$

Since $\partial D = 0$, we have, with evident notations, $\Gamma''_r = \Gamma_r = \Gamma'_r$ for $r = 1,\ldots,q$. Moreover :

$$\Gamma''_{q+1} - \Gamma_{q+1} = - G_{q+1}$$

where :

$$3!\; G_{q+1}(u,v) = D\, S^3_{\Gamma}(u,v) - S^3_{\Gamma}(Du,v) - S^3_{\Gamma}(u,Dv)$$

It follows from the lemma :

$$3!\; G_{q+1} = 2\, k(Z)\; S^3_{\Gamma} + A_{T,\Gamma}$$

choose $Z \in L^c$ so that $k(Z) = -3h$. We obtain :

$$\Gamma''_{q+1} - \Gamma'_{q+1} = -\,\partial\left(\frac{1}{6} A_{T,\Gamma} + B\right)$$

A new ordinary transformation gives $\Gamma''_{q+1} = \Gamma'_{q+1}$.

(22-7) is the unique transformation, in this argument, which is not null on the constants. We have proved by induction that all the considered Lie algebras are weakly equivalent.

23 - Weak twisted product.

Consider the associative deformation of (N,.) of the form :

$$u *_\nu v = u.v + \nu P(u,v) + \sum_{r=2}^{\infty} \nu^r C_r(u,v) \qquad (23\text{-}1)$$

where the differential two-cochains C_r' satisfy always the parity assumption, *the* C_{2r+1} *being null on the constants, but the* C_{2r} *not necessarily.* If such is the case, we say that (23-1) defines a *weak twisted product.* By antisymmetrization, (23-1) generates a formal Lie algebra :

$$[u,v]_\lambda = P(u,v) + \sum_{r=1}^{\infty} \lambda^r C_{2r+1}(u,v) \qquad (\lambda = \nu^2) \qquad (23\text{-}2)$$

with cochains which are null on the constants, that is of the type (14-3). We note that $\tilde{E}_{2t+1} = \tilde{\partial} C_{2t+1}$ is null on the constants for $t = 1,2,\ldots$

a) Study the weak twisted products. The following lemma is useful.

Lemma - Let C *be a differential Hochschild two-cochain such that* $\tilde{\partial}C = E$ *is null on the constants. If* C *is odd, it is null on the constants. If* C *is even, it has the following form :*

$$C(u,v) = \bar{C}(u,v) + a\,u\,v \qquad a \in N \qquad (23\text{-}3)$$

where $\bar{C}$ *is null on the constants. Note that the coboundary of the term* $a\,u\,v$ *is null.*

In fact :

$$\tilde{\partial}C(u,v,w) = u\,C(v,w) - C(uv,w) + C(u,vw) - C(u,v)w$$

If $\tilde{\partial}C = E$ is null on the constants, we have (for $v,w = 1$) $C(u,1) = au$, where $a = C(1,1)$.

If C *is odd,* $C(1,1) = 0$ and we obtain $C(u,1) = 0$; therefore C is null on the constants.

If C *is even,* we can write locally on the domain of a chart $\{x^i\}$:

$$C(u,v) = \bar{C}(u,v) + \sum C^{i_1..i_t}(u\,\partial_{i_1..i_t} v + v\,\partial_{i_1..i_t} u) + a\,u\,v$$

where $\bar{C}$ is null on the constants and $a = C(1,1)$. It follows that :

$$C(u,1) = \sum C^{i_1..i_t}\,\partial_{i_1..i_t} u + au$$

and the first term of the right side vanishs necessarily ; we obtain

(23-3).

b) For the weak twisted product (23-1), consider $\tilde{E}_2$ which is null on the constants. It follows that :

$$C_2(u,v) = \bar{C}_2(u,v) + a_2\ uv \qquad (a_2 \in N)$$

where $\bar{C}_2$ is null on the constants. We have :

$$\tilde{E}_3(u,v,w) = C_2(P(u,v),w) + P(C_2(u,v),w) - C_2(u,P(v,w))$$
$$- P(u,C_2(v,w))$$

Since $\tilde{E}_3$ is null on the constants, it is the same for :

$$a_2\ P(u,v).w + P(a_2uv,w) - a_2\ u\ P(v,w) - P(u,a_2vw)$$

If $u = 1$, we see that necessarily $P(a_2v,w) = a_2\ P(v,w)$ for all $v,w \in N$; this relation implies $a_2 = \text{const.} = k_2 \in \mathbb{R}$.

Suppose that we have :

$$C_{2r}(u,v) = \bar{C}_{2r}(u,v) + k_{2r}uv \qquad (k_{2r} \in \mathbb{R})$$

where $\bar{C}_{2r}$ is null on the constants, for $r = 1,\dots,t-1$ $(t \geqslant 2)$. It is easy to see on the expression of $\tilde{E}_{2t}$ that this three-cochain is null on the constants. It follows from $\tilde{\partial} C_{2t} = \tilde{E}_{2t}$ that C_{2t} has the form :

$$C_{2t}(u,v) = \bar{C}_{2t}(u,v) + a_{2t}uv \qquad (a_{2t} \in \mathbb{R})$$

Similarly $\tilde{E}_{2t+1}$ being null on the constants, we see easily that it is the same for the three-cochain

$$C_{2t}(P(u,v),w) + P(C_{2t}(u,v),w) - C_{2t}(u,P(v,w)) - P(u,C_{2t}(v,w))$$

and the same argument gives $a_{2t} = \text{const.} = k_{2t} \in \mathbb{R}$.

We have proved :

Proposition - The even two-cochains of a weak twisted product (23-1) *have the following form :*

$$C_2(u,v) = \bar{C}_2(u,v) + k_{2r}\ uv \qquad (k_{2r} \in \mathbb{R}) \qquad (23\text{-}4)$$

where the $\bar{C}_{2r}$ *'s are null on the constants.*

c) We shall prove :

Proposition - Each weak twisted product on the symplectic manifold

(W,F) *is weakly equivalent to a twisted product.*

Proof : Consider a weak twisted product (23-1) and suppose that we have obtained by transformation $k_{2r} = 0$ for $0 \leqslant 2r \leqslant 2(q-1)$. We have :

$$C_{2q}(u,v) = \bar{C}_{2q}(u,v) + k_{2q}\, uv$$

Introduce the transformation :

$$T_\nu = \mathrm{Id} - \nu^{2q} k_{2q} \tag{23-5}$$

We deduce from the formal Lie algebra (23-2) by means of (23-5) a new formal Lie algebra for which the two cochains are always null on the constants. It follows that we deduce from (23-1) by means of (23-5) a new weak twisted product denoted by $*'_\nu$. We have, with evident notations, $C'_r = C_r$ for $r = 1,\ldots,2q-1$. Moreover :

$$C'_{2q} = C_{2q} + \tilde{\partial} T_{2q}$$

that is :

$$C'_{2q}(u,v) = C_{2q}(u,v) - k_2 uv = \bar{C}_{2q}(u,v)$$

The proposition is proved by induction.

d) The uniqueness theorem of §14 can be adapted to the weak products. We will use the following ([9]) :

Lemma (S. Gutt) - There does not exist an even differential Hochschild two-cocycle M *such that :*

$$S\, \tilde{\partial}\, Q^3(u,v,w) = S\,\{M(P(u,v),w + P(M(u,v),w) \tag{23-6}$$

$$- M(u,P(v,w)) - P(u,M(v,w))\}$$

where S *is the summation after cyclic permutation on* u,v,w.

The proof consists in a study of the maximum bidifferential type of M. It is possible to prove that the formula (23-6) implies that this type is necessarily exactly (2,2). But there does not exist an even two-cocycle admitting such a type.

We have :

Uniqueness theorem - If two weak twisted products generate the same formal Lie algebra (23-2) *and if the term in* ν^3 *has* $Q^3/3!$ *as coefficient, these products coincide.*

Proof : Consider (23-1) with :

$$C_{2r}(u,v) = \bar{C}_{2r}(u,v) + k_{2r}uv$$

and another weak twisted product :

$$u *'_\nu v = \sum_{r=0}^{\infty} \nu^r C'_r(u,v) = uv + \nu P(u,v) + \sum_{r=2}^{\infty} \nu^r C'_r(u,v)$$

with

$$C'_{2r}(u,v) = \bar{C}'_{2r}(u,v) + k'_{2r}\, uv$$

We have supposed $C'_3 = C_3 = Q^3/3!$. We have :

$$\tilde{\partial} C_2 = \tilde{\partial} C'_2 = \tilde{E}_2$$

and it follows from the assumption that there exists Q^2 such that :

$$\frac{1}{2!}\ \tilde{\partial}\, Q^2 = \tilde{E}_2 \qquad\qquad \frac{1}{3!}\ \tilde{\partial}\, Q^3 = \tilde{E}_3$$

We deduce from these relations :

$$\bar{C}'_2 = \bar{C}_2 = Q^2/2!$$

For $2r = 4$, we obtain by a calculus similar to the calculus of § 14,c :

$$\tilde{E}'_4 - \tilde{E}_4 = -\,h_2\ \tilde{\partial}\ C_2$$

and if we put $D_4 = C'_4 - C_4$, it follows that :

$$D_4 = -\,h_2\ C_2 + M_4 \qquad (23\text{-}7)$$

where M_4 is *an even Hochschild two-cocycle*. We denote by $\bar{M}_4$ the part of M_4 which is null on the constants. We have :

$$\tilde{E}'_5 - \tilde{E}_5 = D_4(P(u,v),w) + P(D_4(u,v),w) - D_4(u,P(v,w)) - P(u,D_4(v,w)) \qquad (23\text{-}8)$$

But :

$$C_2(P(u,v),w + P(C_2(u,v),w) - C_2(u,P(v,w)) - P(u,C_2(v,w)) = \tilde{E}_3(u,v,w)$$

and (23-8) can be written :

$$\tilde{E}'_5 - \tilde{E}_5 = \overline{M}_4(P(u,v),w) + P(\overline{M}_4(u,v),w) - \overline{M}_4(u,P(v,w)) \quad (23\text{-}9)$$

$$- P(u,\overline{M}_4(v,w)) - 2h_2 \tilde{\partial} C_3(u,v,w)$$

e) If the two Lie algebras generated by $\star_\nu$ and $\star'_\nu$ coincide, we have $C'_5 = C_5$ and $\tilde{E}'_5 = \tilde{E}_5$. It follows from (23-9) and from the previous lemma that we have necessarily $h_2 = 0$. Moreover the lemma of §14,d gives $\overline{M}_4 = 0$. We obtain :

$$C'_2 = C_2 \qquad\qquad \overline{C}'_4 = \overline{C}_4$$

and we can continue the proof by induction. The two weak twisted products coincide.

f) We say that a weak twisted product is *a weak Vey twisted product if we have :*

$$C_{2r}(u,v) = Q^{2r}(u,v)/(2r)! + k_{2r}uv \quad (23\text{-}10)$$

$$C_{2r+1}(u,v) = Q^{2r+1}(u,v)/(2r+1)!$$

The argument of §21,a and b, shows that we have the following :

Theorem - Each weak twisted product of (W,F) *is equivalent to a weak Vey twisted product.*

24 - Lie algebras generated by a weak twisted product and Vey Lie algebras.

a) Consider, on an arbitrary symplectic manifold, a formal Lie algebra :

$$[u,v]_{\nu^2} = P(u,v) + \sum_{r=1} \nu^{2r} C_{2r+1}(u,v) \quad (24\text{-}1)$$

where the C_{2r+1} are null on the constants. *Suppose that* $C_3 = Q^3/3!$.

Let U be a contractible domain ($b_2(U)=0$) of W and introduce the restriction to U of (24-1). It follows from the proposition of §22,b that this Lie algebra on $(U,F|_U)$ is weakly equivalent to a formal Lie algebra on U (Moyal Lie algebra) generated by the Moyal product on U defined by means of a flat symplectic connection Γ_U. It is easy to verify that each transformation (22-7) transforms a twisted product in a weak twisted product. Our Lie algebra $(C_{2r+1}|_U)$ $(r = 0,1,\ldots)$ on U is generated by a weak product $(C_{2r(U)}, C_{2r+1}|_U)$ on U. We set :

$$C_{2r(U)}(u,v) = \overline{C}_{2r(U)}(u,v) + k_{2r(U)}uv \qquad k_{2r(U)} \in \mathbb{R}$$

where $\overline{C}_{2r(U)}$ is null on the constants. Consider two contractible domains U,V of W with $U \cap V \neq \emptyset$. We deduce from the uniqueness theorem of §23,d that we have on $U \cap V$:

$$\overline{C}_{2r(U)}(u,v) + k_{2r(U)}uv = \overline{C}_{2r(V)}(u,v) + k_{2r(V)}uv$$

We see that $k_{2r(U)} = k_{2r(V)} = k_{2r}$ and that $\overline{C}_{2r(V)} = \overline{C}_{2r(U)}$ on $U \cap V$. Therefore there exist bidifferential operators $\overline{C}_{2r}$ null on the constants such that $\overline{C}_{2r(U)} = \overline{C}_{2r}|_U$. The k_{2r} and the $\overline{C}_{2r}$, C_{2r+1} define on W a weak twisted product that generates the given Lie algebra.

The weak twisted product being equivalent to a weak Vey twisted product, the formal Lie algebra (24-1) where $C_3 = Q^3/3!$ is equivalent to a Vey Lie algebra. It is the same for a Lie algebra (24-1) such that :

$$C_3 = Q^3/3! + \partial A$$

where A is a differential operator null on the constants. We have :

Theorem - 1) A formal Lie algebra (24-1) *is equivalent to a Vey Lie algebra if and only if the corresponding Chevalley two-cocycle* C_3 *is equivalent to a two-cocycle of the form* $Q^3/3!$

2) A formal Lie algebra (24-1) *is generated by a weak twisted product if and only if it is equivalent to a Vey Lie algebra.*

It follows from the argument of §21,b.

Corollary - A Vey Lie algebra is generated by a weak Vey twisted product.

b) Consider on (W,F) a weak twisted product (23-1) with :

$$C_{2r}(u,v) = \overline{C}_{2r}(u,v) + k_{2r}\,uv \qquad C_{2r+1}(u,v) = \overline{C}_{2r+1}(u,v)$$

where the $\overline{C}_r$ are null on the constants. Such a weak twisted product can be written :

$$u *_\nu v = k_{\nu^2}\,uv + \nu P(u,v) + \sum_{r=2}^{\infty} \nu^r\,\overline{C}_r(u,v)$$

where

$$k_{\nu^2} = 1 + \sum_{r=1}^{\infty} \nu^{2r}\,k_{2r}$$

Introduce the formal series :

$$h_{\nu^2} = + \sum_{r=1} \nu^{2r} h_{2r} \qquad (h_{2r} \in \mathbb{R})$$

If we transform (23-1) by means of the weak equivalence $T_{\nu^2}u = h_{\nu^2}u$, we obtain a new twisted product $*'_\nu$ given by :

$$u *'_\nu v = h_{\nu^2}(u *_\nu v)$$

We say that $*'_\nu$ is deduced from $*_\nu$ by the product by the constants h_{ν^2}. If $h_{\nu^2} k_{\nu^2} = 1$, we obtain a new twisted product. We have :

Proposition - A weak twisted product $*_\nu$ *being given, there is a unique twisted product deduced from* $*_\nu$ *by product by a constant.*

We have similarly for a Vey Lie algebra :

Proposition - A Vey Lie algebra being given, there is a unique Vey Lie algebra deduced by product by a constant which is generated by a Vey twisted product.

V - AUTOMORPHISMS OF A TWISTED PRODUCT.

The study of the problems of invariance of Quantum Mechanics is strictly connected with the study of the group $Aut(*_\nu)$ of the automorphisms of a twisted product (see §15). This study uses the study of a class of automorphisms of the corresponding Lie algebra.

25 - Derivations null on the constants for a formal Lie algebra. Application.

Let (W,F) be a symplectic manifold *admitting a formal Lie algebra null on the constants :*

$$[u,v]_\lambda = P(u,v) + \sum_{r=1}^{\infty} \lambda^r C_{2r+1}(u,v) \qquad (25\text{-}1)$$

a) Consider a symplectic vector field $Z \in L$; if U is a contractible domain of W, there is $w_U \in C^\infty(U;R)$, defined up to an additive constant, such that $Z_{|U} = \mu^{-1}(dw_U)$; C_{2r+1} being null on the constants we define on N a differential operator C^Z_{2r+1} by :

$$C^Z_{2r+1}(u)_{|U} = C_{2r+1}(w_U, u_{|U}) \qquad (u \in N)$$

Let $E(L;\lambda)$ (resp. $E(L^*;\lambda)$) be the space of formal functions in λ with coefficients in L (resp. L^*). These spaces admit Lie algebra structures. We can prove by induction [11] :

Proposition - The Lie algebra of the derivations of a formal Lie algebra (25-1) which are null on the constants is isomorphic to the Lie algebra $E(L;\lambda)$ *by the isomorphism* $\rho : Z_\lambda = \Sigma\, \lambda^s Z_s \to D_\lambda$, *where :*

$$D_\lambda = \sum_{r,s=0}^{\infty} \lambda^{r+s}\, C_{2r+1}^{Z_s} \qquad (25\text{-}2)$$

The Lie algebra of the inner derivations is isomorphic to $E(L^*;\lambda)$.

b) An automorphism of the space $E(N;\lambda)$

$$A_\lambda = A_o + \sum_{s=1}^{\infty} \lambda^s A_s \qquad (25\text{-}3)$$

where A_o is an automorphism of the space N and the A_s $(s \geqslant 1)$ are endomorphisms of N, is an automorphism of the Lie algebra (25-1) iff :

$$A_\lambda [u,v]_\lambda = [A_\lambda u, A_\lambda v]_\lambda \qquad (25\text{-}4)$$

A *symplectomorphism* of (W,F) is a diffeomorphism of W preserving F. Let *Symp*(W,F) be the group of all the symplectomorphisms of (W,F) and $Symp_c(W,F)$ its component connected to the identity by differentiable arcs. If $\sigma \in Symp_c(W,F)$, we denote by $\sigma(t)$ $(0 \leqslant t \leqslant 1)$ a symplectic isotopy such that $\sigma(0)$: Id., $\sigma(1) = \sigma$. For each t, $\dot{\sigma}(t) = (d\sigma(t)/dt)\, \sigma(t)^{-1}$ defines a symplectic vector field; according to (25-2), $\dot{\sigma}(t)$ determines for each t a derivation :

$$D_\lambda(t) = \sum_{r=0}^{\infty} \lambda^r\, C_{2r+1}^{\dot{\sigma}(t)}$$

Consider the differential equation :

$$dA_\lambda(t)/dt = D_\lambda(t)\, A_\lambda(t)$$

where $A_\lambda(t)$ is an automorphism of the space $E(N;\lambda)$ and suppose $A_\lambda(0) = Id$. Studying the solution of this equation, we can construct a one-parameter differentiable family of automorphisms of (25-1) of the form :

$$A_\lambda(t) = (Id + \sum_{s=1}^{\infty} \lambda^s B_s(t))\, \sigma^*(t)$$

We denote by K *the space of the differential operators on* N *which are null on the constants.* The $B_s(t) \in K$ **depend** differentiably upon t. It follows that [11] :

Proposition - For each element σ *of* $Symp_c(W,F)$, *there exists an automorphism of the formal Lie algebra (25-1) of the form :*

$$A_\lambda = (Id + \sum_{s=1}^{\infty} \lambda^s B_s)\,\sigma^* \qquad B_s \in K$$

26 - Derivations and properties of the automorphisms of a twisted product.

We suppose that (W,F) admits a twisted product (14-1) and so a corresponding Lie algebra (14-3) which can be transported to $E(N;\nu)$

$$[u,v]_{\nu^2} = P(u,v) + \sum_{r=1}^{\infty} \nu^{2r} C_{2r+1}(u,v) \qquad (26\text{-}1)$$

a) It follows from the proposition of §25,a that the Lie algebra of the derivations $D_\nu = \Sigma\, \nu^s D_s$ of (26-1) which are null on the constants, is isomorphic to the Lie algebra $E(L;\nu)$ by the isomorphism $Z_\nu = \Sigma\, \nu^s Z_s \to D_\nu$, with :

$$D_\nu = \sum_{r,s \geqslant 0} \nu^{2r+s} C^{Z_s}_{2r+1} \qquad (26\text{-}2)$$

Each derivation D_ν of the twisted product is a derivation of the Lie algebra (26-1) which is null on the constants, according to the proposition of §10,a ; D_ν necessarily has the form (26-2). Conversely it is clear that each derivation (26-2) of (26-1) is a derivation of the twisted product.

Theorem - The derivations of a twisted product and the derivations of the corresponding Lie algebra that are null on the constants coincide and are given by (26-2) . *If* $b_1(W) = 0$, *all the derivations of the twisted product are inner derivations.*

b) Study the automorphisms $A_\nu = A_o + \Sigma\, \nu^s A_s$ of the twisted product. For order 0, we have $A_o(uv) = A_o u \,.\, A_o v$. It follows classically that there is a diffeomorphism σ of W such that $A_o = \sigma^*$. We obtain for order 1 :

$$\sigma^* P(u,v) = P(\sigma^* u, \sigma^* v)$$

and σ is necessarily a *symplectomorphism.* We deduce from §10,a :

Proposition - Each automorphism A_ν *of a twisted product has the form :*

$$A_\nu = (Id + \sum_{s=1}^{\infty} \nu^s B_s)\,\sigma^* \qquad (B_s \in K) \qquad (26\text{-}3)$$

where σ *is a symplectomorphism.*

c) Consider for a moment the automorphism A_ν of *the Lie algebra* (26-1) which are *even with respect to* ν (or *even automorphisms*) and null on the constants; A_ν transforms $*_\nu$ in $*'_\nu$ satisfying

$$u *'_{-\nu} v = v *'_\nu u$$

$*'_\nu$ being a twisted product generating the same Lie algebra as $*_\nu$, we have $*'_\nu = *_\nu$.

Proposition - The group of the even automorphisms of a twisted product coincides with the group of the even automorphisms of the corresponding Lie algebra that are null on the constants.

27 - The group $\mathrm{Aut}_t(*_\nu)$.

Study the automorphisms of our $*_\nu$-product which have a trivial main part. According to the proposition of §26,b such an automorphism is given by :

$$A_\nu = \mathrm{Id.} + \sum_{s=1}^{\infty} \nu^s A_s \qquad (A_s \in K) \qquad (27\text{-}1)$$

a) *Suppose* $b_1(W) = 0$. Let A_ν be an automorphism of *the Lie algebra* (26-1) of the form (27-1). It follows from (12-9) that we have for $t = 1,2, \ldots$

$$A_t(u,v) = \sum_{2r+s=t} A_s\, C_{2r+1}(u,v) - \sum_{r+r'=t} P(A_r u, A_{r'} v) \qquad (27\text{-}2)$$

$$- \sum_{2r+s+s'=t} (A_s u, A_{s'} v)$$

where $r,r' \geqslant 1$, $s,s' \geqslant 0$. If $t=1$, we have $\partial A_1 = 0$ and $A_1 = \mathcal{L}(Z_o)$ where $Z_o \in L = L^*$. There is $b_o \in N$ such that $A_1 u = P(b_o,u)$. Associate with $a_o = \exp(-b_o/2) > 0$ the automorphisms :

$$T_\nu^{(1)} u = a_o *_\nu u *_\nu a_o^{(*)-1} \qquad A_\nu^{(1)} = T_\nu^{(1)} A_\nu = \sum \nu^s A_s^{(1)}$$

$A_\nu^{(1)}$ is an element of $\mathrm{Aut}_t(*_\nu)$ such that $A_1^{(1)} = 0$ and $A_2^{(1)} = 0$. There is $b_1 \in N$ such that $A_2^{(1)} = P(b_1,u)$. Introduce for $a_1 = -b_1/2$:

$$T_\nu^{(1)} u = (1 + \nu a_1) *_\nu u *_\nu (1 + \nu a_1)^{(*)-1}$$

$$A_\nu^{(2)} = T_\nu^{(2)} A_\nu^{(1)} = \sum \nu^s A_s^{(2)}$$

where $A_1^{(2)} = A_2^{(2)} = 0$. If we proceed by induction, we see that there is :

$$c_\nu = \left(\prod_{s=1}^{\infty}{}_{*} (1 + \nu^s a_s)\right) *_\nu a_o \in \hat{E}(N;\nu)$$

such that $A_\nu u = c_\nu^{(*)-1} *_\nu u *_\nu c_\nu$ and A_ν is necessarily *an inner automorphism* of $*_\nu$.

b) If $b_1(W) \neq 0$, we introduce the universal covering of W and we obtain :

Theorem - The group $Aut_t(*_\nu)$ *of a twisted product coincides with the group of the automorphisms of the corresponding Lie algebra which have the form* (27-1). *If* $b_1(W) = 0$, $Aut_t(*_\nu)$ *coincides with the group* $Aut_i(*_\nu)$ *of the inner automorphisms of the* $*_\nu$*-product.*

28 - The main theorem.

a) Consider now the automorphisms of the Lie algebra (26-1) which have the form (26-3). By an argument similar to the argument of §15,c, it is possible to prove that such an automorphism A_ν admits the decomposition :

$$A_\nu = T_\nu A_\nu^{(\ell)}$$

where $A_\nu^{(\ell)}$ is an even automorphism of the Lie algebra and where T_ν is an automorphism of $*_\nu$, element of $Aut_t(*_\nu)$. It follows from the proposition of §26,c, that $A_\nu^{(\ell)}$ is also an automorphism of $*_\nu$. We have [11] :

Main theorem - The group $Aut(*_\nu)$ *of a twisted product coincides with the group of the automorphisms of the corresponding Lie algebra that have the form :*

$$A_\nu = (Id + \sum_{s=1} \nu^s B_s)\, \sigma^* \qquad (B_s \in K)$$

where σ *is a symplectomorphism.*

The group $Aut(*_\nu)/Aut_t(*_\nu)$ is thus isomorphic to a subgroup G of the group *Symp*(W,F). It follows from §25,b that :

$$Symp_c(W,F) \subset G$$

It is possible to conjecture that G = Symp(W,F).

c) Consider in particular a Vey twisted product. We have (18-5) that is :

$$Q^3 = S_\Gamma^3 + M + 3\,\partial H$$

where Γ is a suitable symplectic connection.

The study of the automorphisms independent of ν of this twisted product shows that the symplectic connection Γ is necessarily invariant under the symplectomorphism of (26-3). We have [11] :

Proposition - The group of the automorphisms independent of ν *of a Vey twisted product is isomorphic to a subgroup of the group of the symplectomorphisms affine for the connection* Γ *defined by* Q^3. *This group is necessarily finite dimensional.*

In particular if (W,F,Γ) is a flat symplectic manifold, the group of the automorphisms independent of ν of the Moyal product is isomorphic to the group of the affine symplectic transformations of (W,F,Γ) .

REFERENCES

(1) Avez, A. and Lichnerowicz, A.: 1972, C.R. Acad. Sci. Paris, 275, A, pp. 113. Avez, A., Lichnerowicz, A., Diaz-Miranda, A.: 1974, J. Diff. Geom., 9, pp. 1-40.

(2) Bayen, F., Flato, M., Fronsdal, C., Lichnerowicz, A. and Sternheimer, D.: 1978, Ann. of Phys., 111, pp. 61-110 and pp. 111-151.

(3) Chevalley, C. and Eilenberg, S.: 1948, Trans. Amer. Math. Soc., 63, pp. 85-124.

(4) Flato, M., Lichnerowicz, A., Sternheimer, D.: 1975, Compos. Matem., 31, pp. 47-82.

(5) Gerstenhaber, M.: 1964, Ann. Math., 79, pp. 59-90.

(6) Godbillon, G.,: 1969, Géométrie différentielle et mécanique analytique. Hermann, Paris.

(7) De Groot, S.R.,: 1974, La transformation de Weyl et la fonction de Wigner; une forme alternative de la mécanique quantique. Presses Univ. de Montréal.

(8) Guillemin, V. and Sternberg, S.: 1966, Deformation theory of pseudogroup structures. Mem. Amer. Math. Soc. n°65.

(9) Gutt, S.: 1979, Lett. in Math. Phys., 3, pp. 297-310.

(10) Lichnerowicz, A.: 1977, Lett. in Math. Phys., 2, pp. 133-143.

(11) Lichnerowicz, A.: 1951, C.R. Acad. Sci. Paris, 233, pp. 723-726; 1978, 286, A, pp. 49-53; 1978, 287, A, pp. 1029-1033; Automorphisms of Formal Algebras; Essays in General Relativity, a volume in honor of Taub, A.H., Academ. Press 1980; 1980, Algèbres formelles associées par déformation à une variété symplectique, Ann. di Matem., 123, pp. 287-330.

(12) Lichnerowicz, A.: 1979, Lett. in Math. Phys., 3, pp. 495-502.

(13) Marle, C.: 1977, in Differential Geometry and Relativity, volume dedicated to Lichnerowicz, A., Cahen, M. and Flato, M. Eds, Reidel.

(14) Moyal, J.E.: 1949, Proc. Cambridge Philos. Soc., 45, pp. 49-124.

(15) Neroslavsky, O.M. and Vlassov, A.T., (to appear).
(16) Nijenhuis, A.: 1955, Indag. Math., 17, pp. 390.
(17) Kostant, B.: 1970, Quantization and unitary representations. Lect. Notes Math., 170, Springer.
(18) Schouten, J.A.: 1954, Conv. Int. Geom. Diff., Crémonèse, Roma, pp. 1-7.
(19) Segal, I.E.: 1974, Sympos. Mathem., 14, pp. 99-117. Academic Press and references quoted here.
(20) Souriau, J.M.: 1970, Structure des systèmes dynamiques, Dunod, Paris.
(21) Weyl, H.: 1932, The theory of Groups and Quantum Mechanics, Dover, New York.
(22) Vey, J.: 1975, Comm. Math. Helv., 50, pp. 421-454.
(23) Wigner, E.P.: 1932, Phys. Rev., 40, pp. 749.
(24) Lichnerowicz, A.: 1973, C.R. Acad. Sci. Paris, 277, A, pp. 215-219; 1975, 280, A, pp. 37-40; 1975, 280, A, pp. 1217-1220; 1977, J. of Diff. Geom., 12, pp. 253-300 and Dirac, P.A.: 1959, Canad. J. Math., 2, pp. 129-148.
(25) Abraham, R. and Marsden, J.E.: 1978, Foundations of Mechanics, Second Edition, Benjamin, New York.

QUANTIZATION AS MAPPING AND AS DEFORMATION [1/]

J. Niederle

International Centre for Theoretical Physics, Trieste, Italy; permanent address: Institute of Physics, Czechosl.Academy of Sciences, 180 40 Prague 8, Czechoslovakia

Quantizations are considered as mapping from the algebra of classical polynomial observables into the algebra of quantum polynomial observables satisfying a minimal set of natural requirements. Physically important subclasses of quantizations are specified by further symmetry assumptions.

In the second part quantizations defined as deformations of the classical algebra of polynomial observables are reviewed. The relation of these deformations to the quantization mappings is clarified.

1. INTRODUCTION

In this lecture some results obtained by Tolar and myself [1 , 2] will be reviewed which concern classes of quantization from classical to quantum observables.

First, we shall discuss classes of quantizations as classes of maps from classical to quantum observables. More precisely we shall determine classes of all quantizations of polynomial observables for a system of one degree of freedom. Let us note, however, that the derived results can be generalized to several degrees of freedom in a straightforward way and that for an irreducible representation of Q and P by self-adjoint operators $\hat{Q}$ and $\hat{P}$ in $L^2(R)$ any self-adjoint operator in $L^2(R)$ can be approximated (in the sense of a strong resolvent convergence) by the self-adjoint closures of polynomials in $\hat{Q}$ and $\hat{P}$ [3]. [2]

Secondly, we shall summarize some results obtained in the approach where quantization is defined as deformation of the algebra of classical observables into the algebra of quantum observables. Finally, a brief comparison of both quantization approaches will be given.

2. QUANTIZATION AS MAPPING

Let us first specify our notation. We shall denote by $S[q,p]$ the algebra of classical polynomial observables in commuting variables q,p and by $W[Q,P]$ the algebra of quantum polynomial observables in Q,P, i.e. the complex algebra of polynomial observables in Q,P which fulfil the Heisenberg relation

$$QP - PQ = i\hbar \mathbb{1} \quad ,$$

equipped with the unique involution under which Q,P are self-adjoint. [3]

As far as the bases of these algebras are concerned we use the basis in $S[q,p]$ formed by all monomials $q^j p^k$ which span a $(j+k+1)$-dimensional vector space. In $W[Q,P]$ the basis is formed (in accordance with the Poincaré-Birkhoff-Witt theorem [5]) by standard monomials $Q^j P^k$. We prefer, however, the basis formed by symmetrized monomials according to the Weyl-McCoy rule, i.e. the monomials

$$\left\langle Q^j P^k \right\rangle_W = \frac{1}{2^j} \sum_{a=0}^{j} \binom{j}{a} Q^{j-a} P^k Q^a \quad .$$

They span a complex $(j+k+1)$-dimensional vector space. Let us now define a quantization.

<u>Definition:</u> A quantization is said to be a map

$$\Omega \; : \; S[q,p] \longrightarrow W[Q,P]$$

satisfying the following conditions:

(I) Ω is linear,
(II) Ω fulfils the classical limit, i.e.

$$\lim_{\hbar \to 0} \Omega(x)\Big|_{\substack{Q=q\\P=p}} = x \quad , \quad x \in S[q,p] \; ,$$

(III) Ω preserves the physical dimension of any $x \in S[q,p]$. In this connection let us remark that any mechanical quantity x has the dimension

$$[x] = M^{\alpha} L^{\beta} T^{\gamma} \sim (\alpha, \beta, \gamma)$$

where M,L,T is a unit of mass, length and time respectively. Consequently, property (III) means that

$$[q^j p^k] = [\langle Q^j P^k \rangle_W] \sim (k, j+k, -k) .$$

(IV) $\Omega(x)$ is self-adjoint for each $x \in S[q,p]$.

In many classes the quantization as map Ω with properties (I) - (IV) appears to be too general. One, therefore, imposes some further restrictions on Ω. Namely,

(V) the only dimensional parameter in the theory is $\hbar$

or

(V′) besides $\hbar$ there is also another dimensional parameter in the theory, say μ, with the dimension $[\mu] \sim (1,0,-1)$. 4/

(VI) Ω is translationally invariant

and finally

(VII) Ω is symplectically invariant.

The last two properties express the G-equivariance of quantization Ω with respect to group $G = T_2$ or $Sp(2,R)$ respectively, i.e. the commutativity of the diagram

$$\begin{array}{cccc} & S[q,p] & \xrightarrow{\Omega} & W[Q,P] & \\ T^s(g) & \downarrow & & \downarrow & T^w(g) \\ & S[q,p] & \xrightarrow{\Omega} & W[Q,P] & \end{array}$$

where $g \mapsto T^s(g)$ and $g \mapsto T^w(g)$ are representations of group G. Now, we are well prepared to specify all classes of quantizations Ω satisfying properties (I) - (IV) and some additional ones from (V) - (VII).

<u>Theorem:</u> Ω is a bijective linear map from $S[q,p]$

real subspace of $W[Q,P]$ spanned by the Weyl basis and its most general form is given by

$$\Omega(q^j p^k) = \langle Q^j P^k \rangle_W + \sum_{a=1}^{\min(j,k)} A_a^{j,k} \hbar^a \langle Q^{j-a} P^{k-a} \rangle_W , \qquad A_a^{j,k} \in \mathbb{R},$$

provided (V) holds;

$$\Omega(q^j p^k) = \langle Q^j P^k \rangle_W + \sum_{a=-k+1}^{j} \sum_{b=-a+1}^{k} B_{a,b}^{j,k} \hbar^{\frac{a+b}{2}} \mu^{\frac{b-a}{2}} \langle Q^{j-a} P^{k-b} \rangle_W ,$$
$$B_{a,b}^{j,k} \in \mathbb{R},$$

provided (V′) holds;

$$\Omega(q^j p^k) = \sum_{a=0}^{\min(j,k)} C_a \hbar^a \frac{j!}{(j-a)!} \frac{k!}{(k-a)!} \langle Q^{j-a} P^{k-a} \rangle_W , C_a \in \mathbb{R}, C_0 = 1,$$

provided (V) + (VI) hold;

$$\Omega(q^j p^k) = \sum_{a=0}^{j} \sum_{b=0}^{k} D_{ab} \hbar^{\frac{a+b}{2}} \mu^{\frac{b-a}{2}} \frac{j!}{(j-a)!} \frac{k!}{(k-b)!} \langle Q^{j-a} P^{k-b} \rangle_W ,$$
$$D_{ab} \in \mathbb{R}, D_{00} = 1,$$

provided (V′) + (VI) hold;

$$\Omega(q^j p^k) = \langle Q^j P^k \rangle_W$$

provided (V) or (V′) + (VII) hold. In this case Ω is just the Weyl quantization map Ω_W generated by the map

$$e^{i(\xi q + \eta p)} \longmapsto e^{i(\xi Q + \eta P)}$$

The classical limit of the commutator of two quantum observables is equal to the Poisson bracket of the corresponding classical observables, namely

$$\lim_{\hbar \to 0} \Omega^{-1}\left(\frac{1}{i\hbar}\left[\Omega(x), \Omega(y)\right]\right) = \mathcal{P}(x,y), \quad \forall x,y \in S[q,p].$$

The proof of this theorem can be found in [1]. Here, let us only note: first, that bijectivity of Ω follows from its linearity and the existence of the classical limit (since then $\Omega(x) = 0$, iff $x = 0$). Second, when Ω is translationally invariant, its general form follows from the fact that any translationally equivariant quantization $\hat{\Omega}$ can be written as

$$\tilde{\Omega} = \Omega_W \circ F$$

where Ω_W is the (translationally invariant) Weyl map specified above and

$$F = \sum_{a,b} f_{ab} \frac{\partial^a}{\partial q^a} \frac{\partial^b}{\partial p^b}$$

(since $\Omega_W^{-1} \tilde{\Omega}$ is a functional on $S[q,p]$ and thus can be written in terms of the basis in the dual of $S[q,p]$).

By comparing formal power series expansions

$$\Omega(q^j p^k) = i^{-j-k} \frac{\partial^{j+k}}{\partial \xi^j \partial \eta^k} \left[\omega(\xi,\eta)\, e^{i(\xi Q + \eta P)} \right] \Big|_{\xi=\eta=0}$$

and

$$\Omega(q^j p^k) = \sum_{a,b=0}^{j,k} \frac{\omega_{ab}}{i^{a+b}} \frac{j!}{(j-a)!} \frac{k!}{(k-b)!} \left\langle Q^{j-a} P^{k-b} \right\rangle_W$$

where

$$\omega(\xi,\eta) = \sum_{a,b=0}^{\infty} \omega_{ab}\, \xi^a \eta^b \quad \text{and} \quad \frac{\omega_{ab}}{i^{a+b}} = D_{ab}\, \hbar^{\frac{a+b}{2}} \mu^{\frac{b-a}{2}},$$

one easily sees that any translationally invariant quantization can be generated by the map

$$\Omega : e^{i(\xi q + \eta p)} \longmapsto \omega(\xi,\eta)\, e^{i(\xi Q + \eta P)}$$

where

$$\omega(\xi,\eta) = \sum_{a,b=0}^{\infty} \omega_{ab}\, \xi^a \eta^b \quad \text{and} \quad \omega_{00} = 1,\ \overline{\omega(\xi,\eta)} = \omega(-\xi,-\eta).$$

In fact, all frequently used symmetrization rules are special cases of translationally invariant quantizations as indicated in Table I.

Symmetrization rule	$\omega(\xi,\eta)$
Weyl-McCoy [6]	1
Born-Jordan [7]	$\frac{\sin(\frac{\hbar}{2}\xi\eta)}{\frac{\hbar}{2}\xi\eta}$
Rivier [8]	$\cos(\frac{\hbar}{2}\xi\eta)$
normal [9]	$e^{\frac{\hbar}{4}(\xi^2+\eta^2)}$
antinormal [9]	$e^{-\frac{\hbar}{4}(\xi^2+\eta^2)}$

Tab. I.

3. QUANTIZATION AS DEFORMATION

Recently, it has been proved [10] that quantum mechanics is a non-trivial deformation of classical mechanics with respect to parameter $\hbar$. It is unique provided one restricts oneself to the flat phase space R^{2n} and infinitesimal deformations expressible in terms of the Poisson bracket. 5/ In order to appreciate this result and to compare it with the quantization procedures studied in the previous section let us recall a few facts about the deformation theory [12] and about the Weyl-Wigner-Moyal formulation of quantum mechanics [13].

Let $\mathfrak{A}$ be an algebra, i.e. a vector space with a binary operation (multiplication)

$$\mu : \mathfrak{A} \times \mathfrak{A} \rightarrow \mathfrak{A} .$$

By a <u>deformation</u> of the multiplication μ a continuous change of μ is meant, i.e. a one-parameter family $\{\mu_\lambda\}$ which can be written in the form of the power series in a continuous parameter

$$\mu_\lambda(a,b) = \mu(a,b) + \sum_{r=1}^{\infty} \lambda^r C_r(a,b)$$

where

$$\lim_{\lambda \to 0} \mu_\lambda(a,b) = \mu(a,b)$$

and $C_r(a,b)$ fulfil relations which follow from the required properties of μ_λ . For example, for associative algebras these relations follow from the associativity condition

$$\mu_\lambda(a, \mu_\lambda(b,c)) = \mu_\lambda(\mu_\lambda(a,b), c) .$$

A deformation is non-trivial whenever μ_λ is not equivalent to for all $\lambda \neq 0$ (i.e. the corresponding algebras are not isomorphic for all $\lambda \neq 0$).

Since the usual probabilistic interpretation of quantum mechanics contrasts with the deterministic character of classical mechanics and the usual mathematical settings of the two types of mechanics look rather disjoint, [67] one may wonder how quantization can be understood as a deformation of the structure of the classical observables rather than as a radical change in the nature of the observables. In this connection let us remark that such a view of quantization is possible because there exists a framework for the description of both classical and quantum mechanics, within which the continuity of the quantization process is brought out. It is the framework in which both mechanics are formulated as statistical theories of functions or distributions on phase space with products and brackets connected by deformations. This framework was worked out by many people (see [13]). A phase space distribution function was introduced by Wigner [13] in 1932. Most interesting developments are due to Moyal [13], who introduced the "sine-Poisson" bracket, now called Moyal bracket, for functions on phase space (that corresponds to the commutator bracket of quantum mechanics) and to Bayen, Flato, Fronsdal, Lichnerowicz and Sternheimer [10],[13] who examined in detail quantization as a deformation of the algebra of observables and determined the spectra of some important physical observables by direct phase space methods.

The main features of the Wigner-Moyal approach in a flat phase space and its comparison with the usual formulation of quantum mechanics can be briefly summarized as follows:

/i/ Observables

In the usual formulation of quantum mechanics, to any classical observable f there corresponds a self-adjoint operator $\hat{F}$ in a Hilbert space $\mathcal{H}$

$$f \xrightarrow{\Omega} \hat{F} .$$

In the Wigner-Moyal formulation of quantum mechanics, to any classical observable f there corresponds a symbol (a certain function or distribution) $F(q,p)$ on phase space R^{2n}. Moreover:

- Instead of the product $\hat{F}.\hat{G}$ of two operators there will be a $\star$-product $F\star G$ (a non-commutative binary operation) of the corresponding symbols F and G defined by

$$\begin{aligned}(F \star G)(q,p) &= e^{\frac{i\hbar}{2}\mathcal{P}(F,G)} \\ &= F\,G + \sum_{k=1}^{\infty} \frac{1}{k!} \left(\frac{i\hbar}{2}\right)^k \mathcal{P}^k(F,G)\end{aligned}$$

where $F\,G$ is the usual product of functions and $\mathcal{P}^k(F,G)$ is a "k-th power of the Poisson bracket"

$$\mathcal{P}(F,G) = \sum_{i=1}^{n} \left(\frac{\partial F}{\partial q_i}\frac{\partial G}{\partial p_i} - \frac{\partial F}{\partial p_i}\frac{\partial G}{\partial q_i} \right).$$

More precisely, $\mathcal{P}^k(F,G)$ is a bidifferential operator given by

$$\mathcal{P}^k(F,G) = \omega^{\alpha_1\beta_1} \cdots \omega^{\alpha_k\beta_k} (\partial_{\alpha_1} \cdots \partial_{\alpha_k} F)(\partial_{\beta_1} \cdots \partial_{\beta_k} G)$$

where $\alpha, \beta = 1,2,\ldots 2n$, $\partial_i = \frac{\partial}{\partial q_i}$, $\partial_{i+n} = \frac{\partial}{\partial p_i}$,

$i = 1,2,\ldots,n$, $q,p \in R^n$, and $\omega^{i,i+n} = -\omega^{i+n,i} = 1$ and the other $\omega^{\alpha\beta}$ are equal to zero.

- Instead of the commutator $[\hat{F},\hat{G}]$ there will be the Moyal bracket $\mathcal{M}(F,G)$ defined by

$$\begin{aligned}\mathcal{M}(F,G) &= \frac{1}{\hbar}(F\star G - G\star F) \\ &= \frac{2}{\hbar}\sin\left(\frac{\hbar}{2}\mathcal{P}(F,G)\right) \\ &= \sum_{k=0}^{\infty} \frac{(-1)^k}{(2k+1)!} \left(\frac{\hbar}{2}\right)^k \mathcal{P}^{2k+1}(F,G)\end{aligned}$$

/ii/States

In the Wigner-Moyal approach instead of a density matrix $\hat{\rho}$ there will be a quasi-distribution function $\rho(p,q)$ so that the result of a measurement of the observable $\hat{F}$ at the time t on the state $\hat{\rho}$

$$\langle \hat{F} \rangle_{\hat{\rho}} = \mathrm{Tr}\,(\hat{F}\hat{\rho})$$

is expressed in the form

$$\langle F \rangle_{\rho} = \int F(q,p) \star \rho\,(q,p)\; d^n q\; d^n p .$$

Here, quasi-distribution $\rho(q,p)$ is not necessarily non-negative.

/iii/ Equations of motion

In the Schrödinger picture, the equation of motion

$$\frac{\partial \hat{\rho}}{\partial t} = -\frac{1}{i\hbar}\left[\hat{\rho}, \hat{H}\right]$$

, $\hat{H}$ - the Hamiltonian of the system,

has the form in the Wigner-Moyal approach

$$\frac{\partial \rho}{\partial t} = -\mathfrak{M}(\rho, H)$$

from which one-parameter family $\{\rho_t\}$, $t \geqq 0$ is derived provided $\rho_{t=0}$ is given.

In the Heisenberg picture, instead of the usual equation

$$\dot{\hat{F}} = \frac{\partial \hat{F}}{\partial t} + \frac{1}{i\hbar}\left[F, H\right]$$

there will be the equation in the Wigner-Moyal approach of the form

$$\dot{F} = \frac{\partial F}{\partial t} + \mathfrak{M}(F, H).$$

/iv/ Eigenvalue problem

The eigenvalue equation

$$\hat{F}\,\psi_\lambda = \lambda\,\psi_\lambda$$

has the form in the Wigner-Moyal formulation given by 7/

$$F \star P_\lambda = \lambda\, P_\lambda$$

where $P_\lambda(q,p)$ is a quasi-distribution which corresponds to the projector $\hat{P}_\lambda = \hat{\rho}_\lambda = |\psi_\lambda\rangle\langle\psi_\lambda|$.
Let us note that quantization $f \longmapsto F$ in the Wigner-Moyal quantum mechanics is introduced in such a way that the Bohr correspondence principle is fulfilled, namely

$$\lim_{\hbar \to 0} F = f ,$$

$$\lim_{\hbar \to 0} F \star G = \lim_{\hbar \to 0} \exp \left(i \frac{\hbar}{2} \mathcal{P}(F,G)\right) = fg ,$$

$$\lim_{\hbar \to 0} \mathcal{M}(F,G) = \lim_{\hbar \to 0} \frac{2}{\hbar} \sin \left(\frac{\hbar}{2} \mathcal{P}(F,G)\right) = \mathcal{P}(f,g).$$

Finally, let us note that analogously to various quantizations $\Omega : f \to \hat{F}$ discussed in the previous Section 2 one may introduce various $\star$-products and, consequently, various $\star$-formulations of quantum mechanics of which the Wigner-Moyal formulation is just one particular example. What is then the correspondence between classical and $\star$-formulations of quantum mechanics?

From classical observables one may form two algebras. First, the associative algebra, which is the vector space $C^\infty(R^{2n})$ endowed with commutative pointwise multiplication $(f.g)(q,p) = f(g,p)\, g(q,p)$. Second, the Lie algebra with the same vector space but endowed with the Poisson bracket binary operation

$$(f,g) \longmapsto \mathcal{P}(f,g) \quad .$$

The associative algebra of classical observables is deformed into a class of equivalent associative algebra of non-commuting quantum observables, i.e. into the vector space of symbols $F(q,p) \in C^\infty(R^{2n})$ equipped with various $\star$-products

$$(F,G) \longmapsto F \star G \quad . \text{8/}$$

The Lie algebra of classical observable is analogously deformaed into a class of equivalent Lie algebras of quantum observables, i.e. into the same vector space of symbols $C^\infty(R^{2n})$ endowed with various deformed Lie structures

$$(F,G) \longmapsto \frac{1}{i\hbar} (F \star G - G \star F).\text{9/}$$

Thus we see that the quantization via deformation is nothing else than the map from the algebra of classical observables $f(q,p)$ into the algebra of symbols $F(q,p)$ which can be taken such that

$$f \longmapsto F = f \; .$$

Of course, this map is not a homomorphism of algebras.

In order to compare this result with the previous construction of quantization via the bijective maps Ω, one has to take into account that to the algebra of classical observables $S[q,p]$ there correspond both the algebras of non-commutative symbols described above and the algebra of non-commutative variables Q and P discussed in Section 2. Since the connections between the classical observables **x** on the one hand and the symbols or non-commuting quantities $X(Q,P)$ on the other hand are bijective, any algebra of symbols is isomorphic to the algebra of non-commuting quantities $X(Q,P)$.

Due to this property any symmetrization rule discussed in Section 2 can be related with the corresponding $\star$-product. This correspondence is specified for translationally invariant quantization in the following theorem 1.

<u>Theorem:</u> Let the map ϕ from the classical observables into the symbols be taken as

$$\phi : \quad f \longmapsto F = f .$$

Then the $\star$-product corresponding uniquely to the general translationally invariant quantization

$$\Omega : e^{i(\xi q + \eta p)} \longmapsto \omega_\Omega(\xi,\eta)\, e^{i(\xi Q + \eta P)}$$

characterized by the formal power series

$$\omega_\Omega(\xi,\eta) = e^{\chi_\Omega(\xi,\eta)} = \sum_{a,b=0}^{\infty} \omega_{ab}\, \xi^a \eta^b, \quad \chi(0,0)=1,$$

is given by

$$F \star_\Omega G = (e^{\frac{i\hbar}{2}\mathcal{P}_\Omega})(F,G) = F.G + \sum_{k=1}^{\infty} \frac{1}{k!} (i\hbar/2)^k\, \mathcal{P}_\Omega^k(F,G)$$

where $\mathcal{P}_\Omega^k$ is the k-th power of the bidifferential operator $\mathcal{P}_\Omega$ related to the Poisson operator $\mathcal{P}$ by

$$\frac{i\hbar}{2}\mathcal{P}_\Omega = \frac{i\hbar}{2}\mathcal{P} + \delta\chi_\Omega$$

with

$$\delta\chi_\Omega = \chi_\Omega\left(\frac{\partial}{\partial q'}, \frac{\partial}{\partial p'}\right) + \chi_\Omega\left(\frac{\partial}{\partial q}, \frac{\partial}{\partial p}\right) - \chi_\Omega\left(\frac{\partial}{\partial q} + \frac{\partial}{\partial q'}, \frac{\partial}{\partial p} + \frac{\partial}{\partial p'}\right).$$

The proof is given in [1] .

Remarks.

/i/ The Weyl symmetrization rule, i.e. $\omega = \exp \chi = 1$, leads to the $\star$-product of the Wigner-Moyal approach with $\mathcal{P}_\Omega = \mathcal{P}$.
The correspondence with other quantization rules is indicated in Table II.

Symmetrization rule	$\chi(\xi,\eta)$	$\frac{i\hbar}{2}\mathcal{P}_\Omega(\xi,\eta;\xi',\eta')$
Weyl-McCoy	0	$\frac{i\hbar}{2}(\xi\eta' - \eta\xi')$
Born-Jordan	$\ln\left(\frac{\sin\frac{\hbar}{2}\xi\eta}{\frac{\hbar}{2}\xi\eta}\right)$	$\frac{i\hbar}{2}(\xi\eta'-\eta\xi') + \ln\frac{\sin\frac{\hbar}{2}\xi\eta\sin\frac{\hbar}{2}\xi'\eta'(\xi+\xi')(\eta+\eta')}{\sin\frac{\hbar}{2}(\xi+\xi')(\eta+\eta')\frac{\hbar}{2}\xi\eta\xi'\eta'}$
Rivier	$\ln(\cos\frac{\hbar}{2}\xi\eta)$	$\frac{i\hbar}{2}(\xi\eta'-\eta\xi') + \ln\frac{\cos\frac{\hbar}{2}\xi\eta\cos\frac{\hbar}{2}\xi'\eta'}{\cos\frac{\hbar}{2}(\xi+\xi')(\eta+\eta')}$
normal	$\frac{\hbar}{4}(\xi^2+\eta^2)$	$-\frac{\hbar}{2}(\eta-i\xi)(\eta'+i\xi')$
antinormal	$-\frac{\hbar}{4}(\xi^2+\eta^2)$	$\frac{\hbar}{2}(\eta+i\xi)(\eta'-i\xi')$

/ii/Different Ω , Ω' may yield the same $\star$-product. This happens whenever χ_Ω and $\chi_{\Omega'}$ differ by $i(a\xi + b\eta)$.

/iii/
Quantization via deformation is very useful when one studies systems which have classical and quantum subsystems (e.g. when one tries to quantize solitons, to study scatterings of heavy ions, to calculate quantum corrections to classical radiation etc.).

/iv/For practical purposes, it will be useful to clarify the way of determining the most natural $\star$-product for symmetries of the Hamiltonian of the considered system other than translation ones studied in the Theorem.

Acknowledgments

The author would like to thank Professors Abdus Salam and Paolo Budinich as well as the International Atomic Energy Agency and UNESCO for hospitality at the International Centre for Theoretical Physics, Trieste, where the work was done. The author is also grateful to Professors A.O.Barut and E.Inönö for hospitality at the Bogaziçi University during the Colloquium.

Note added in proof.

For a more recent development of the deformation theory see a review by M.Flato, Czech.J.Phys. B 32 (1982), 472.

FOOTNOTES

1/ Invited talk at the Colloquium on Mathematical Physics, Bogaziçi University, Istanbul, Turkey.

2/ It is clear that this is not the whole story since not every symmetric operator in the enveloping algebra has a self-adjoint extension [4].

3/ In a concrete representation of W the self-adjoint variables Q,P will be represented by symmetric operators.

4/ One can consider μ to be for instance the term $m\omega$ of a harmonic oscilator.

5/ The result was generalized for global deformations in [11].

6/ Classical observables form an algebra of function on the corresponding phase space whereas quantum observables form an algebra of self-adjoint operators on a Hilbert space of state vectors.

7/ This follows from the derivative $i\hbar \frac{d}{dt}$ Exp(Ft) of the formal series

$\mathrm{Exp}(Ft) = \sum_{n=0}^{\infty} \frac{1}{n!} \left(\frac{t}{i\hbar}\right)^n (F \ast)^n$ and the Fourier-

Dirichlet expansion $\text{Exp}\,(Ft) = \sum_{\lambda \in I} P_\lambda \; e^{\lambda t / i\hbar}$

where I is the spectrum of F.

8/ For instance, in the Wigner-Moyal approach

$$F \star G = \exp\left(\frac{i\hbar}{2}\mathcal{P}\right)(F,G)$$

so that the corresponding deformation parameter is $\lambda = i\hbar/2$.

9/ For instance, in the Wigner-Moyal approach

$$\frac{1}{i\hbar}(F \star G - G \star F) = \frac{2}{\hbar}\sin\left(\frac{\hbar}{2}\mathcal{P}\right)(F,G) = \mathcal{M}(F,G)$$

so that the deformation parameter $\lambda = -\hbar^2/4$.

REFERENCES

[1] Niederle, J., Tolar, J.: 1979, Czech.J.Phys.B 29, p. 1358.

[2] Niederle, J., Tolar, J.: 1978, Čs.čas.fyz. A 28, p. 258 (in Czech).

[3] Waniewski, J.: 1977, Repts. Math. Phys. 11, p.331

[4] Arnal, D.: 1976, A.Funct.Analysis 21, p.432

[5] Jacobson, N.: 1961, Lie Algebra. Interscience Publishers, New York.

[6] McCoy, N.H.: 1932, Proc.Nat.Acad.Sci. (US) 18, p. 674;
Weyl, H.: 1927, Z.Physik 46, p. 1;
Weyl, H.: 1931, Gruppentheorie und Quantenmechanik, Leipzig;

[7] Born, M., Jordan, P.: 1925, Z.Physik 34, p.858.

[8] Rivier, D.C.: 1951, Phys.Rev. 83, p.862.

[9] Agarwal, G.S., Wolf, E.: 1970, Phys.Rev. D2, p.2161.

[10] Bayen, F. et al.: 1977, Letters Math.Phys. 1, p. 521;
Bayen, F. et al.: 1978, Ann.Phys.(US) 111, p.61.

[11] Guth, S.: 1979, Letters Math.Phys. 3, p.297.

[12] Gerstenhaber, M.: 1964, Ann.Math. 79, p.59.

[13] Weyl, H.: 1931, Gruppentheorie und Quantenmechanik, Leipzig.
(The Theory of Groups and Quantum Mechanics, Dover, New York, 1931);
Wigner, E.P.: 1932, Phys.Rev.40, p.749;
Moyal, J.E.: 1949, Proc. Cambridge Phil. Soc. 45, p. 99;
See also [2],[9],[10] and Groenewold, H.J.: 1946, Physica 12, p.405;
Rivier, D.C.: 1951, Phys.Rev. 83, 862;
Jordan, T.F., Sudarshan, E.C.G.: 1961, Rev.Mod. Phys. 33, p.515;
Cohen, L.: 1966, J.Math.Phys. 7, p.781;
Cahill, K.E., Glauber, R.J.: 1969, Phys.Rev.177 p. 1857, p.1882;
Flato, M., Lichnerowicz, A., Sternheimer, D.: 1974, C.R.Acad.Sci., Paris A 279, p.877; 1976, A 283, p.19;
Vey, J.: 1975, Comment.Math.Helvet. 50, p.421;
Flato, M., Lichnerowicz, A., Sternheimer, D.: 1976, J.Math.Phys. 17, p.1754;
Shirokov, Yu.M.: 1979, in "Physics of Elementary Particles and of Atomic Nuclei" Vol.10, No.1,p.5.

Part II

Group Representations and Gauge Theories

$\overline{SO}_o(3,2)$: LINEAR AND UNITARY IRREDUCIBLE REPRESENTATIONS

E. ANGELOPOULOS

Physique - Mathématique
Université de Dijon
BP. 138
F - 21004 DIJON (FRANCE)

Abstract :

Complete results on : linear irreducible representations of $\overline{SO}_o(3,2)$ on Banach spaces, up to infinitesimal equivalence ; unitary irreducible representations of the same group ; reduction up to multiplicity of the restriction on the covering of SO(3) x SO(2); are obtained by using universal enveloping algebra methods.

FOREWORD

The goal of this work was the determination of the unitary dual of G = $\overline{SO}_o(3,2)$.The guideline of the method used is the following : Let $\mathfrak{g}$ be the Lie algebra of G, $\underline{k} \oplus \underline{p}$ a Cartan decomposition, $\mathcal{U}(\mathfrak{g})$, $\mathcal{U}(\underline{k})$ their enveloping algebras, $\mathcal{E}$ a simple $\mathfrak{g}$-module, $\mathcal{E} = \oplus \mathcal{E}^\lambda$ its reduction to $\underline{k}$-isotypic components. If there is a nondegenerate, $\mathfrak{g}$-invariant, sesquilinear form (s.f.) on $\mathcal{E}$, then the unitarizability problem is reduced to the positive definiteness of this s.f.. On the other hand, the reduction of the tensor product $\underline{p} \otimes \mathcal{E}^\lambda$, (or, more generally of $x \otimes \mathcal{E}^\lambda$, with x any $\underline{k}$-submodule of $\mathcal{U}(\mathfrak{g})$ under ad) establishes relations among the restrictions of the s.f. in $\mathcal{E}^\lambda$ and in the components of $\underline{p} \otimes \mathcal{E}^\lambda$,through $(X\varphi|\psi) = (\varphi|X^*\psi)$. This implies that elements of the form Y*Y, with $Y \in \underline{p}.\mathcal{U}(k)$, must be represented by positive linear operators on $\mathcal{E}$. Using the fact $[\underline{p},\underline{p}] \subset \underline{k}$ one can establish equalities involving elements of this form, or, equivalently, squared vectors $(Y\varphi|Y\varphi)$.

Suppose now that there exists a (partial) order relation on $\text{Range}_{\mathcal{E}}(\lambda)$ (which is a subset of the unitary dual of $\underline{k}$), such that, for every nonminimal λ, and for every φ_λ in $\mathcal{E}^\lambda$, the above relations yield equalities of the form $(\varphi_\lambda|\varphi_\lambda) = \sum_{\mu<\lambda}(\varphi_\mu|\varphi_\mu) + C_\lambda$,

A. O. Barut (ed.), Quantum Theory, Groups, Fields and Particles, 101–148.

with $\varphi_\mu \in \mathcal{E}^\mu$, and C_λ some positive scalar. If this happens, the positive-definiteness of the s.f. is the consequence of the positive-definiteness of its restriction to $\oplus \mathcal{E}^\lambda$ with summation over minimal λ's only. If the order relation is adequate enough, minimality of λ will be translated algebraically by constraints on the corresponding representation of the commutant of k inside $\mathcal{U}(\mathfrak{g})$. Using these supplementary constraints, one can try to iterate the procedure by looking for a finer order relation inside the subset of minimal λ's , and so on, up to the complete characterization of the signature of the s.f. .

Of course, there is no guarantee that this approach works a priori. When $\mathfrak{g}$ = so(2,1) it does : the partial order in question is roughly the order of increasing values of $m^2+|m|$, im running over the spectrum of the compact generator. It also works for $\mathfrak{g}$ = so(3,2), although the nontrivial multiplicity of $\mathcal{E}^\lambda$ yields a complication which does not appear in so(2,1). To overcome it, positive-definiteness is established (modulo an induction hypothesis) separately for two orthocomplementary subspaces of $\mathcal{E}^\lambda$, which are respectively the image of some operator and the kernel of its adjoint. (Section 5.1, lemma 8, formulas 101-102, and proposition 5). Investigation of minimal λ subspaces (called in this paper lowest-quadratic-weight subspaces) presents no major difficulties.

This study is the startpoint of a more ambitions work : the case $\mathfrak{g}$ = so(p,q). For q = 2, the problem has been completely solved (the so(4,2) case is actually a preprint of the Université de Dijon, Math.Phys.Dept, while the general p case has been published in C.R.Acad.Sc. Paris, I, 292, 1981, p.469, both by the author), while for general q it has not yet been solved. The major innovation of these further developments consists on computational tricks to yield formulas inside the commutant of k :they are obtained by using polynomials with coefficients in $\mathcal{U}(\mathfrak{g})$, establishing their general properties, and giving particular values to the variables when examining what happens on $\mathcal{E}^\lambda$.

In view of these more global results, the case p=2, q=3 may appear as a mere application of more general formulas. However, the choice to publish this paper in its original form is due to several reasons : first of all, it is quite interesting from the physical point of view, to dispose of a list of representations of the de Sitter group, both unitary and nonunitary, as well as of properties and explicit formulas on the enveloping algebra. Next, the generalizations just mentioned rely on the introduction of quite elaborate formalism and calculating devices, not yet published, and the so(3,2) case may be treated (in fact has been treated) independently of their introduction.

Indeed, if I was to rewrite this paper today instead of 1979, I would have done it differently, insisting more on generalizable features (like, say, eq. (59)), and less on particularities, so that it could serve as an introduction to more general cases. But when it was effectively treated, I had no idea about the generalization, except the vague thought that this method would also work elsewhere : and this proved to be the case. Consequently, when this paper was written it reflected the scope I had at that time. By modifying towards a more "elegant" presentation I would accredit the idea that mathematics is as smooth as it is usually presented ; and I don't think so.

0. INTRODUCTION

The de Sitter group $G=SO_o(3,2)$ is the "motion of curved space-time with constant negative curvature - the de Sitter universe. It plays exactly the same role in that universe, as the Poincaré group P does in flat space-time, that is, physical laws are invariant under its action. Because of its relation to the Poincaré group (which is a "limit" of the de Sitter group through Wigner-Inonü contraction), the group G, as well the other noncompact real form $SO_o(4,1)$, has become a rather familiar object in mathematical physics, so that the "public relations" part of this introduction may end here with no harm.

The object of this paper is the exhaustive study of the irreducible continuous ray-representations of G ; or, equivalently, true representations of its universal covering, $\bar{G}$.

What is meant by "exhaustive" needs a little clarification. Indeed, many representations of $\bar{G}$ are known explicitely, either constructed by hand, like the Majorana-Dirac representations which remain irreducible when restricted to the Lorentz group (in the notations of Theorem 1 these are the four representations $(j;[\pm(j+\frac{1}{2}),\pm(j+\frac{1}{2})])$, with $2j=0,1$), or obtained by standard techniques, like representations induced by a subgroup AN appearing in the Iwasawa decomposition. But a complete list of all linear irreducible representations, or a list of all unitary ones have not yet been given, at least up to the author's knowledge. Both these lists are given here, in Theorem 1 (sec.3) and Table 1 (sec.5). Moreover, the reduction of the restriction to the maximal compact subgroup $K=SO(3)xSO(2)$ is carried on for every irreducible representation of $\bar{G}$, up to multiplicity coefficients for each isotypic component. There are also a lot of by-products, like some generalized matrix elements or formulas leading to them, but they are not systematically presented ; the interested reader is invited to work out explicit formulas by himself, with little or no algebraic calculations. The reason for this situation is that the methods used are constructive and based on explicit formulas inside the enveloping algebra, which are developed here ; but the calculations have been carried on only so far as to establish the desired results.

The irreducible representations are constructed by infinitesimal means, i.e. Lie algebra and eveloping algebra properties. The maximal compact subgroup K and its universal covering $\overline{K}$ play an important role. Indeed, the choice of a Banach space for the representation space is dictated by the fact that the general properties of compact groups' representations (especially discrete reduction to inequivalent isotypic components) are established in Banach spaces. These properties may be extended to $\overline{K}=SU(2)\times\mathbb{R}$, though this group is no more compact (the universal covering $\overline{G}$ has infinite center), provided one considers only restrictions to $\overline{K}$ of irreducible representations of $\overline{G}$: this follows from the fact that $\overline{K}/K$ must be represented by scalar multiples of the identity because of Schur irreducibility.

The existence of a discrete $\overline{K}$-reduction is a crucial point : indeed, i.r.'s of $\overline{K}$ can be ordered with respect to their SU(2) content. Calling a quadratic weight the number j which determines the eigenvalue j(j+1) of the Casimir operator of SU(2), there always exists a lowest quadratic weight (LQW). It appears that the action of the Lie algebra $\underline{g}$ (of G) on a LQW eigenspace determines the representation in question ; in particular, any i.r. (j,n) of $\overline{K}$ appears with multiplicity one, if j is the LQW. In that sense, LQW's appear as the generalization of ordinary weights, which are defined relatively to a Cartan subalgebra. The main difference between ordinary and quadratic weights is that the linearity is lost : the ladder operators are polynomial expressions on the Lie algebra generators for quadratic weights. This is due to the fact that $\overline{K}$ is non-abelian.

The multiplicity-one property for LQW spaces implies that the multiplicity is finite for every i.r. of $\overline{K}$, and that there exists a dense space of K-finite vectors, on which the Lie algebra representation of $\underline{g}$ can be realized. The classification problem reduces to a much easier problem, which is purely algebraic : to classify all irreducible representations of the associative algebra $\mathcal{C}_3$ which satisfy the LQW constraints, $\mathcal{C}_3$ being the commutant of SO(3) inside the enveloping algebra.

The unitarity list is established by requiring positive-definiteness for the (unique) sesquilinear form with respect to which the Lie algebra generators are skew-adjoint. The properties of the ladder operators are used to establish an induction on the quadratic weight j ; it results that positivity on the LQW space, together with a condition involving only j and the Casimir operators eigenvalues, is a necessary and sufficient condition for unitary.

This paper has been written to be as self-sufficient as possible. Some proofs have been omitted, but the hints given are sufficient to guide the reader who wants to reconstruct them, since the techniques used are quite standard.

The paper's plan is the following :

Section 1 contains all algebraic preliminaries : notations and formulas for the group, Lie algebra, enveloping algebra, important compact subalgebras and their commutants as well as ladder operators relative to them.

Section 2 contains the general considerations which lead to the one-to-one correspondence between i.r.'s of $\bar{G}$ and LQW i.r.'s of $\mathcal{C}_3$ (proposition 4). The classification of i.r.'s of $\bar{G}$ is carried on in Section 3, Theorem 1 ; this theorem is established in three pieces, Theorems 1A+1B+1C , corresponding to the different possible choices of the LQW.

In Section 4 the K-content of each i.r. R of $\bar{G}$ is determined : every i.r. of $\bar{K}$ appearing in the reduction of R may be labelled (j',n'), with $j' \in j+\mathbb{N}$, $n' \in n+\mathbb{Z}$, where j is the LQW and n is some complex number ; the hypothesis that the a priori possible range of (j',n') does not contain an element (j",n") leads to a complete classification.

Section 5 deals with the unitarity problem, using the approach sketched above. The complete list of u.i.r.'s is given in Table 1. Section 6, finally, contains concluding remarks.

1. THE DE SITTER GROUP AND LIE ALGEBRA

1.1 Presentation of the structures involved

In what follows G will denote the (connected, centerless) Lie group $SO_o(3,2)$, $\bar{G}$ its universal covering group, $\underline{g}$ its Lie algebra and $\mathcal{U}_g$ its universal enveloping algebra.

We shall use tensor notations to describe elements of $\mathcal{U}_g$ and relations between them. Define the pseudometric tensors associated to the de Sitter 3-2 bilinear product in $\mathbb{R}^5$ as $g^{\lambda\mu}$ and $g_{\lambda\mu}$, with $\lambda,\mu \in \{1,2,3,1',2'\}$: we have $g^{\lambda\mu} = c(\lambda)\ \delta_{\lambda\mu}$, $g_{\lambda\mu} = c(\lambda)\ \delta_{\lambda\mu}$, with $c(1)=c(2)=c(3)=1=-c(1')=-c(2')$. Let also $\varepsilon_{\lambda\mu\nu\rho\sigma}$ denote the completely skew-symmetric tensor on 5 coordinates, with $\varepsilon_{1'2'123}=1$. We shall use greek letters to represent any of the five coördinates, with the summation convention between identical upper and lower indices ; latin letters will be used to represent coordinates of a definite signature : unprimed letters for 1,2,3 and primed one for 1',2' . Summation convention will be more loose for latin letters : identical indices will be summed even if they are both lower or both upper indices. We shall also make use of the correspondence between tensors of rank one and skew-symmetric tensors of rank two in $\mathbb{R}^3$, with the help of ε_{ijk}, the completely skew-symmetric tensor in three coordinates.

The generators of $\underline{g}$ will be denoted by $M_{\mu\nu}$ or $M^{\mu\nu} = g^{\mu\nu}\ g^{\nu\rho}\ M_{\lambda\rho}$, and they satisfy $M^{\mu\nu} = -M^{\nu\mu}$. The commutation relations are given by :

$$(1)\quad [M^{\mu\nu},M^{\lambda\rho}] = g^{\nu\lambda}\ M^{\mu\rho} + g^{\mu\rho}\ M^{\nu\lambda} - g^{\nu\rho}\ M^{\mu\lambda} - g^{\mu\lambda}\ M^{\nu\rho}$$

The maximal compact subalgebra $\underline{k}$ $\underline{so}(3) \oplus \underline{so}(2)$ is generated by the elements $U=M_{1'2'}$ and $J_k=\frac{1}{2}\ \varepsilon_{kij}\ M_{ji}$, so that the nonzero brackets in $\underline{k}$ are given by :

$$(2)\quad [J_i,J_j] = \varepsilon_{ijk}\ J_k = M_{ji}$$

Let now introduce some important elements of $\mathcal{U}_g$. The two-degree polynomials defined by :

$$(3)\quad F_\mu = \frac{1}{8}\ \varepsilon_{\mu\nu\rho\sigma\tau}\ M^{\nu\rho}\ M^{\sigma\tau}$$

generate an irreducible five-dimensional $\underline{g}$-module (under the adjoint action of $\underline{g}$ on $\mathcal{U}_g$), isomorphic to the de Sitter 3-2 space, so that :

$$(4)\quad [M_{\mu\nu},F_\rho] = g_{\nu\rho}\ F_\mu - g_{\mu\rho}\ F_\nu$$

One also easily establishes :

(5) $M_{\mu\nu} F^{\nu} = 2 F_{\mu} = - F^{\nu} M_{\mu\nu}$

The bracket of two F's is given by :

(6) $[F_{\mu} , F_{\nu}] = - \frac{1}{2} \varepsilon_{\mu\nu\lambda\rho\sigma} M^{\lambda\rho} F^{\sigma}$

Distinguishing primed and unprimed indices one gets :

(7) $F_k = - U J_k + \varepsilon_{kij} M_{1'j} M_{2'i}$

(8) $F_{k'} = \varepsilon_{k'h'} M_{h'i} J_i \qquad (\varepsilon_{1'2'} = - \varepsilon_{2'1'} = 1)$

We denote by UF the left ideal $\oplus_{\mu} \mathcal{U}_g F_{\mu}$ of $\mathcal{U}_g$; it follows from (4) that

(9) $F_{\mu} X \in UF$, for every $X \in \mathcal{U}_g$

The center $\mathcal{C}_o$ of $\mathcal{U}_g$ is generated by the elements

(10) $C_2 = \frac{1}{2} M^{\mu\nu} M_{\mu\nu} \quad ; C_4 = F^{\mu} F_{\mu} = - \frac{1}{4} M_{\mu\nu} \quad [F^{\mu}, F^{\nu}]$

1.2. Interesting so(3)-submodules of the so(3)-module $\mathcal{U}_g$

We are interested here in exhibiting a set of so(3)-scalars which span algebraically $\mathcal{C}_3$, the commutant of so(3) in $\mathcal{U}_g$, as well as studying some formulas involving 3-vectors - that is elements of $\mathcal{U}_g$ which span three-dimensional so(3)-modules isomorphic to the adjoint module of so(3).

Considered as an so(3)-module, $\mathcal{U}_g$ behaves exactly like a vector space of polynomials in ten commutative variables, as it easily follows from Poincaré-Birkhoff-Witt theorem. Under the action of so(3) the ten generators of g separate into three 3-vectors, namely J_i, $M_{1'i}$, $M_{2'i}$, and one so(3)-scalar, U. The condition for an element to be in $\mathcal{C}_3$ is to contain only contracted unprimed latin indices, so that every such element will contain as a (say, left) factor of every highest-degree term one of the following so(3)-scalars of minimal degree : U ; $W = -J_k J_k$; $F_{k'}$; $\{M_{h'i}.M_{k'i}\}$; $\varepsilon_{ijk} M_{1'i} M_{2'i} J_k$.

Remark : The notation $\{\Phi.\Psi\}$ means $\Phi\Psi + \Psi\Phi$ hereafter.

The last of the polynomials above lies in the algebraic span of the ones that precede it ; indeed, from (6) and (7) one gets :

(11) $V = [F_{1'} , F_{2'}] = J_i F_i = UW - \varepsilon_{ijk} M_{1'i} M_{2'j} J_k$

Thus, the six elements of degree 2 above and U form a set which spans $\mathscr{C}_3$. Notice that :

(12) $M_{k'i}\ M_{k'i} = -\ C_2 - W + U^2$

so that C_2 may replace $M_{k'i}\ M_{k'i}$ as a spanning element.

Consider now 3-vectors. These are polynomials in the generators, which have either a single noncontracted latin index, or only two ones with respect to which they are skew-symmetric. Nondecomposable 3-vectors (i.e. ones which are not sums of terms $X_i B$, B being a nonscalar elements of $\mathscr{C}_3$), are either the generators (except U) or their "vector products". Elementary properties of double vector products show that these are the only ones.

For every 3-vector A_k one has

(13) $[J_j\ ,\ A_k] = \varepsilon_{ijk}\ A_i$

Defining A'_k by

(14) $A'_k = \frac{1}{2}\ \varepsilon_{ijk}\ \ \{J_i . A_j\}$

one has

(15) $[\frac{1}{2}\ W, A_k] = A'_k = \varepsilon_{kmn}\ J_m\ A_n - A_k = \varepsilon_{kmn}\ A_n\ J_m + A_k$

(16) $A''_k = (A'_k)' = A_k\ W + A'_k + J_k(A_i\ J_i)$

Concerning the 3-vector F_i one easily establishes from (4) , (5) :

(17) $F'_i = -\ M_{k'i}\ F_{k'} + F_i = -\ F_{k'}\ M_{k'i} - F_i$

(18) $F_i\ F'_i = F_i\ F_i + U\ V + F_{k'}\ F_{k'}$

1.3. <u>The commutant of the maximal compact subalgebra</u>

Let $M_{\pm k} = (iM_{1'k} \pm M_{2'k})/\sqrt{2}$ be the eigenvectors of ad U corresponding to the eigenvalues $\pm i$. These elements belong to the complexified vector space $\underline{g}^c$ as well as in $\mathcal{U}_g$ and not to $\underline{g}$ itself. Rewriting the nonzero commutation relations in terms of the generators U , J_k , $M_{\ k}$ one has

(19) $[U, M_{\pm k}] = \ i\ M_{\pm k}\ ;\ [J_j, M_{\pm k}] = \varepsilon_{ijk}\ M_{\pm i}\ ;\ [J_j, J_k] = \varepsilon_{jkh}\ J_h\ ;$

$$[M_{\pm j}, M_{\mp k}] = M_{kj}\ \ \pm i\ U\ \delta_{kj} = \varepsilon_{hjk}\ J_h \pm i\ U\ \delta_{jk}$$

Extending this notation to the F's we shall write :

(20a) $F_k = - U J_k + \frac{1}{2} i \varepsilon_{kmn} \{M_{-m}.M_{+n}\}$

(20b) $F_\pm = i M_{\pm k} J_k = (F_2, \pm i F_1,)/\sqrt{2}$

(20c) $F'_k = M_{-k} F_+ - M_{+k} F_- + F_k = F_+ M_{-k} - F_- M_{+k} - F_k$

and we get from (4) and (11)

(21) $[M_{\pm k}, F_\mp] = \pm F_k \; ; \; [M_{\pm k}, F_h] = \pm \delta_{kh} F_\pm \; ; \; [F_+, F_-] = i V$

It immediately follows from (19) that $\mathcal{U}_g$ is isomorphic (as an so(2)-module) to the direct sum of eigenspaces X^m of ad U, with corresponding eigenvalue im ($m \in \mathbb{Z}$). This is also true of $\mathcal{C}_3$ which is also an so(2)-module. So, the commutant $\mathcal{C}$ of the maximal compact subalgebra $\underline{k}$ = so(2) $\oplus$ so(3) is $X^o \cap \mathcal{C}_3$.

We are interested in finding a set of elements which span $\mathcal{C}$. Examining the set of seven elements spanning $\mathcal{C}_3$ we search for eigenvectors of ad U in their linear span : There are three independent ones for the eigenvalue 0, namely U, W, C_2, and one for each of the eigenvalues $\pm i$, $\pm 2i$, that is, respectively, $F_\pm$ and $\omega_\pm = M_{\pm k} M_{\pm k}$.

Every (noncommutative) monomial in F_+, F_-, ω_+, ω_-, with respective degrees a,b,c,d, is an eigenvector of ad U with eigenvalue i(a-b+2c-2d). To find a spanning set we have to consider products inside the associative algebra $\mathcal{C}$. Observe first that $[M_{\pm h}, M_{\pm k}]=0$, so that :

(22) $[M_{\pm k}, F_\pm] = [M_{\pm k}, \omega_\pm] = [F_\pm, \omega_\pm] = 0$

One has next :

(23) $[\omega_\pm, M_{\mp k}] = 2M'_{\pm k} \pm i \{M_{\pm k}.U\}$

so that

(24a) $[\omega_\pm, F_\mp] = \pm i \{U.F_\pm\} = F_\pm(\pm 2iU - 1) = (\pm 2iU + 1) F$

(24b) $[\omega_\pm, F_k] = \pm 2 F_\pm M_{\pm k}$

(25) $[\omega_\pm, \omega_\mp] = \pm i (4V + (2C_2 - 2W - 2U^2 - 3) U)$

Introduce now $\Psi = \{F_+.F_-\}$, so that $C_4 = F_k F_k - \Psi$. Since $F_k + UJ_k$ is the vector product of the 3-vectors M_{+k} and M_{-k}, by using the double vector product formulas and the commutation relations one obtains :

(26) $2\{\omega_+ . \omega_-\} = -4\Psi - 8UV - 4C_4 + (C_2 - U^2 + W)^2 + 4U^2W - 2C_2 - 6W - U^2$

By more elaborate calculations, which we do not reproduce, one also establishes that the terms $\omega_\mp F_\pm F_\pm$ lie in the algebraic span of the elements U, W, C_2, C_4, V, Ψ . In order to prove that these elements span $\mathcal{C}$, one has to show that, if X is anyone of them and AXB belongs to $\mathcal{C}$ with $[U,B] \neq 0$ (A and B being so(3)-scalars), then AXB belongs to the algebraic span. It is sufficient to take for X any linear combination of V, Ψ alone.
From (24) one obtains

(27a) $[\omega_\pm, F_\mp F_\pm] = F_\pm F_\pm(\pm 2iU-3)$; $[\omega_\pm, F_\pm F_\mp] = F_\pm F_\pm(\pm 2iU-1)$

Since one obviously has $(F_+F_-)F_+ = F_+(F_-F_+)$, it suffices to examine $V F_\pm$; from (24),(25) one gets :

(27b) $2[iV, F_\pm] = \pm F_\pm(4U^2+1) \mp 4F_\mp\omega_\pm$

An induction on the degree of B shows that AXB is inside the algebraic span. Thus one has :
Proposition 1 . The commutant $\mathcal{C}$ of k in $\mathcal{U}_g$ is spanned by the six elements U, W, C_2, C_4, V, Ψ , or, equivalently to U, W, C_2, C_4, $\omega_+\omega_-$ and $\omega_-\omega_+$.

1.4. Ladder operators for so(3)

Let j be a complex number, and let A_i be a 3-vector. In what follows, and till lemma 1, we shall systematically drop the index i from all 3-vectors. From A and j define the following 3-vectors :

(28) $A(j)\uparrow = A'' + jA'$; $A(j)\downarrow = A(-j-1)\uparrow = A'' - (j+1)A'$

Consider now a representation of $\mathcal{U}_g$ on a space $\mathcal{E}$, and assume that $\mathcal{E}$ is the algebraic direct sum $\oplus_j \mathcal{E}(j)$, such that $W = j(j+1)$ when restricted on $\mathcal{E}(j)$; if the representation is integrable to a ray-representation of SO(3) when restricted to so(3), then $\mathcal{E}(j)$ is a multiple of the 2j+1-dimensional irreducible so(3)-module and j takes only positive values, half-integer or integer ones. Define then the two operators $A\uparrow$ and $A\downarrow$ acting on $\mathcal{E}$ by :

(29a) $(2j+1)(j+1)A\uparrow f = A(j)\uparrow f$, $f \in \mathcal{E}(j)$

(29b) $(2j+1)\, j\, A\downarrow f = A(j)\downarrow f$, $j \neq 0$, $f \in \mathcal{E}(j)$

(29c) $A\downarrow f = 0$, $f \in \mathcal{E}(0)$

With these notations one has, for every 3-vector A :

(30a) $Af = A\uparrow f + A\downarrow f - (j^2+j)^{-1}(A_i J_i)J f$, for $j \neq 0$, $f \in \mathcal{E}(j)$

(30b) $A'f = (j+1)A\uparrow f - j\, A\downarrow f$, for $f \in \mathcal{E}(j)$

(30c) $A f = A\uparrow f$ for $f \in \mathcal{E}(0)$

The interest of these operators is that they shift from $\mathcal{E}(j)$ to $\mathcal{E}(j\pm1)$. For every f in $\mathcal{E}(j)$ one has :

(31) $W A\uparrow f = (j+1)(j+2)A\uparrow f$; $W A\downarrow f = (j-1) j A\downarrow f$

Remark : These 3-vector operators can be defined independently of any representation, as elements of a quadratic extension of $\mathcal{U}_g$. However, we avoided such a definition, since we want to use them precisely for the study of representations of g. Notice also that what precedes in "independent" of g, in the sense that these operators can be defined with every associative algebra which contains so(3).

One can also prove the following lemma :

Lemma 1 : Let $\mathcal{E}(j) = \mathbb{C}^{2j+1} \otimes \mathcal{F}(j)$ be an so(3)-module, multiple of the 2j+1-dimensional irreducible one. Every linear operator X defined on $\mathcal{E}(j)$ factorizes into the finite sum $X = \Sigma_i U_i \otimes X_i$, where X_i is a linear operator on $\mathcal{F}(j)$, commuting with so(3), and U_i is a polynomial expression on the generators of so(3), belonging to $gl(2j+1,\mathbb{C})$.

The proof of this lemma is based on the existence of interwinning operators between equivalent representations of so(3) and on the fact that the representation of so(3) on $\mathbb{C}^{2j+1}$ realizes a surjective mapping from the enveloping algebra of so(3) on $gl(2j+1,\mathbb{C})$.

This result applied to $\mathcal{U}_g$ enables to focus attention on $\mathcal{E}_3$, since any operator commuting with W can be related to elements of $\mathcal{E}_3$. In particular, if A,B,C, are 3-vectors acting on $\mathcal{E}(j)$, one has :

(32) $j(2j-1)A\uparrow_m B\downarrow_n = (j J_n J_m - (j-1)J_m J_n + j^2 \delta_{mn})(A\uparrow_i B\downarrow_i)$

(33) $A_i B\downarrow_i = A\uparrow_i B\uparrow_i$; $A_i B\uparrow_i = A\downarrow_i B\uparrow_i$

(34a) $(2j-1)(A\downarrow_i B\uparrow_i)C\downarrow_k = (2j+1)A\downarrow_k(B\uparrow_i C\downarrow_i)$

(34b) $(2j+3)(A\uparrow_i B\downarrow_i)C\uparrow_k = (2j+1)A\uparrow_k(B\downarrow_i C\uparrow_i)$

The arrowed 3-vectors will be called ladder operators with respect to so(3) because they shift from one u.i.r. to another. For every 3-vector A, the so(3)-module spanned by the expression $A_i f$ with $f \in \mathbb{C}^{2j+1}$ is $\mathbb{C}^3 \otimes \mathbb{C}^{2j+1}$; the arrowed operators provide its reduction and eqs. 32-34 are closely related to some three j symbols.

Applying these formulas to 3-vectors of $\mathcal{U}_g$ one gets, using (20c), (21) :

$$(35a)\quad j\,F{\uparrow}_k - M{\uparrow}_{-k}F_+ + M{\uparrow}_{+k}F_- = (j+2)F{\uparrow}_k - F_+M{\uparrow}_{-k} + F_-\,M{\uparrow}_{+k} = 0$$

$$(35b)\quad (j+1)F{\downarrow}_k + M{\downarrow}_{-k}F_+ - M{\downarrow}_{+k}F_- = (j-1)F{\downarrow}_k + F_+M{\downarrow}_{-k} - F_-M{\downarrow}_{+k} = 0$$

$$(36a)\quad (j+1)M{\uparrow}_{\pm k}F_{\mp} - M{\uparrow}_{\mp k}F_{\pm} - jF_{\mp}M{\uparrow}_{\pm k} = (j+1)F_{\pm}M{\uparrow}_{\mp k} + F_{\mp}M{\uparrow}_{\pm k} - (j+2)M{\uparrow}_{\mp k}F_{\pm} = 0$$

$$(36b)\quad jM{\downarrow}_{\pm k}F_{\mp} + M{\downarrow}_{\mp k}F_{\pm} - (j+1)F_{\mp}M{\downarrow}_{\pm k} = jF_{\pm}M{\downarrow}_{\mp k} - F_{\mp}M{\downarrow}_{\pm k} - (j-1)M{\downarrow}_{\mp k}F_{\pm} = 0$$

Remark : Formulas containing j and arrowed 3-vectors remain true if the direction of the arrows is reversed and j is changed to -j-1.

1.5. More formulas in $\mathcal{U}_g$

The following formulas concern elements of $\mathcal{C}$ and $\mathcal{C}_3$, mainly contracted (in latin indices) products of ladder operators. In order to avoid duplication of these formulas later we have systematically substituted the eigenvalues j^2+j and i n for the operators W and U respectively. Consequently, the equalities below must be interpreted as ones between operators acting on adequate eigenspaces of W and U.

In view of 6,19,30, one immediately obtains :

$$(37)\quad j(2j+1)M{\uparrow}_{\pm k}M{\downarrow}_{\mp k} = (j-1)F_{\pm}F_{\mp} - jF_{\mp}F_{\pm} + \frac{1}{2}\,j^2(C_2+(j+1\pm n)(j-2\pm n))$$

$$(38)\quad (j+1)(2j+1)M{\downarrow}_{\mp k}M{\uparrow}_{\pm k} = -(j+2)F_{\mp}F_{\pm} + (j+1)F_{\pm}F_{\mp} + \frac{1}{2}(j+1)^2(C_2+(j+3\pm n)(j\pm n))$$

$$(39)\quad j(2j+1)F{\uparrow}_kF{\downarrow}_k = j^2\,C_4 + (j^2-j)\Psi - j\,n(iV) - (iV)^2$$

Introducing

$$(40)\quad S = \Sigma_{\pm}\,M{\uparrow}_{\pm k}\,M{\downarrow}_{\mp k} \quad ; \quad \Delta = \Sigma_{\pm}\,\pm\,M{\uparrow}_{\pm k}\,M{\downarrow}_{\mp k}$$

one obtains through 37 :

$$(41)\quad (2j-1)(iV + j^2n) = j(2j+1)\,\Delta$$

$$(42)\quad \Psi = -j(2j+1)S + j^2(C_2+n^2+(j+1)(j-2))$$

$$(43)\quad (2j-1)F_{\mp}F_{\pm} = -j(2j+1)(jM{\uparrow}_{\pm k}M{\downarrow}_{\mp k} + (j-1)M{\uparrow}_{\mp k}M{\downarrow}_{\pm k}) + {} + j^2(j-\tfrac{1}{2})(C+n^2\pm n+j^2-j-2)$$

Combining 37 and 38 one gets :

(44) $j^2(j+1)(2j+1)M{\downarrow}_{\mp k}M{\uparrow}_{\pm k}=j(j+1)^2(2j+1)M{\uparrow}_{\pm k}M{\downarrow}_{\mp k}+(j+1)F_{\pm}F_{\mp}+$
$+jF_{\mp}F_{\pm}+2(j^2+j)^2(j+\tfrac{1}{2}\pm n)$

Introducing the quantity

(45) $P_j = (j^2-j) + (C_2-2)\ (j^2-j) + C_4$

one obtains through 39,41,42 :

(46) $j^2P_j - (iV + j^2n)^2 = j(2j+1)(F{\uparrow}_kF{\downarrow}_k + (j^2-j)S - jn\Delta\Delta)$

Finally one obtains from 19, 30, 20c :

(47) $j(2j+1)M{\uparrow}_{\pm k}\ M{\downarrow}_{\pm k} = j^2\omega_{\pm} - F_{\pm}F_{\pm}$

(48) $(j+1)(2j+1)M{\downarrow}_{\pm k}\ M{\uparrow}_{\pm k} = (j+1)^2\ \omega_{\pm} - F_{\pm}F_{\pm}$

(49) $\omega_{\pm} = (j+1)M{\downarrow}_{\pm k}\ M{\uparrow}_{\pm k} - j\ M{\uparrow}_{\pm k}\ M{\downarrow}_{\pm k}$

(50) $F_{\pm}\ F_{\pm} = j^2(j+1)M{\downarrow}_{\pm k}\ M{\uparrow}_{\pm k} - j(j+1)^2\ M{\uparrow}_{\pm k}\ M{\downarrow}_{\pm k}$

2. BANACH REPRESENTATIONS OF $\bar{G}$ AND HOW TO HANDLE THEM

Let R be a continuous irreducible representation (i.r.) of $\bar{G}$ on a Banach space H, dR being the derived representation of g, defined on the maximal analytic domain. We are interested in classifying i.r.'s of $\bar{G}$ up to infinitesimal (or H-C , from Harish-Chandra) equivalence :

Definition 1 : Two i.r.'s R and R' on H and H' are H-C equivalent iff there are dense analytic subspaces E in H and E' in H' and an invertible operator T from E to E', such that $T^{-1}dR'(X)T=dR(X)$ for every X in g.

A first consequence of the irreducibility of R is that the center Z of $\bar{G}$ must be represented by scalars. By general theory it is known that Z is also the center of the universal covering $\bar{K}$ of the maximal compact subgroup K = SO(3) x SO(2), so that $Z=\mathbb{Z}_2 \times \mathbb{Z}$. This means that there is a complex number N such that $R(\exp 2i\pi U) = \exp 2i\pi N$, U being the generator of the subalgebra so(2). An immediate consequence of this is :

Lemma 2 . If the i.r. R satisfies $R(\exp 2i\pi U)=\exp 2i\pi N$, then H is the completed direct sum $\bar{\oplus}_n$ H(n) of subspaces H(n), the range of n being a subset of $N+\mathbb{Z}$, H(n) being the completion of a subspace A(n) on which dR(U) is well-defined and scalar with eigenvalue i n.

The proof of this lemma is based on the existence of a dense analytic subspace in H and on standard integration theory in Banach spaces. The detailed proof is left to the reader.

From lemma 2 and from general properties of compact groups we have :

Proposition 2 . The restriction to $\bar{K} = SU(2) \times \mathbb{R}$ of an i.r. R of $\bar{G}$ is the discrete direct sum of representations (j,n) of $\bar{K}$, with Range(n)$\subset N+\mathbb{Z}$, Range(j)$\subset j_{min} + \mathbb{N}$, $N \in \mathbb{C}$, $2j_{min} \in \mathbb{N}$, such that (j,n) is a (possibly infinite, a priori) multiple of $D_j \otimes \chi_n$, where $d\chi_n(U) = i\, n$ and D_j is the 2j+1-dimensional i.r. of SU(2).

Notice that, since Z is scalarly represented, the SU(2) label j may be either integer or half-integer, but cannot take both kinds of values.

Since j by convention is a positive number, we introduce :

Definition 2 . For every i.r. R, the smallest element in Range j will be called the lowest quadratic weight (LQW) of R.

Let now H(j,n) be the carrier (closed) subspace of the representation (j,n) of $\bar{K}$, such that H is the completed direct sum$\bar{\oplus}$H(j,n) Let A be the analyticity domain of dR(g). Let f be a vector inside A$\cap$H(j,n) ; we shall write j = j(f) and n = n(f) to specify these values. The space E(f) = {dR(X)f ; $X \in \mathcal{U}_g$} is a g-invariant subspaces of A dense in H. Let I_f be the left ideal of $\mathcal{U}_g$ which annihilates f, i.e. I_f = {X ; dR(X)f=0}. The left g-module $\mathcal{U}_g$ (defined by associative multiplication on the left) has a factor g-module $\mathcal{U}_g/I_f$, which is isomorphic to E(f) by construction.

For arbitrary j,n, define the spaces E(j,n;f) = E(f)$\cap$H(j,n) ; E(j;f) = $\oplus_n$ E(j,n;f) ; E(-,n;f) = $\oplus_j$ E(j,n;f) , the sums being algebraic (finite linear combinations only). Clearly, E(j,n;f) must be dense in H(j,n), E(j;f) in H(j) = $\bar{\oplus}_n$ H(j,n) and E(-,n;f) in H(-,n) = $\bar{\oplus}_j$ H(j,n).

Now one has the factorizations :

$$H(j,n) = \mathbb{C}^{2j+1} \otimes B(j,n) \quad ; \quad H(j) = \mathbb{C}^{2j+1} \otimes B(j) \ ;$$

$$E(j,n;f) = \mathbb{C}^{2j+1} \otimes D(j,n;f) \quad ; \quad E(j;f) = \mathbb{C}^{2j+1} \otimes D(j;f) \ :$$

so that D(j;f) (resp. D(j,n;f)) is dense in B(j) (resp: B(j,n)), both being $\mathcal{C}_3$-modules (resp. $\mathcal{C}$-modules) on which W = j(j+1) (resp. W = j(j+1)and U = i n).

Take now j = j(f), n = n(f) ; by construction of E(j,n;f) and E(j,f) , and using lemma 1, it follows that D(j;f) and $\mathcal{C}_3/\mathcal{C}_3 \cap I_f$

are isomorphic $\mathcal{C}_3$-modules and that D(j,n;f) and $\mathcal{C}/\mathcal{C}\cap I_f$ are isomorphic $\mathcal{C}$-modules.

Since R is irreducible it follows easily that D(j,n;f) (resp: D(j;f)) must also be an irreducible $\mathcal{C}$-module (resp. $\mathcal{C}_3$-module) for every choice of f. If in particular one chooses f to be a LQW vector, that is, j(f) is the LQW j_{min} of R, there is much information about the ideal I_f : indeed, for any 3-vector X,Y, the elements $X\downarrow_k$ and $X\uparrow_k\ Y\downarrow_k$ must vanish on f. This is a quite strong condition, because of :

<u>Proposition 3</u> : Let j be the LQW of the i.r. R. Then, for every n inside Range(n), the representation (j,n) of $\bar{K}$ is irreducible and the space D(j,n;f) is one-dimensional and independent of the choice of f in $A\cap H(j,n)$.

<u>Proof</u> : Taking S=0 in (42), one has $\Psi = j^2(C_2+n^2+j^2-j-2)$ on $\overline{D(j,n;f)}$, so that the image $dR(\mathcal{C})$ is an abelian associative algebra ; by (41) one sees that V is also scalarly represented, except when 2j=1, in which case it is a diagonalizable operator which may take two eigenvalues because, by (46), $(4V-U)^2$ must be a scalar.Because of irreductibility the above result follows.

Proposition 3 ensures that E=E(f) will not depend on the choice of the LQW vector f, and that the multiplicity of (j,n) will be finite for every j,n i.e. E will be a Harish-Chandra module, or, in other words, an analytic subspace of <u>k</u>-finite vectors. Since the ideal I_f will be a maximal one satisfying the LQW constraints, the representation of <u>g</u> on E is completely determined ; such a representation is integrable to the group $\bar{G}$ and it can be realized on a Banach (even Hilbert) space of functions on $\bar{K}$ by means of harmonic analysis.

Obviously, any i.r. of $\bar{G}$ defines a unique representation of $\mathcal{C}_3$ on the space $D(j_{min})=D(j_{min};f)$ for any choice of the LQW vector f. The converse is also true, since $E(j_{min})=E(j_{min};f)$ is spanned by the action of $\mathcal{C}_3$ on $E(j_{min},n)=E(j_{min},n;f)$, so that the classification of LQW i.r.'s of $\mathcal{C}_3$ provides that of i.r.'s of $\bar{G}$. On the other hand, a LQW i.r. of $\mathcal{C}_3$ contains in general many inequivalent LQW i.r.'s of $\mathcal{C}$ which depend on the choice of n inside Range(n). That is why we shall focus on LQW i.r.'s of $\mathcal{C}_3$. The above discussion is summarized in :

<u>Proposition 4</u> . The i.r.'s of $\bar{G}$ on a Banach space are characterized, up to Harish-Chandra equivalence, by the LQW i.r. of $\mathcal{C}_3$ acting on the space $D(j_{min}) = \oplus_n D(j_{min},n)$ with one-dimensional $D(j_{min},n)$, such that the restriction of the given i.r. to SU(2) and to the LQW j_{min} is realized on the completion of the analytic subspace of <u>k</u>-finite vectors $E(j_{min}) = \mathbb{C}^{2jmin+1} \otimes D(j_{min})$.

3. IRREDUCIBLE LQW $\mathcal{C}_3$ MODULES

In this section we shall classify all irreducible LQW $\mathcal{C}_3$-modules, with the help of relations (37) to (5o). Through out this section j will denote a LQW. From now-on we shall call $D(j)_\nu = \oplus_{n\in\nu+Z} D(j,n)$ any LQW $\mathcal{C}_3$-module, irreducible or not, such that the action of $\mathcal{C}_3$ on it obeys the following constraints :

1) $W = j(j+1)$; 2) $U = i\,n$ on $D(j,n)$; 3) $D(j,n)$

is one dimensional. The label ν may be thought of eigher as a complex number, with $D(j)_\nu = D(j)_{\nu+1}$ or as an element of $\mathcal{N} = \mathbb{C}/\mathbb{Z}$. $D(j)_\nu$ is an infinite string of eigenspaces of U, and relations (37) to (50) will define the action of $\mathcal{C}_3$ in it. To obtain irreducible modules we shall determine the minimal invariant submodules or factor modules of $D(j)_\nu$ - if any : they will correspond to the splitting of the infinite string to substrings, that is to a partition of Range(n) into at most five parts. A representation of $\bar{G}$ which yields by restriction $D(j)_\nu$ as its LQW $\mathcal{C}_3$-module will be denoted by $\{j,\nu\}$.

First of all, irreducibility says that C_2 and C_4 are represented by scalar operators throughout $\{j,\nu\}$; we shall not distinguish between the operators and their eigenvalues from now on.

Next, we obtain from (47) :

(51) $$j^2\,\omega_\pm = F_\pm\,F_\pm$$

so that, for $j\neq 0$, $\mathcal{C}_3$ is (projected on) the algebra spanned by $\{F_+,F_-,U\}$; and, for $j=0$, $F_\pm$ vanishes because so(3) is trivially represented, so that $\mathcal{C}_3$ is (projected on) the algebra spanned by $\{\omega_+;\omega_-,U\}$.

The operator Ψ and V, or, equivalently, $F_\pm\,F_\mp$ (or $\omega_\pm\,\omega_\mp$) must be scalar on every D(j,n), so we shall not distinguish between them and their eigenvalues in what follows. Relations (41) , (42), (46) (or (25) , (26) for j=0) yield relations among the eigenvalues of spanning elements of $\mathcal{C}$.
Let us first observe that from (41) one gets

(52) $$[F_+,F_-] = iV = -j^2 n \qquad \text{for } j \neq \tfrac{1}{2}$$

so that the Lie algebra spanned by F_+,F_-,U is isomorphic to sl(2) in LQW modules with $j > \frac{1}{2}$. This is not the case for j=0 or $\frac{1}{2}$, so that a direct study of the associative algebra which realizes $\mathcal{C}_3$ is necessary in these cases.

We shall next parametrize C_2 and C_4 by four complex numbers $\pm x_1$,

$\pm x_2$, which are the roots of the biquadratic equation :

(53) $$x^4 + (2C_2-5)x^2 + (C_2-2)^2 - 4C_4 = 0$$

so that

(54) $$C_2 = -\frac{1}{2}(x_1^2 + x_2^2 - 5)$$

(55) $$16C_4 = (x_1+x_2-1)(x_1-x_2-1)(x_1+x_2+1)(x_1-x_2+1)$$

The four roots are distinct in general. There are only 3 roots if, say, $x_2=0$, so that :

(56) $$16C_4 = 4(C_2-2)^2 = (x_1^2-1)^2$$

and there are 2 roots if $x_1^2 = x_2^2$, so that

(57) $$16C_4 = 4C_2-9 = 1-4x_1^2$$

There is only one root $x_1 = x_2 = 0$ iff

(58) $$16C_4 = 1 \quad \text{and} \quad 2C_2 = 5$$

The sums and differences of the roots, $\pm(x_1+x_2)$, $\pm(x_1-x_2)$ are, as easily seen, roots of the biquadratic equation :

(59) $$t^4 + 2(2C_2-5)t^2 + 16C_4 - 4C_2 + 9 = 0$$

which has a unique double root if (57) holds, two distinct opposite double roots if (56) holds and a unique root equal to 0 if (58) holds. The importance of the parameters x_i^2 and of the equations involving them will appear in what follows.

Using now (42) and (54) we obtain :

(60) $$F_+F_-+F_-F_+ = \Psi = \frac{1}{4}\, j^2(4n^2+(2j-1)^2+1-2x_1^2 - 2x_2^2)$$

Let us introduce the operator $\Xi = -j^{-2}\, iV + i\, U$, for $j \neq 0$.
By (52), $\Xi = 0$ if $j > \frac{1}{2}$, so that (45) can be written :

(61) $$((x_1+x_2)^2 - (2j-1)^2).((x_1-x_2)^2 - (2j-1)^2) = 16j^2\, \Xi^2$$

For $j > \frac{1}{2}$ this means nothing else but that $\pm(2j-1)$ are a couple of opposite roots of (59), since Ξ^2 vanishes. The eigenvalues of the two Casimirs cannot be arbitrary but they are related by (59), in which $t^2 = (2j-1)^2$, for a representation with LQW > $\frac{1}{2}$.

<u>Remark</u> : The reversal of ladder operators by $j \leftrightarrow -j-1$ shows that if one looks for representations having a highest quadratic weight j for so(3), then 2j+3 will be a solution of (62). For finite

dimensional representations — which are representations of SO(5) as well — such a j is the dominant weight of the representation. This case is discussed in Sec.4.4 and 4.5.

For $j = \frac{1}{2}$, (61) relates the eigenvalue of Ξ^2, hence the two possible eigenvalues of i V, to x_1^2 and x_2^2 ; it becomes :

(62) $\quad (x_1^2 - x_2^2)^2 = 4\Xi^2$

Now, using (24) , (51) one obtains

(63) $\quad 0 = [F_{\mp}, \omega_{\pm} - 4F_{\pm} F_{\pm}] = \pm(-\Xi F_{\pm} - F_{\pm}\Xi)$

so that Ξ anticommute with $F_{\pm}$. This means that if $2\xi_n = x_1^2 - x_2^2$ is the eigenvalue of 2Ξ on $D(j,n)$ (which involves a choice between the two roots of (53)), then $\xi_{n+k} = (-1)^k \xi_n$ for every $k \in \mathbb{Z}$.

We are now going to construct all possible modules $D(j)_\nu$ with Range (n) = $\nu+\mathbb{Z}$, and one-dimensional subspaces D(j,n). Let first $\varphi_\nu \in D(j,\nu)$ be a nonzero vector and $j \neq 0$. Put, for $k \in \mathbb{N}$, $\varphi_{\nu\pm k\pm 1} = F_{\pm} \varphi_{\nu\pm k}$. If x_1^2, x_2^2 (and, for $j = \frac{1}{2}$, the eigenvalue ξ_ν of Ξ on D(j,n)) are fixed, obeying (61), the quantities $F_{\mp} F_{\pm}$ are fixed on every D(j,n), so that $F_{\mp} \varphi_{\nu\pm k\pm 1} = (F_{\mp} F_{\pm})\varphi_{\nu\pm k} \in D(j,\bar{\nu}\pm k)$.

Clearly, $D(j)_\nu = \oplus_{h\in\mathbb{Z}} D(j,\nu+h)$ acquires a structure of LQW $\mathcal{C}_3$-module, since all three operators F_+, F_- , U are well defined on it, satisfying all constraints. On the other hand, $D(j)_\nu = \mathcal{C}_3 \varphi_\nu$ is monogeneous, and the representation obtained is indecomposable.

To examine irreducibility we see that $D(j)_\nu$ is irreducible iff it is generated by the action of $\mathcal{C}_3$ on any vector φ_n, $n \in \nu+\mathbb{Z}$; otherwise there exist invariant submodules. Now, it will appear on the explicit formulas that the restriction of F_+F_- on D(j,n) and of F_-F_+ on D(j,n-1) are equal, so that irreducibility is equivalent to the condition that $F_{\pm}F_{\mp}$ does not have the eigenvalue zero on any D(j,n), and so that both F_+, F_- are invertible operators. It is clear that the construction of $D(j)_\nu$ starting with any other eigenvector $\varphi_{\nu+h}$ yields an equivalent $\mathcal{C}_3$-module if $D(j)_\nu$ is irreducible. If it is reducible, the new module may be nonequivalent to the initial one, though all irreducible factors of the two indecomposable representations will be the same if the new construction is nonequivalent to the previous one, its only difference will be the conversion of some invariant submodules to factor modules and vice-versa. Notice that, if the representation splits, there will be a finite number of components only : indeed, it will appear that setting $F_{\mp} F_{\pm} = 0$ yields a finite number of finite degree equations in n.

The same discussion and results hold for j=0, with the following changes : $\mathcal{C}_3$ is spanned no more by F_+, F_- but by ω_+, ω_-.

Thus range(n) will be $\nu+2\mathbb{Z}$ on $\mathcal{C}_3\,\varphi_\nu$, since $[-i\,U, \omega_\pm] = \pm 2\,\omega_\pm$. So, for fixed C_2, C_4 we have two distinct inequivalent modules, $\mathcal{C}_3\,\varphi_\nu$ and $\mathcal{C}_3\varphi_{\nu+1}$, into which every module with Range(n) = $\nu+\mathbb{Z}$ splits ; and the preceding discussion holds for each of $\mathcal{C}_3\,\varphi_\nu$ $\mathcal{C}_3\,\varphi_{\nu+1}$, replacing in what precedes $F_\pm$ by $\omega_\pm$.

We are now going to study each case j=0, $j=\frac{1}{2}$, $j>\frac{1}{2}$ explicitely.

A) $j > \frac{1}{2}$

From (52) we obtain

(64) $\quad [j^{-1}\,F_+\,,\,j^{-1}\,F_-] = j^{-2}\,i\,V = -n$

so that the operators $j^{-1}\,F_+$, $j^{-1}\,F_-$, U satisfy the commutation relations of sl(2). All IR's of sl(2) for which the regular element U has an eigenvector are known [1]. They depend on two parameters, the range of n and the eigenvalue of the Casimir operator.

To describe them in our notations we first impose a choice of x_1, x_2, using (61) :

(65) $\quad x_1 = j - \frac{1}{2} + y\ ,\ x_2 = j - \frac{1}{2} - y$ so that $C_2 = -y^2 - (j+\frac{1}{2})^2 + \frac{5}{2}$

and denote by $(j,\nu;y^2)$ the representation on $D(j)_\nu$ characterized by these values of x_i^2.

Then we have, by (52) and (60) on D(j,n) :

(66) $\quad F_\pm\,F_\mp = \frac{1}{2}\,j^2\,[(n \mp \frac{1}{2})^2 - y^2]$

Setting this expression to zero we see that $(j,\nu;y^2)$ has 1,2 or 3 irreducible factors, if and only if 0,1 or 2 respectively among the square roots +y and -y belong to the set $\frac{1}{2}+\nu+\mathbb{Z}$. More precisely we have :

Theorem 1.A : There are the following IR's of $\mathfrak{g}$ corresponding to LQW $j > \frac{1}{2}$, $2j \in 2 + \mathbb{N}$:

a) $(j;\nu;y^2)$ with $\nu \in \mathcal{N}$, $y^2 \in \mathbb{C}$, $\pm y \notin (\nu+\frac{1}{2}+\mathbb{Z})$

b) $(j;[-m,\ldots[)$ with $m \in \mathbb{C}$, $2m \notin \mathbb{N}$.

c) $(j;]\ldots,m])$ with $m \in \mathbb{C}$, $2m \notin \mathbb{N}$.

d) $(j;[-m,m])$ with $2m \in \mathbb{N}$.

IR's of types b,c,d are factors of $(j;\nu;(m+\frac{1}{2})^2)$, with $\nu \in \mathcal{N}$ and $m \in \nu+\mathbb{Z}$ (for c,d) or $-m \in \nu+\mathbb{Z}$ (for b,d). Range(n) in D(j) is $\nu+\mathbb{Z}$ for a) and for the other IR's it is the set indicated by the brackets with step $\Delta n = 1$.

Remark : The 4 roots of (59) are, in this case : $\pm(2j-1)$, $\pm 2y$.

B) $j = \frac{1}{2}$

Using what precedes we obtain on $D(\frac{1}{2},n)$:

$$8F_{\pm} F_{\mp} = (n \mp \tfrac{1}{2})^2 - \tfrac{1}{2}(x_1^2 + x_2^2) \mp \xi_n$$

Recalling that $\xi_n = -\xi_{n+1}$ and that $\xi_n^2 = \xi^2 = \frac{1}{4}(x_1^2 - x_2^2)^2$ from (62) and (63), we first see that exchange of x_1^2 and x_2^2 is equivalent to $\Xi \leftrightarrow -\Xi$ and leads to inequivalent representations. To obtain simpler notations, let $x_1 = x_-$, $x_2 = x_+$, and for any complex ν denote by $(\frac{1}{2};\nu;x_-^2,x_+^2)$ the representation defined on $D(\frac{1}{2})_\nu$ by the eigenvalues of $F_{\pm}F_{\mp}$ on $D(j,n)$:

(67a) $\quad 8F_{\mp} F_{\pm} = (\pm n + \frac{1}{2})^2 - x_{\pm}^2 \qquad$ iff $\; n \in \nu+2\mathbb{Z}$

(67b) $\quad 8F_{\mp} F_{\pm} = (\pm n + \frac{1}{2})^2 - x_{\mp}^2 \qquad$ iff $\; n \in \nu+2\mathbb{Z}+1$

One obviously has $(\frac{1}{2};\nu;x_-^2,x_+^2) = (\frac{1}{2};\nu+1;x_+^2, x_-^2)$.

When $x_+^2 = x_-^2$ we obtain again the commutation relations of sl(2) and the situation is identical to that of $j > \frac{1}{2}$. One easily checks that F_+F_- and $D(\frac{1}{2},n)$ and F_-F_+ on $D(\frac{1}{2},n-1)$ are equal. Setting $F_+F_-=0$ it appears that $\nu+\mathbb{Z}$ splits at as many places as there are distinct elements in the set $S= (\{-x_-,x_-\}\cap\{\nu-\frac{1}{2}+2\mathbb{Z}\})\cup(\{-x_+x_+\}\cap(\nu+\frac{1}{2}+2$

So, suppose there is a splitting between m and m+1 ; modifying by translations the initial labelling of the representation, one can write

$$(\tfrac{1}{2};m;x_-^2,(m+\tfrac{1}{2})^2) = (\tfrac{1}{2};]\ldots,m];x_-^2) + (\tfrac{1}{2};[m+1,\ldots[;x_-^2)$$

The two summands denoting the representations obtained on the corresponding semi-bounded strings $m-\mathbb{N}$ and $m+1+\mathbb{N}$. Let now examine if such a semi-bounded string, denoted in what follows $n_{\pm} \mp \mathbb{N}$ to treat both directions simulatenously, admits further splitting or not ; one has, x^2 being the remaining parameter :

(68a) $\quad 8F_{\mp}F_{\pm}= (\pm n+\frac{1}{2})^2-(\pm n_{\pm}+\frac{1}{2})^2 = (n-n_{\pm})(n+n_{\pm}\pm 1) \quad$ iff $-n \in -n_{\pm}\pm 2\mathbb{N}$

(68b) $\quad 8F_{\mp}F_{\pm}= (\pm n+\frac{1}{2})^2 - x^2 \quad$ iff $-n \in -n_{\pm}\pm 1\pm 2\mathbb{N}$

The expression of (68a) cannot vanish for $n\neq n_{\pm}$ unless $\pm n_{\pm} \in \frac{1}{2}+\mathbb{N}$. The finite string obtained in that case is even dimensional and symmetric with respect to n=0. Putting $n_+=m$ the representation obtained will be denoted by $(\frac{1}{2};[-m,m];x^2)$, the remaining parameter x^2 being arbitrary.

The expression of (68b) can vanish only once if $n_{\pm} \notin \frac{1}{2}+\mathbb{Z}$. The finite string obtained that way must be odd dimensional ; it will be denoted $(\frac{1}{2};[m_1,m_2])$, the infinite string to which it belongs being $(\frac{1}{2};m_1;(m_1-\frac{1}{2})^2, (m_2+\frac{1}{2})^2)$. Notice that this implies $m_2-m_1 \in 2\mathbb{N}$.

Singletons, that is one-dimensional $\mathcal{C}_3$-modules $D(\frac{1}{2})$ are obtained when $m_1=m_2$ the difference (or sum) between the two x's being ± 1. The representations $(\frac{1}{2};[m_1,m_2])$ are irreducible if $m_1 \notin \frac{1}{2}+\mathbb{Z}$; nevertheless, they are still well defined if $m_1 \in \frac{1}{2}+\mathbb{Z}$.

Finally, if $x_+ \in 2\mathbb{Z}+1$ and $x_- \in 2\mathbb{Z}$, the representation $(\frac{1}{2};\frac{1}{2};x_-^2,x_+^2)$ splits into 5 irreducible factors, or 4 ones if $x_-=0$.
Let $x_{\pm}^2 > x_{\mp}^2$ and $m_{\pm} = |x_{\pm}| - \frac{1}{2}$, $m_{\mp} = |x_{\mp}| + \frac{1}{2}$; with the notations already introduced these factors are $(\frac{1}{2};[m_{\pm}+1,\ldots[;x_{\mp}^2)$; $(\frac{1}{2};]\ldots,-m_{\pm}-1];x_{\mp}^2)$; $(\frac{1}{2};[m_{\mp},m_{\pm}])$; $(\frac{1}{2};[-m_{\pm},-m_{\mp}])$, and, except in case $x_{\mp} = m_{\mp} -\frac{1}{2} = 0$, $(\frac{1}{2};[-m_{\mp}+1,m_{\mp}-1];x_{\pm}^2)$. Notice that the last factor acts on an even-dimensional space.

Summarizing what precedes we obtain :

<u>Theorem 1B</u> : There are the following IR's of $\mathfrak{g}$ corresponding to LQW $\frac{1}{2}$:

a) $(\frac{1}{2};\nu;x_-^2,x_+^2) \sim (\frac{1}{2};\nu\pm 1;x_+^2,x_-^2)$ with ν, $x_{\pm}^2 \in \mathbb{C}$ and $x_{\pm},-x_{\pm} \notin \nu\pm\frac{1}{2}+\mathbb{Z}$

b) $(\frac{1}{2};[-m,\ldots[;x^2)$ with $m \in \mathbb{C}$, $m \notin \frac{1}{2}+\mathbb{N}$, $\pm x \notin -m+\frac{1}{2}+2\mathbb{N}$

c) $(\frac{1}{2};]\ldots,m];x^2)$ with same constraints as b)

d) $(\frac{1}{2};[-m,m];x^2)$ with $m \in \frac{1}{2}+\mathbb{N}$, $x \notin (-m+\frac{1}{2}+2\mathbb{N}) \cap (m-\frac{1}{2}-2\mathbb{N})$

e) $(\frac{1}{2};[m_1,m_2])$ with $m_i \in \mathbb{C}$, $m_2-m_1 \in 2\mathbb{N}$ and if $m_i \in \frac{1}{2}+\mathbb{Z}$ then $m_1.m_2 > 0$.

The range of n in $D(\frac{1}{2})$ is $\nu+\mathbb{Z}$ for type a), and it is the string indicated by the brackets for the other types : infinite dimensional semi-bounded for types b,c ; even dimensional for type d, odd dimensional for type e.

C) <u>j = 0</u>

Using (25), (26) we obtain on D(o,n) :

$$(69) \qquad 4\omega_{\pm}\omega_{\mp} = ((n\mp 1)^2 + C_2-2)^2 - (n\mp 1)^2 - 4C_4$$

so that $\omega_+\omega_-(n) = \omega_+\omega_-(2-n) = \omega_-\omega_+(n-2) = \omega_-\omega_+(-n)$.

Formula (69) can be used to define an action of $\mathcal{C}_3$ on $D(0)_\nu$, but it will never become a monogeneous $\mathcal{C}_3$-module since there is no means to go to D(0,n+1) from D(0,n). So, from now on, $D(0)_\nu$ <u>will denote the string $\nu+2\mathbb{Z}$</u> which is irreducible in general, unless $\omega_{\pm}\omega_{\mp}$ vanishes somewhere.

Introducing the roots x_1^2, x_2^2 of (53) one obtains

$$(70) \quad 4\omega_{\pm}\omega_{\mp} = ((n\mp 1)^2 - x_1^2).((n\mp 1)^2 - x_2^2)$$

Any representation defined on $D(0)_\nu$ and satisfying (70) will be designed by $(0;\nu;x_1^2;x_2^2) = (0;\nu;x_2^2;x_1^2) = (0;\nu\pm 2;x_2^2,x_1^2)$. (We take again the liberty of equalling representations which may be nonequivalent because their irreducible factors are equivalent). There will be k+1 irreducible factors in $(0;\nu;x_1^2,x_2^2)$ if k is the number of elements of the set $\{x_1,-x_1, x_2,-x_2\}\cap(\nu+1+2\mathbb{Z})$.

So, suppose, as in the previous case that $(0;n;(n+1)^2,x^2)$ is reducible to $(0;[m+2,\ldots[;x^2) + (0;]\ldots,m];x^2)$.

The study of the irreducible components is quite similar to the case $j=\frac{1}{2}$, and even simpler. Let a splitting occur because of x_1 ; we shall write $(0;x_1+1;x_1^2;x_2^2)$ as the sum $(o;[x_1+1,\ldots[;x_2^2) + (0;]\ldots,x_1-1],x_2^2)$ the corresponding ranges of n being $x_1+1+2\mathbb{N}$, $x_1-1-2\mathbb{N}$.

Next we search for splittings on a string such that Range(n)$\subset n_{\pm}-(\pm 2\mathbb{N})$. One must have

$$0 = (n-n_{\pm})(n+n_{\pm}\pm 2)((n\pm 1)^2 - x^2)$$

There is a splitting either if $\pm n_{\pm}\in\mathbb{N}$, the finite part being denoted by $(0;[\mp n_{\pm},\pm n_{\pm}];x^2)$ its dimension can be odd or even (the step is 2) its ends are opposite integers ([0,0] is not excluded) and the remaining parameter x^2 can be any number ; or if $\sqrt{x^2}\in\pm n_{\pm}+1-2\mathbb{N}$, which can occur only for one square root of x^2 unless $n_{\pm}\in\mathbb{Z}$. Such a representation will be denoted by $(0;[m_1,m_2])$ and it is a factor of $(0;m_1;(m_1-1)^2, (m_2+1)^2)$. The dimension can be any number, and x_1-x_2 is an even integer, different from zero.

When n takes integer values multiple splittings may occur. Indeed, $(0;\alpha,(2k+1-\alpha)^2, (2h+1-\alpha)^2)$ with $\alpha=0$ or 1, $k<h$ both integers, contains the irreducible factors labelled

$(0;[2h+2-\alpha,\ldots[;(2k+1-\alpha)^2)$; $(0;]\ldots,-2h-2+\alpha]; (2k+1-\alpha)^2)$;

$(0;[2k+2-\alpha,2h-\alpha])$; $(0;[-2h+\alpha,-2k-2+\alpha])$; and (unless $x_1=k=0$, $\alpha=1$),

$(0;[-2k+\alpha,2k-\alpha]; (2h+1-\alpha)^2)$.

Summarizing we have :

<u>Theorem 1C</u> : There are the following IR's of $\mathfrak{g}$ having LQW 0 :

a) $(0;\nu;x_1^2,x_2^2) \sim (0;\nu+2k;x_2^2,x_1^2)$, $\forall k \in \mathbb{Z}$, with $\nu, x_i^2 \in \mathbb{C}$ and $\pm x_i \notin (\nu+1+2\mathbb{Z})$

b) $(0;[-m,\ldots[;x^2)$ with $m \in \mathbb{C}$, $m \notin \mathbb{N}$, $x^2 \in \mathbb{C}$, $\pm x \notin -m+1+2\mathbb{N}$.

c) $(0;]\ldots,m];x^2)$ with same constraints as b)

d) $(0;[-m,m];x^2)$ with $m \in \mathbb{N}$, $x^2 \in \mathbb{C}$, $x \notin (-m+1+2\mathbb{N}) \cap (m-1-2\mathbb{N})$.

e) $(0;[m_1,m_2])$ with $m_1 \in \mathbb{C}$, $m_2 \in m_1+2\mathbb{N}$ and $(m_1 \in \mathbb{Z} \Longrightarrow m_1.m_2 > 0)$.

IR's of types b,c,d are factors of a unique reducible representation $(0;\nu;(m-1)^2,x^2)$ with $\nu \in -m+2\mathbb{Z}$ for b,d and $\nu \in m+2\mathbb{Z}$ for c,d. IR's of type e are factors of $(0;m_1;(m_1-1)^2, (m_2+1)^2)$. Range(n) in D(0) is $\nu+2\mathbb{Z}$ for type a, and it is the set indicated by the brackets for the other types , with step $\Delta n=2$.

4. REDUCTION TO $\bar{K}$ OF AN i.r. OF $\bar{G}$ (UP TO MULTIPLICITY)

4.1. General considerations

The object of this section is to study Range(j,n) for a given i.r. R characterized by the LQW j_{min}, by $\nu \in \mathbb{C}/\mathbb{Z}$, and by the other parameters just studied. From what precedes we know that Range(j,n) will be a subset of the set $(j_{min} + \mathbb{N}) \times (\nu+\mathbb{Z})$. We have also seen that restrictions of Range(j,n) $\cap \{j=j_{min}\}$ either fix one of the two parameters characterizing R (if $j_{min} = 0,1/2$), or they completely characterize R if $j_{min} > \frac{1}{2}$. We have also seen that if $j_{min}=0$, Range(0,n) $\subset (0,\nu+\alpha+2\mathbb{Z})$ with $\alpha = 0$ or 1.

Here we shall examine what happens if there is a hole somewhere, that is, if for some $(j,n) \in (j_{min} + \mathbb{N}) \times (\nu+\mathbb{Z})$ the space D(j,n) is {0}.
We shall first prove for any i.r. :

<u>Lemme 3</u> : Let $\varphi \in D_n^j$ such that $\omega_\pm \varphi = F_\pm \varphi = 0 \neq \varphi$ then $(j,n\pm k)$ does not belong to Range(j,n) for any integer $k > 0$.

<u>Proof</u> : Fixing $\pm 1=-1$ for a better visualisation, we see from (24) that $\omega_- F_+ \varphi = 0$ as well ; for (27a), (27b) we obtain that $\omega_- D(j,n) = F_- D(j,n) = \{0\}$ since $D(j,n) = \mathcal{C}\varphi$, so $\omega_- F_- D(j,n) = \{0\}$ also. Hence $D(j,n-1) = \{0\}$ and $D(j,n-k) \subset \mathcal{C}_3\, \omega_- D(j,n) + \mathcal{C}_3 D(j,n-1) = \{0\}$ for every k.
Next, we shall prove :

<u>Lemma 4</u> : If $D(j,n\pm 1) = \{0\} \neq D(j,n)$, then eigher $D(j,n\pm k) = \{0\}$ for every integer $k > 0$, or

$$D(j,n+1) = D(j,n-1) = 0 .$$

Proof : Suppose again $\pm 1=-1$. If F_-F_+ does not vanish identically on $D(j,n)$, take $\varphi = F_-F_+\psi \neq 0$ in $\bar{D}(j,n)$.

Then apply lemma 3 on φ : one has $F_-\varphi = 0$ and $\omega_-\varphi = F_-\omega_-F_+\psi \in F_-D(j,n-1) = \{0\}$, so that the condition of lemma 3 are satisfied. Now if $F_-F_+ D(j,n) = \{0\}$, we have $D(j,n+1) \subset F_+D(j,n) + \omega_+F_-D(j,n)$ so it is sufficient to show that $F_+\psi = 0$ for every ψ in $D(j,n)$. If this were not the case , take $\varphi = F_+\psi \neq 0$. One has $F_-\varphi = \omega_-\varphi = 0$, so that $\mathcal{C}_3$ does not contain $D(j,n)$ (by lemma 3), hence $\mathcal{C}_3\varphi$ is a proper invariant $\mathcal{C}_3$-submodule of $D(j)$, hence $\{0\}$ because of irreducibility.

4.2. The case $F_+ = F_- = 0$.

We shall study the two cases of Lemma 4 separately. Suppose first that both $D(j,n\pm 1) = \{0\} \neq D(j,n)$, and $j \neq 0$ (to avoid already seen situations). Then $F_\pm = 0$ on $D(j,n)$, hence, by (35), $jF_k\uparrow = (j+1)F_k\downarrow = (F_kJ_k)^- = 0$, to that $F_k = 0$ as well and the ideal I_φ which annihilates a nonzero $\varphi \in E(j,n)$ contains UF. From (9) we conclude that F vanishes all over $E = \mathcal{U}_g\varphi$, so that $\mathcal{C}$ is abelian, $C_4=0$, and every nonzero $D(j,n)$ is one dimensional. Moreover one has

$$(71) \quad M_{\pm k}\, E(j,n) \subset E(j+1,n\pm 1) \oplus E(j-1,n\pm 1)$$

so that if $(j_o,n_o) \in \text{Range}(j,n)$, every element (j,n) of this set must satisfy :

$$(72) \quad j+n-j_o-n_o \in 2\mathbb{Z}$$

What can be the LQW restriction of such an i.r. ? If $j_{min} > \frac{1}{2}$ is the LQW, one must have $i\,V = 0$, so that $(j_{min})^2.n = 0$. It follows that n=0 for $j=j_{min}$, and we have i.r.'s of type Ad, labelled $(j_{min};[0,0])$ according to the classification of theorem 1A. If $j_{min} = \frac{1}{2}$, the $\mathcal{C}_3$-module must again be a singleton since $\mathcal{C}_3$ is still spanned by $\{U,F_+,F_-\}$; hence it has to be a representation of type B.e, labelled $(\frac{1}{2},[m,m])$. Such singletons may exist for any complex value m. The Casimir values are then

$$(73) \quad C_4 = 0,\ C_2 = \frac{9}{4} - m^2 \ , \text{ with } x_1 = m - \frac{1}{2} \ , \ x_2 = m + \frac{1}{2}$$

while in the $j_{min} > \frac{1}{2}$ case they are :

$$(74) \quad C_4 = 0,\ C_2 = \frac{9}{4} - (j_{min}-\tfrac{1}{2})^2; \text{ with } x_1=j_{min} \ , \ x_2 = j_{min}-1.$$

On the other hand, if $j_{min} = 0$, then $D(0)$ need not be a singleton. The condition $C_4=0$ is sufficient ; for if $C_4\varphi = 0$ and $F_k\varphi \neq 0$ then

$F_k\varphi$ is a LQW vector for the invariant proper $\underline{g}$-submodule $\mathcal{U}_g F_k\varphi$, with LQW equal to 1, which is contradictory to irreducibility. If $C_4=0$ put $x_1=x$, $x_2=x-1$ and $C_2=\frac{9}{4}-(x-\frac{1}{2})^2$. Then (70) becomes :

$$\text{(75)} \qquad \omega_{\pm}\omega = ((n\mp 1)^2 - x^2)((n\mp 1)^2-(x-1)^2)$$

which shows that if $D(0)_\nu$ is reducible then $D(0)_{\nu+1}$ is also reducilbe and that there cannot be more than a double splitting.

The different types of i.r.'s which satisfy $C_4=0$ are :

$$\text{(76)} \begin{cases} \text{Type Ca}:(0;\nu;x^2,(x-1)^2) \text{ with } \pm x \notin \nu+\mathbb{Z}, C_2=2+x-x^2 \\ \text{Type Cb}:(0;[-m,\ldots[;(m-1+\varepsilon)^2) \text{with } \varepsilon=\pm 1, 2m\notin \mathbb{N}, C_2=-m^2+(2-\varepsilon)m+1+\varepsilon \\ \text{Type Cc}:(0;]\ldots,m];(m-1+\varepsilon)^2) \text{with } \varepsilon=\pm 1, 2m\notin \mathbb{N}, C_2=-m^2+(2-\varepsilon)m+1+\varepsilon \\ \text{Type Cd}:(0;[-m,\ldots,m];(m-1+\varepsilon)^2) \text{with } \varepsilon=\pm 1, m\in \mathbb{N}, C_2=-m^2+(2-\varepsilon)m+1+\varepsilon \\ \text{Type Ce}:(0;[-m+\tfrac{1}{2}\varepsilon,m+\tfrac{1}{2}\varepsilon]) \text{ with } m\in\mathbb{N}, \quad \varepsilon=\pm 1, \ C_2=-m^2+\frac{9}{4} \end{cases}$$

Another manner to describe D(0) is to form the direct sum $D = D(0)_\nu \oplus D(0)_{\nu+1}$ taking x to have the same value on both, with $C_4=0$. Grant D with a structure of $\underline{sl}(2)$-module where $\{T_+,T_-,U=in\}$ are the generators of $\underline{sl}(2)$, represented by

$$\text{(77)} \qquad T_{\pm}\, T_{\mp} = (n_{\mp}\, 1\pm x)\ (n\mp x)$$

Then $\omega_{\pm}$ can be identified to $T_{\pm}\, T_{\pm}$ through (75) ; determine first the $\underline{sl}(2)$ splitting using the Miller classification : then each irreducible $\underline{sl}(2)$-module (except the one-dimensional one) yields two $\mathcal{C}_3$-irreducible components corresponding to the choice ν or $\nu+1$.

On the other hand, the expression of $M^{\uparrow}_{\mp k}\ M^{\downarrow}_{\pm k}$ and $M^{\downarrow}_{\mp k}\ M^{\uparrow}_{\pm k}$ are directly available from 35, 36. Substituting $C_2=2+x-x^2$ we obtain on D(j,n) :

$$\text{(78)} \qquad 2(2j+1)\ M^{\uparrow}_{\pm k}\ M^{\downarrow}_{\mp k} = j(\pm n+j-x)(\pm n+j-1+x)$$

$$\text{(79)} \qquad 2(2j+1)\ M^{\downarrow}_{\pm k}\ M^{\uparrow}_{\mp k} = (j+1)(\pm n-j-1-x)(\pm n-j-2+x)$$

These equations are valid for every j and n and for every one of the i.r.'s studied here, provided x is conveniently interpreted in the different cases. We summarize all this discussion in the following theorem, and we leave the complete characterization of Range(j,n) to later.

<u>Theorem 2</u> : The i.r.'s for which $F_\mu = 0$ all over the space are multiplicity free and $j_1+n_1-j_2-n_2 \in 2\mathbb{Z}$ for any couples (j_i,n_i) in

Range(j,n). The LQW characterizations of such i.r.'s are given by (76) for $j_{min} = 0$; for $j_{min} > \frac{1}{2}$ only $(j_{min};[0,0])$ is of that type, and for $j_{min} = \frac{1}{2}$ every singleton $(\frac{1}{2};[m,m])$ is of that type. The expressions of $M^{\uparrow}_{\mp k} M^{\downarrow}_{\pm k}$ and $M^{\downarrow}_{\mp k} M^{\uparrow}_{\pm k}$ on every D(j,n) are given by (78), (79).

4.3. The case where Re(±n) has a lower bound in D(j)

Let study now the other case of Lemma 4. We suppose $D(j,n\mp k)=0\neq D(j,n)$ for every $k>0$ and a definite choice of ±. We are going to prove :

Lemma 5 : If $F_{\pm}=\omega_{\pm}=0$ on D(j,n) , then at least one of the spaces $M^{\uparrow}_{\pm k}$ E(j,n) or $\bar{M}^{\downarrow}_{\pm k}$ E(j,n) is {0}.

Proof : Suppose $\omega_{-} = F_{-} = 0$. Then $(M^{\uparrow}_{+i} M^{\downarrow}_{+i})(M^{\downarrow}_{-k} M^{\uparrow}_{-k}) = 0$ as well. But $M^{\downarrow}_{+i} M^{\downarrow}_{-k} - M^{\downarrow}_{-i} M^{\downarrow}_{+k}$ is an eigenvector of $Ad\bar{W}$, acting on $M_{+i} M_{-k} - M_{-i} M_{+k} = [M_{+i},M_{-k}]+ \varepsilon_{hki}(\varepsilon_{hu\upsilon} M_{-u} M_{+\upsilon})$, corresponding to the eigenvalue $-j(j+1) + (j-2)(j-1)$. Since the non arrowed quantity is a combination of so(3)-scalars and 3-vectors the arrowed quantity vanishes and

$$(80) \quad (M^{\uparrow}_{+i} M^{\downarrow}_{-i})(M^{\downarrow}_{+k} M^{\uparrow}_{-k}) = 0 = (M^{\downarrow}_{+k} M^{\uparrow}_{-k})(M^{\uparrow}_{+i} M^{\downarrow}_{-i})$$

(the second equality is obtained by reversing the direction of the arrows).

An argument similar to that of Lemma 3 shows that either $M^{\downarrow}_{+k} M^{\uparrow}_{-k}$ or $M^{\uparrow}_{+k} M^{\downarrow}_{-k}$ vanishes on D(j,n).

Suppose that $M^{\uparrow}_{+k} M^{\downarrow}_{-k} = 0$. If there is a nonzero φ in $M^{\downarrow}_{-k}$ E(j,n) (we remind that $E(j,n) = \mathbb{C}^{2j+1} \otimes D(j,n)$), its annihilator I_{φ} will contain $M^{\uparrow}_{\pm k}$ and $F^{\uparrow}_{k}$. But the invariant g-module $\mathcal{U}_{g}\varphi$ will not contain $E(j,n')$ for any n', because from (23), (36) one gets :

$$(81a) \quad M^{\uparrow}_{\pm k} \; \mathcal{C}_3 \; E(j,n) \subset \sum_{\varepsilon=\pm 1} \mathcal{C}_3 \; M^{\uparrow}_{\varepsilon k} \; E(j,n) \text{ for every } j,n.$$

Because of irreducibility $\mathcal{U}_{g}\varphi$ must vanish, hence $\varphi = 0$.

The case $M^{\downarrow}_{+k} M^{\uparrow}_{-k} = 0$ is similar, with the exception j=0. Indeed, from (23), (36) one gets :

$$(81b) \quad M^{\downarrow}_{\pm k} \; \mathcal{C}_3 \; E(j,n) \subset \sum_{\varepsilon=\pm 1} \mathcal{C}_3 \; M^{\downarrow}_{\varepsilon k} \; E(j,n) \text{ for every } j\neq 1, \text{ every } n.$$

so that the case $\varphi \in M^{\uparrow}_{-k}$ E(0,n) may a priori be an exception Though $M^{\downarrow}_{k}$ vanishes identically for j=0, it is interesting to see that $\omega_{-} = M_{+k} M_{-k} = 0$ on E(0,n) = D(0,n) imply also $M_{-k}=0$ (we may drop the arrows since $A=A^{\uparrow}$ on E(0) for every 3-vector). We simply have to show that, on E(j,n) :

$$A_k^{\downarrow} X = X_1 M_{+k}^{\downarrow} + X_2 M_{-k}^{\downarrow} + X_3 F_k^{\downarrow}$$

with $X_i \in \mathcal{C}_3$, X being any generator of $\mathcal{C}_3$ and $A_k \in \{M_{+k}, M_{-k}, F_k\}$. This can be immediately derived from (21), (22), (23) if $A_k = M_{\pm k}$. Putting A = F one has from (21) .

$$F_k F_+ = [M_{+k}, F_- F_+]$$

and since $F_- F_+$ is expressible linearly by means of $\omega_\pm \omega_\mp$ and the center of $\mathcal{C}$, $F_k^{\downarrow} F_+$ will take the desired form by using (23). The case A=F, $X = \omega_\pm$ is covered by 24b .

We conclude that $\mathcal{C}_3 M_{-k}^{\uparrow} \dot{E}(0,n)$ is a LQW $\mathcal{C}_3$-module as well, and we may apply the argument of irreducibility, which implies $M_{-k}^{\uparrow} = 0$ on E(0,n).

Space E(j,n), D(j,n) and vectors in them for which $F_\pm = \omega_\pm = 0$ and $M_{\pm k}^{\downarrow} = 0$ (resp. $M_{\pm k}^{\uparrow} = 0$) will be called $\pm\downarrow$ (resp. $\pm\uparrow$) <u>boundary</u> spaces or vectors ;
indeed, if one represents by a dot in the (Re(n),j) plane every element of Range(j,n), such points will be extremal on the line j = constant, minimal (left boundary) if $F_- = \omega_- = 0$ and maximal (right boundary) if $F_+ = \omega_+ = 0$.
We next have, for $\varepsilon = \pm$, $\varepsilon' = -\varepsilon$:

<u>Lemma 6</u> : Let $\varphi \in E(j,n)$ be an $\varepsilon\uparrow$-boundary (resp. $\varepsilon\downarrow$-boundary) vector. Then $M_{\varepsilon k}^{\downarrow}\varphi$ and $M_{\varepsilon' k}^{\uparrow}\varphi$ (resp. $M_{\varepsilon k}^{\uparrow}\varphi$ and $M_{\varepsilon' k}^{\downarrow}\varphi$) either vanish or they are also $\varepsilon\uparrow$ (resp. $\varepsilon\downarrow$) boundary vectors.

<u>Proof</u> : One has just to prove that F_ε and $M_{\varepsilon\ell}^{\uparrow}$ (resp. $M_{\varepsilon\ell}^{\downarrow}$) are zero on the vectors in question : then ω_ε will vanish as well, by (45) (46). Since $0 = [F_\varepsilon, M_{\varepsilon k}] = [M_{\varepsilon\ell}^{\uparrow}, M_{\varepsilon' k}^{\uparrow}] = [M_{\varepsilon\ell}^{\downarrow}, M_{\varepsilon' k}^{\downarrow}]$, and since both $M_{\varepsilon\ell}^{\uparrow} M_{\varepsilon k}^{\downarrow}\varphi$, $M_{\varepsilon\ell}^{\downarrow} M_{\varepsilon k}^{\uparrow}\varphi$ are in $E(j,n-2\varepsilon) = \{0\}$, we have to check $F_\varepsilon M_{\varepsilon' k}^{\uparrow}$ (resp. $F_\varepsilon M_{\varepsilon' k}^{\downarrow}$). From (36a), (36b) one gets :

$$(j+1) F_\varepsilon M_{\varepsilon' k}^{\uparrow} = (j+2) M_{\varepsilon' k}^{\uparrow} F_\varepsilon - F_{\varepsilon'} M_{\varepsilon k}^{\uparrow}$$

$$j F_\varepsilon M_{\varepsilon' k}^{\downarrow} = (j-1) M_{\varepsilon' k}^{\downarrow} F_\varepsilon + F_{\varepsilon'} M_{\varepsilon k}^{\downarrow}$$

and the right members vanish on φ under the respective hypotheses (if j=0, $M_{\varepsilon' k}^{\downarrow} = 0$ any how).

Now $\Psi = \varepsilon i V$ on every ε boundary $\mathcal{C}$-module, so the representation of $\mathcal{C}$ is abelian on it, and the multiplicity of (j,n) must be one. Thus by Lemma 6 we have established the existence of a whole string of irreducible so(3)-submodules which are all $\varepsilon\uparrow$ (resp. $\varepsilon\downarrow$) boundaries, obtained from one another by action of $M_{\varepsilon k}^{\downarrow}$ and $M_{\varepsilon' k}^{\uparrow}$ (resp. $M_{\varepsilon k}^{\uparrow}, M_{\varepsilon' k}^{\downarrow}$).

There is a big difference between i.r.'s containing $\varepsilon\uparrow$ boundaries and those containing $\varepsilon\downarrow$ boundaries : if φ is an $\varepsilon\uparrow$ boundary vector, one cannot indefinitely iterate the procedure $\varphi \rightarrow M^{\downarrow}_{\varepsilon k}\varphi$ and obtain nonzero results. At some step (the latest when reaching the LQW subspace) one will obtain a space E(j,n) on which both $M^{\downarrow}_{\varepsilon k}$ and $M^{\uparrow}_{\varepsilon k}$ will be zero. Thus i.r.'s with $\varepsilon\uparrow$ boundaries form a subset of i.r.'s with $\varepsilon\downarrow$ boundaries. This subset may be characterized, as we shall soon see, by the existence of an upper or lower bound for Re(n) which does not depend on j.

4.4. The $\varepsilon\downarrow$ and $\varepsilon\uparrow$ boundaries of an i.r.

Except types Aa, Ba, Ca, all other types of i.r.'s classified in Theorem 1 contain either one $\varepsilon\downarrow$ boundary $\mathcal{C}$-submodule (ε=-1 for types Ab,Bb,Cb and ε=+1 for types Ac,Bc,Cc) in the LQW $\mathcal{C}_3$-submodule ; or they contain two boundaries, one for each choice of ε, for the remaining types.

Let j_{min} be the LQW and suppose (j_{min},n_ε) are the characteristics of the ε boundary in the LQW space. Define the quantity $s = j_{min} - \varepsilon n_\varepsilon - 1$; it is closely related to the Casimir values : indeed $\pm$s is one couple of the roots of the biquadratic equation (53) , which have been labelled x_1,x_2 previously ; a glance to (66), (68), (70) is sufficient to prove this assertion.

Now, for $j=j_{min}$, the value $Re(\varepsilon n_\varepsilon)$ is the maximal value taken by $Re(\varepsilon n)$. If (j,n) are the characteristics of any other $\varepsilon\downarrow$ boundary vector, generated by repeated action of $M^{\uparrow}_{\varepsilon k}$ on $E(j_{min},n_\varepsilon)$ one has $j-j_{min} = \varepsilon n - \varepsilon n_\varepsilon$, so that

(82) $\quad j-\varepsilon n = s+1$

for every $\varepsilon\downarrow$ boundary E(j,n). It follows that :

(83) $\quad Re(\varepsilon n) \leqslant Re(j-s-1) \qquad$ for every (j,n)

so that $Re(\varepsilon n)$ is bounded for fixed j; but its maximul increases with j.

Suppose now that an $\varepsilon\uparrow$ boundary string exists in an i.r., and let

(84) $\quad d = j+\varepsilon n+2$

This quantity is a constant on the string.

As j decreases on the string we shall reach a couple (j_c,n_c), called critical couple, for which $M^{\downarrow}_{\varepsilon k}$ also vanishes, and from which an $\varepsilon\downarrow$ boundary starts downwards towards j_{min}. We of course have

(85) $\quad Re(\varepsilon n) \leqslant d-j-2 \qquad$ for every (j,n) with $j \geqslant j_c$

and this inequality combined with (83) shows that $Re(\varepsilon n_c)$ is an absolute maximum for $Re(\varepsilon n)$.
The quantities s and d are related by

(86) $s+d = 2j_c+1 \quad ; \quad -s+d = 2\varepsilon n_c+3$

Using the fact that $M_{\varepsilon'k} M_{\varepsilon k} = 0$ on a critical vector one easily sees that ± d is the other (than ±s) couple of roots of (53), and $\pm(2j_c+1)$, $\pm(2\varepsilon n_c+3)$ are the 4 roots of (59).

Using equations (37),(38),(43) we get the following two sets of expressions of eigenvalues of some elements of $\mathcal{C}$ acting on each boundary string, in terms of {s,d} or $\{j_c, \varepsilon n_c\}$ (with $\varepsilon' = -\varepsilon$):

On the $\varepsilon\downarrow$ string ($F_\varepsilon = M^\downarrow_{\varepsilon k} = 0$; $\varepsilon n = j-s-1$) :

(87) $\Psi = \varepsilon iV = \frac{1}{4} j(s+d+1)(s-d+1) = -j(j_c+1)(\varepsilon n_c+1)$

(88) $j(2j+1)M^\uparrow_{\varepsilon k}M^\downarrow_{\varepsilon' k} = \frac{1}{4}j(2j-1)(s+d-2j+1)(s-d-2j+1) = j(2j-1)(j-j_c-1)(j+\varepsilon n_c+1)$

(89) $M^\downarrow_{\varepsilon' k}M^\uparrow_{\varepsilon k} = \frac{1}{4}(s+d-2j-1)(s-d-2j-1) = (j-j_c)(j+\varepsilon n_c+2)$

On the $\varepsilon\uparrow$ string ($F_\varepsilon = M^\uparrow_{\varepsilon k} = 0$; $\varepsilon n = -j+d-2$) :

(90) $\Psi = \varepsilon iV = +\frac{1}{4}(j+1)(s+d-1)(s-d+1) = -(j+1).j_c.(\varepsilon n_c+1)$

(91) $(2j+1)M^\downarrow_{\varepsilon k}M^\uparrow_{\varepsilon' k} = \frac{1}{4}(2j+1)(2j+3-s-d)(2j+3+s-d) = (2j+3)(j+1-j_c)(j-\varepsilon n_c)$

(92) $M^\uparrow_{\varepsilon' k}\ M^\downarrow_{\varepsilon k} = \frac{1}{4}(2j+1-s-d)(2j+1+s-d) = (j-j_c)(j-\varepsilon n_c-1)$

<u>Remark</u> : We point out that (87) to (89) are valid even if there is no $\varepsilon\uparrow$ boundary string in the representation. In this case there is of course no critical point and j_c, εn_c are just complex numbers defined by (86), ±d is just the remaining couple of roots of (53), while s conserves its meaning.

The equations above yield the necessary and sufficient conditions for the existence of a critical subspace and they provide information on the extent of the boundary string. We first of all point out that the extent of the $\varepsilon^\downarrow$ string is infinite if no solution of (59) is in the set $2j_{min}+1+2\mathbb{N}$, and there is no critical subspace.

Suppose now that a critical subspace exists with $j_c \geqslant j_{min}$ and take the case $j_{min} > \frac{1}{2}$. Substituting $j=j_{min}$ in (88), the only possibility for the expression to vanish is

(93) $j_{min} + \varepsilon n_c + 1 = 0$

and, since $j-\varepsilon n$ is constant on the boundary, we also have,

(j_{min},n_ε) being the LQW $\varepsilon\downarrow$ boundary :

(94) $\quad j_c + \varepsilon n_\varepsilon + 1 = 0$

This means that the only i.r.'s which have critical subspaces are of type Ab (if $\varepsilon = -1$) or Ac (if $\varepsilon = +1$). For every such i.r. Range n is the set $-\varepsilon(j_{min}+1+\mathbb{N})$, and for its LQW subspace $\text{Range}(n) = -\varepsilon(j_c+1+\mathbb{N})$ with $j_c - j_{min} \in \mathbb{N}$. Substituting(93) in (91) one checks that the $\varepsilon\uparrow$ string has infinite extent because

$$(2j+1)M^{\downarrow}_{\varepsilon k}\, M^{\uparrow}_{\varepsilon' k} = (2j+3)(j+1-j_c)(j+j_{min}+1) > 0$$

The cases $j_{min}=0$ and $j_{min}=\frac{1}{2}$ are quite similar and can be treated together. The condition for the existence of an $\varepsilon\downarrow$ string is that Range n has an upper bound in the LQW subspace for some choice of ε : this is related to the roots of (53). A critical subspace exists iff there is a root of (59) in the set $2j_{min}+1+2\mathbb{N}$.

To characterize completely cases b and c one must write that εn has no lower bound for $j=j_{min}$.

This is the case iff $\varepsilon n_c - j_c = \varepsilon n_\varepsilon - j_{min}$ is not in $\mathbb{N}$.

This condition also shows that $M^{\uparrow}_{\varepsilon' k}$ never vanishes so that the $\varepsilon\uparrow$ string is of infinite extent.

In a representation of type d (53) has one root in $\mathbb{Z}$, and Range(n) $\subset j_{min}+\mathbb{Z}$, so that if there is a critical subspace all roots of (59) are integers. One has $n_+ + n_- = 0$ and $n_+ - j_{min} = -s-1 \in \mathbb{N}$.

Range n is symmetric with respect to n=0 and there are two critical submodules corresponding to opposite values of n, say $-n_c \leqslant 0 \leqslant n_c$. The positive value is on the $\varepsilon=+1$ string since $n_c - j_c = n_+ - j_{min} \geqslant 0$ and it is superior to both n_+ and j_c. It follows that some day j will take the value $n_c \geqslant j_c$ and $M^{\downarrow}_{\varepsilon k}\, M^{\uparrow}_{\varepsilon' k}$ will vanishes. The value of n at that point will be j_c. We leave it to the reader to prove that this i.r. is the finite-dimensional representation with weights n_c, j_c.

For representations of type e, the range of n is $\{m_1, m_1+\Delta n, \ldots, m_2\}$ with $m_2 - m_1 \in 2\mathbb{N}$ and $\Delta n = 2-2j_{min}$. The roots of (59) are $x_1 = m_1 - 1 + j_{min}$, $x_2 = m_2 + 1 - j_{min}$ and their opposites. To identify s and d among $\{x_1, -x_1, x_2, -x_2\}$ one observes that $s+d \in 2j_{min}+1+2\mathbb{N}$ while $x_2 - x_1 \in 2\mathbb{N} + 2 - 2j_{min}$, so that s+d is either $x_1 + x_2 = m_1 + m_2$ or its opposite. This means that m_i is half-integer if $j_{min} = 0$ and integer if $j_{min} = \frac{1}{2}$. One must have

(95) $\quad m_1 = \pm(j_c + \frac{1}{2}) - k \qquad m_2 = \pm(j_c + \frac{1}{2}) + k \qquad k \in \mathbb{N}$

according to whether the sum $m_1 + m_2$ is positive or not.

There is only one critical point, with $\varepsilon = -1$ if $m_1+m_2 > 0$ or $\varepsilon = +1$ if $m_1+m_2 < 0$. One has

(96) $2j_c+1 = -\varepsilon(m_1+m_2)$; $2\varepsilon n_c = m_2-m_1 - 1-2j_{min} = 2k-1-2j_{min}$

and

(97) $2s = (1-\varepsilon)(j_{min} + m_1-1) + (1+\varepsilon)(j_{min}-m_2-1)$

$2d = (1-\varepsilon)(-j_{min}+m_2+1) + (1+\varepsilon)(-j_{min}-m_1+1)$

Both strings are infinite since (91) never vanishes.
For fixed $j \geqslant j_c$ the range of n is finite ; the distance between the extremities in $m_2-m_1+2j_c-2j_{min} = 2(k+j_c-j_m)$.
The distance is zero only if $k = j_c-j_{min} = 0$: there are 4 i.r.'s in this case, satisfying $n = \pm(j+\frac{1}{2})$, two for each choice of j_{min}.

Remark : The i.r.'s of theorem 2 which have an $\varepsilon\downarrow$ boundary (that is all of them except those of type (a) behave as following :
For $j_{min} = 0$ there is always a critical subspace with $j_c=0$;
for $(\frac{1}{2};[m,m])$ there is a critical subspace iff m Z-{0}; for $\{j_{min};[0,0]\}, j_{min} > 0$ there is no critical subspace.

4.5. Finiteness of Range j

The only representations for which Range j is a finite set we have met with are those of types Bd, Cd having critical subspaces. Are there any other ones ?

Notice first that only types Aa,Ba,Ca are candidates, since otherwise we would have discovered them when walking on the ε-boundaries.

Suppose that Range j is bounded and let φ be any vector in the LQW subspace. Then $(M^{\uparrow}_{+k})^y\varphi$ will vanish for some integer y. This leads to

$$M^{\downarrow}_{+k}\dots M^{\downarrow}_{+h}\,(M^{\downarrow}_{+i}\;M^{\uparrow}_{+i})\;M^{\uparrow}_{+h}\dots M^{\uparrow}_{+k}\;\varphi = 0$$

Using (48) and the fact that ω_+, F_+ commute with $M^{\uparrow}_{+k}$ we obtain

(98) $((j_{min}+1)^2\omega_+-F_+F_+)((j_{min}+2)^2\omega_+-F_+F_+)\dots((j_{min}+y)^2\omega_+-F_+F_+)\varphi = 0$

But $j_{min}^2\omega_+-F_+F_+ = 0$ on the LQW subspace hence ω_+ and F_+ must be nilpotent to satisfy (98) , and the same for ω_-, F_-. Thus Range n must be finite on LQW subspace, so the type of the i.r. must be d or e. This settles the question by the negative.

4.6. Summarizing theorem

The following theorem describes completely Range(j,n) for the different types of representations :

Theorem 3 : For every $(j,n)\in(\mathbb{N}\cup\frac{1}{2}+\mathbb{N})\times\mathbb{C}$ define the following sets :

$X(j,n)=\{(j',n');\ j'\in j+\mathbb{N},\ n'\in n+j'-j+2\mathbb{Z}\}$; $T(0,n)=n+2\mathbb{Z}$;

$T(j+\frac{1}{2},n)=n+\mathbb{Z}$; $Y(0,n)=X(0,n)\cup X(1,n)$;

$Y(j+\frac{1}{2},n)=X(j+\frac{1}{2},n+1)\cup X(j+\frac{1}{2},n)$; $Z(j,n;\pm)=Y(j,n)\cap\{(j',n')$; $j'-j\mp n'\pm n\in\mathbb{N}\}$.

Then, for every i.r. of g, characterized by its LQW j_{min}, by the set N, equal to the range of n in the LQW subspace, and by the roots ±s, ±d of (53) (which obey the adequate constraints if necessary), the set Range(j,n) is defined by the following considerations :

1/ If there is $\nu\in\mathbb{C}$ such that $N=T(j_{min},\nu)$ (types Aa,Ba,Ca) then Range $(j,n)=Y(j_{min},\nu)\cap X$, where $X=X(0,\nu)$ iff

$C_4=((s+d)^2-1)((s-d)^2-1)=0=j_{min}$, and $X=\mathbb{C}$ otherwise.

2/ If there is $n_\varepsilon\in\mathbb{C}$, with $\varepsilon\in\{+1,-1\}$, such that $N=(n_\varepsilon-\varepsilon\mathbb{N})\cap T(j_{min},n_\varepsilon)$ in which case the root **s** obeys $s=j_{min}-\varepsilon n_\varepsilon-1$ (types b,c), then :

2.1) If s±d are not in $2j_{min}+1+2\mathbb{N}$, Range(j,n) is the set $Z(j_{min},n_\varepsilon,\varepsilon)\cap X$, where $X=X(0,n_\varepsilon)$ iff $j_{min}=0=C_4$ and $X=\mathbb{C}$ otherwise.

2.2) if $s+d=2j_c+1\in 2j_{min}+1+2\mathbb{N}$ (which implies, for type Ab,Ac, that $-\varepsilon n_\varepsilon\in j_{min}+1+\mathbb{N}$, and with $\varepsilon n_c=\frac{1}{2}(d-s-3)=\varepsilon n_\varepsilon+j_c-j_{min}$, Range(j,n) is $Z(j_{min},n_\varepsilon,\varepsilon)\cap\{(j,n)\ ;\ n\in n_c+\varepsilon(j_c-j)-\varepsilon\mathbb{N}\}\cap X$, where $X=X(0,n_\varepsilon)$ iff $0=j_{min}=j_c(j_c+\varepsilon n_\varepsilon+1)(j_c+\varepsilon n_\varepsilon+2)$ and $X=\mathbb{C}$ otherwise.

3/ If $j_{min}>\frac{1}{2}$ and $N=\{-n_+,-n_++1,\dots,n_+\}$ with $2n_+\in\mathbb{N}$ (type Ad) then Range $(j,n)=Z(j_{min},n_+,+)\cap Z(j_{min},-n_+,-)\cap X$, where $X=X(j_{min},0)$ if $n_+=0$ and $X=\mathbb{C}$ otherwise;

4/ If $j_{min}\in\{0,\frac{1}{2}\}$ and $N=\{n;|n|\leqslant n_+\}\cap T(j_{min},n_+)$, with $n_+\in j_{min}+\mathbb{N}$ (types Bd,Cd), in which case $s=j_{min}-n_+-1$, then :

4.1) if s±d are not in $2j_{min}+1+2\mathbb{N}$, Range(j,n) = $Z(j_{min},n_+,+)\cap$

$Z(j_{min},-n_+,-)\cap X$, where $X = X(0,n_+)$ iff $j_{min} = C_4 = 0$ and $X = \mathbb{C}$ otherwise.

4.2) If $s+d=2j_c+1 \in 2j_{min}+1+2\mathbb{N}$, with $n_c = n_+ + j_c - j_{min}$, then Range(j,n) is $Z(j_{min},n_+,+)\cap Z(j_{min},-n_+,-)\cap\{(j,n)\ ;\ |n| \leqslant j_c+n_c-j,\ j \leqslant n_c\}\cap X$, where $X=X(0,n_+)$ iff $j_{min}=j_c=0$, and $X = \mathbb{C}$ otherwise.

5/ If $j_{min} \in \{0,\frac{1}{2}\}$ and there are n_+,n_- such that $N = (n_+ - \mathbb{N})\cap(n_- + \mathbb{N})\cap T(j_{min},n_+)$, with $n_+ - n_- \in 2\mathbb{N}$ and with $(n_+ \in j_{min} + \mathbb{Z} \Longrightarrow n_- n_+ > 0)$ (types Be,Ce), then :

5.1) if $\pm(n_+ + n_-)$ are not in $2j_{min} + 1 + 2\mathbb{N}$ Range(j,n) = $Z(j_{min},n_+,+)\cap Z(j_{min},n_-,-)\cap X$, where $X=X(j_{min},n_+)$ iff $j_{min} = \frac{1}{2}$ and $n_+ = n_-$, and $X = \mathbb{C}$ otherwise.

5.2) if, for $\varepsilon \in \{+1,-1\}$, $-\varepsilon(n_+ + n_-) = 2j_c+1 \in 2j_{min}+1+2\mathbb{N}$, with $2\varepsilon n_c = (n_+ - n_- - 1 - 2j_{min})$, Range(j,n)=$Z(j_{min},n_+,+)\cap Z(j_{min},n_-,-)\cap\{(j,n);n \in n_c+\varepsilon(j_c-j)-\varepsilon\mathbb{N}\}\cap X$, where $X=X(j_{min},n_+)$ iff either $j_{min}=j_c=0$ or $j_{min}-\frac{1}{2}=n_+-n_- = 0$, and $X = \mathbb{C}$ otherwise.

Some of these sets are illustrated in figures 1 to 15. These figures indicate regions of the (j,Ren) plane, limitated by lines : either straight lines or, for $j_{min} = 0$ a broken line with globally horizontal direction. Lines ending to arrows are supposed to go on indefinitely. Representative points of (j,Ren) are crosses or circles : circles represent points which are not in the spectrum if U F = 0, i.e. cases described in Theorem 2. The value -1 is systematicall given to ε. The label j increases upwards, Ren from left to right.

Table: Unitary Irreducible Representations of $\overline{SO}_0(3.2)$

Type	Label	Constraints	Roots of (53) - on t^2 -	Observations (Critical subspaces and UF=0)
Aa	$(j;\nu;y^2)$	$\frac{1}{2}<j;\ 0\leq\nu<1;\ y^2<(\nu-\frac{1}{2})^2$	$(j-\frac{1}{2}+y)(j-\frac{1}{2}-y)^2$	
Ba	$(\frac{1}{2};\nu;\ x_-^2,x_+^2)$	$-\frac{1}{2}\leq\nu<\frac{1}{2};\ x_\pm^2<(\pm\nu+\frac{1}{2})^2$	$x_+^2\ \ x_-^2$	
Ca	$(0;\nu;x_1^2,x_2^2)$ equivalent to $(0;\nu;x_2^2,x_1^2)$	$-1\leq\nu<1 \begin{cases} x_1^2+x_2^2<\nu^2+(1-\lvert\nu\rvert)^2 \\ (1-(x_1+x_2)^2)(1-(x_1-x_2)^2)\geq 0 \\ ((1-\lvert\nu\rvert)^2-x_1^2)((1-\lvert\nu\rvert)^2-x_2^2)>0 \end{cases}$	$x_1^2\ \ x_2^2$	$UF=0 \Longleftrightarrow (x_1\pm x_2)^2=1$
Ab	$(j;[m,\dots[\,)$	$\frac{1}{2}<j,\ \ 0<m\leq j+1$	$(j+m-1)^2(j-m)^2$	$m=j+1 \Longleftrightarrow \exists j_c=j$
Ac	$(j;]\dots,]m\,)$			
Bb	$(\frac{1}{2};[m,\dots[;x^2)$	$-\frac{1}{2}<m;\ x^2<\mathrm{Inf}_k(m+\frac{1}{2}-k)^2,$ $k\in\{0\}\cup(3+2\mathbb{N})$	$(m-\frac{1}{2})^2,\ x^2$	
Bc	$(\frac{1}{2};],,,.-m];x^2)$			
Cb	$(0;[m,\dots[;x^2)$	$0<m;\ x^2<\mathrm{Inf}(m-2k)^2,\ k\in\mathbb{N}$	$(m-1)^2\ ,\ x^2$	
Cc	$(0;]\dots,-m];x^2)$			
Bb,Cb	$(j;[m,\dots[;\ x^2)$	$\begin{cases} j\in\{0,\frac{1}{2}\} \\ m>\frac{1}{2}+j \end{cases}$ $x=-m+2-j+2j_c$	$(m-1+j)^2,\ x^2$	$j_c\in j+\mathbb{N};\ j_c>j \Longrightarrow 2j_c+1<m+j$
Bc,Cc	$(j;]\dots,-m];x^2)$			$UF=0 \Longleftrightarrow j_c=j=0$
Cb	$(0;[m,\dots[;m^2)$	$0<m<1$	$(m-1)^2,\ m^2$	$UF=0$
Cc	$(0;]\dots,-m];m^2)$			
Ad	$(j;[0,0])$	$1\leq j$	$j^2,\ (j-1)^2$	$UF=0$
Bd	$(\frac{1}{2};[-\frac{1}{2},\frac{1}{2}];x^2)$	$x^2<0$	$1,\ x^2$	
Cd	$(0;[0,0];x^2)$	$x^2\leq 0$	$1,\ x^2$	$UF=0 \Longleftrightarrow x^2=0$
Be	$(\frac{1}{2};[m,m])^{1/2}$	$-1\leq m\leq 1$	$(m-\frac{1}{2})^2,(m+\frac{1}{2})^2$	$UF=0;\ m=\pm 1 \Longleftrightarrow \exists j_c=\frac{1}{2}$
Ce	$(0;\ [\pm 1/2,\pm 1/2])$		$1/4,\ 9/4$	$UF=0\ ;\ \ j_c=0$
Cd	$(0;[0,0];4)$		$1\ ,\ 4$	TRIVIAL

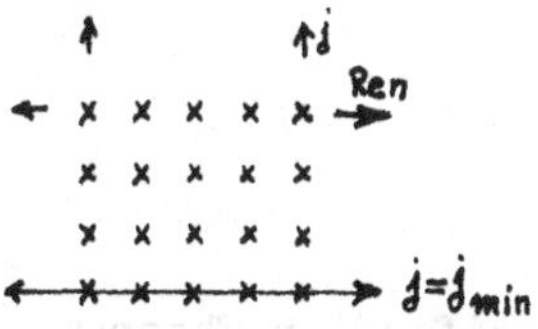

Fig 1: Case 1, $j_{min} > 0$

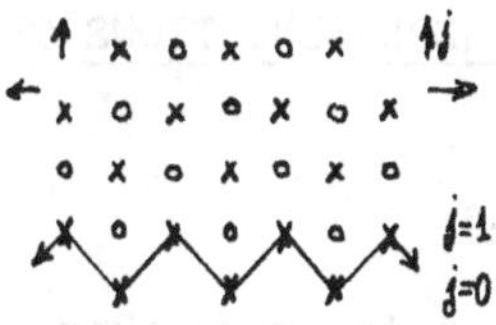

Fig 2: Case 1, $j_{min} = 0$

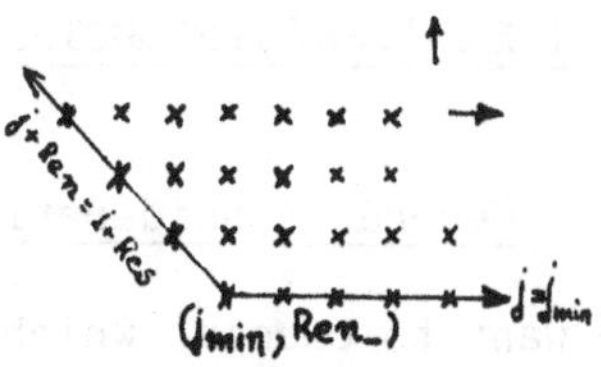

Fig 3: Case 2.1, $j_{min} > 0$

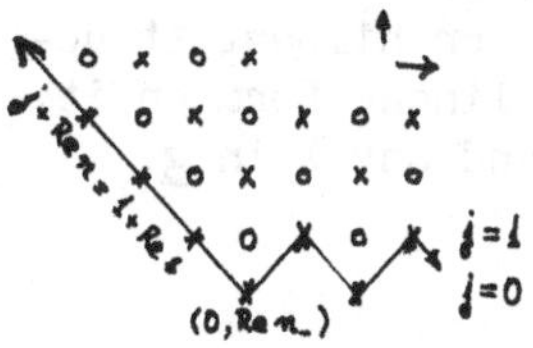

Fig 4: Case 2.1, $j_{min} = 0$

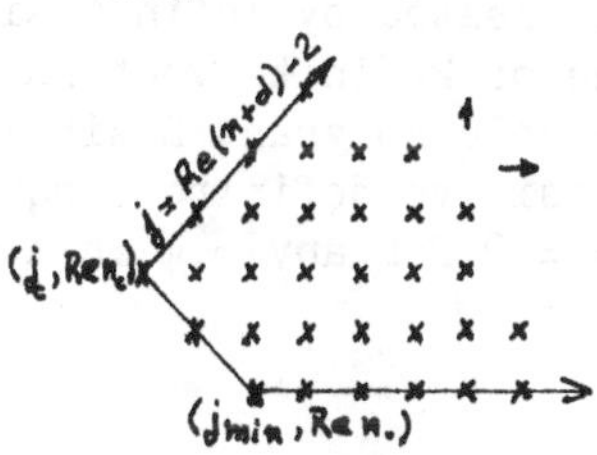

Fig 5: Case 2.2, $j_{min} > 0$ ($j_c = j_{min} + 2$)

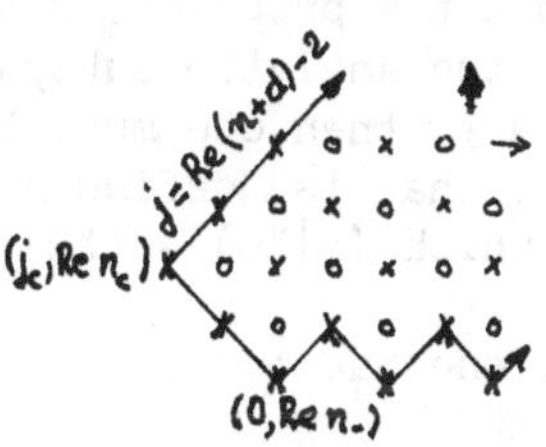

Fig 6: Case 2.2, $j_{min} = 0$ ($j_c = 2$)

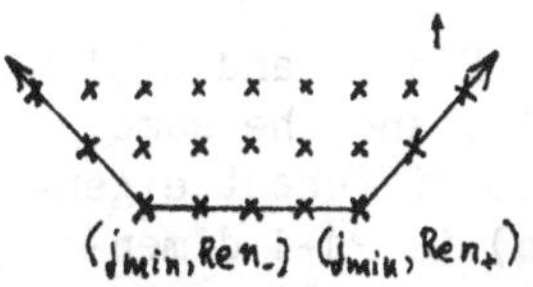

Fig 7: Case 3, $n_+ = -n_- \in 1+\mathbb{N}$ ($n_+=2$) and Case 5.1, $j_{min} = \frac{1}{2}$, $n_+ \neq n_-$ ($n_+ - n_- = 4$)

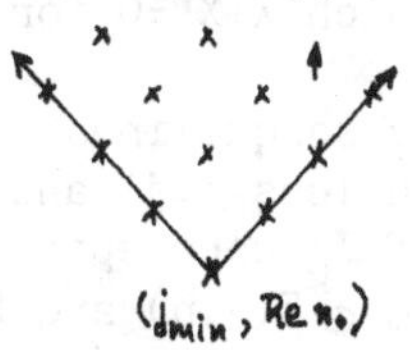

Fig 8: Case 3, $n_+ = 0$ and Case 5.1, $j_{min} = \frac{1}{2}$, $n_+ = n_-$

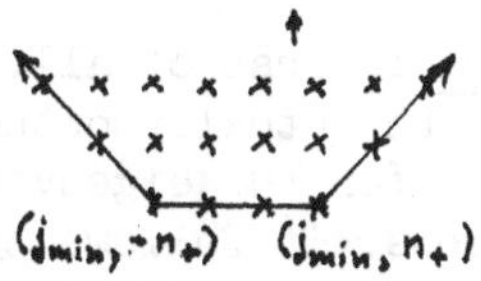

Fig 9: Case 3, $n_+ \in \frac{1}{2} + \mathbb{N}$ ($n_+ = 3/2$) and Case 4.1, $j_{min} = \frac{1}{2}$ ($n_+ = 3/2$)

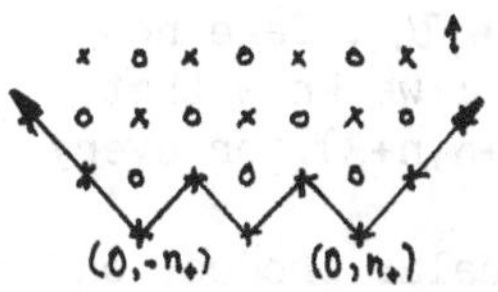

Fig 10: Case 4.1, $j_{min} = 0$ ($n_+ = 2$)

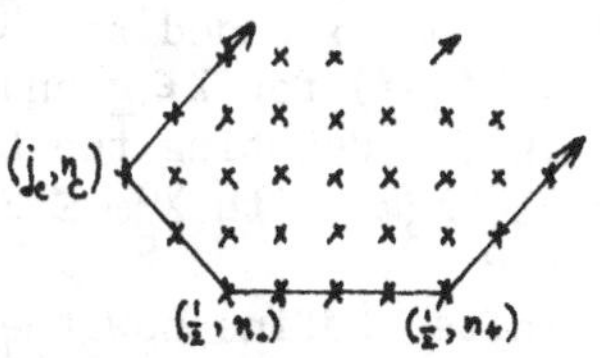

Fig 11: Case 5.2, $j_{min} = \frac{1}{2}$, $n_+ \neq n_-$ ($j_c = \frac{5}{2}$, $n_c = -1$, $n_- = 1$, $n_+ = 5$)

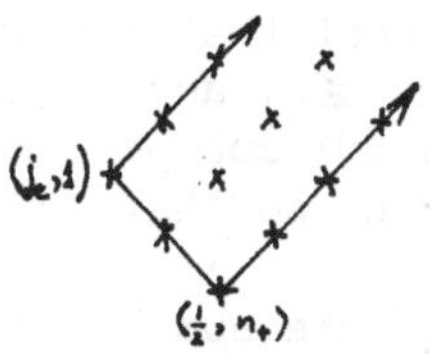

Fig 12: Case 5.2, $j_{min} = \frac{1}{2}$, $n_c = n_-$, $n_c = 1$ ($j_c = \frac{5}{2}$, $n_+ = 3$)

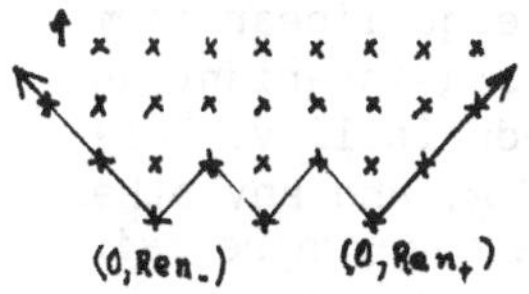

Fig 13: Case 5.1, $j_{min} = 0$ ($n_+ - n_- = 4$)

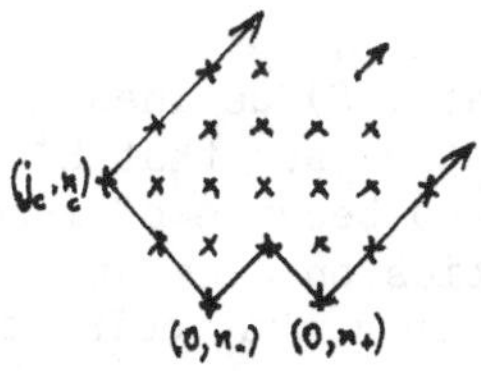

Fig 14: Case 5.2, $j_{min} = 0 \neq j_c$ ($j_c = 2$, $n_- = 3/2$, $n_+ = 7/2$, $n_c = -1/2$)

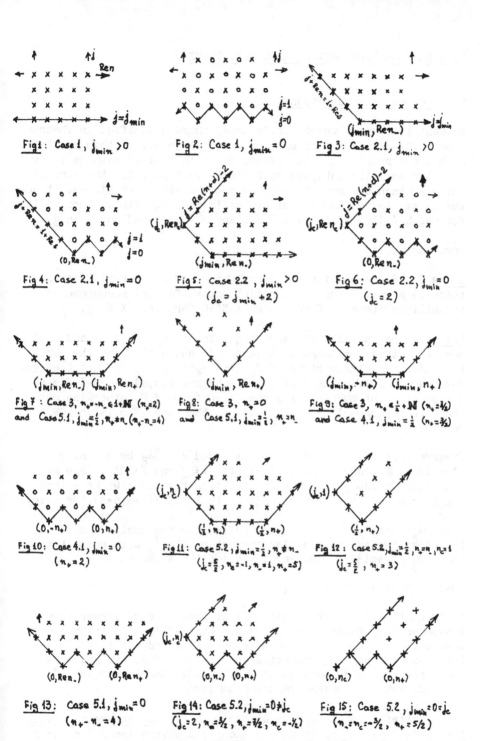

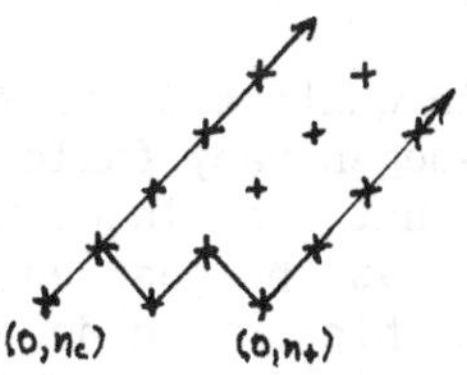

Fig 15: Case 5.2, $j_{min} = 0 = j_c$ ($n_- = n_c = -3/2$, $n_+ = 5/2$)

5. THE UNITARY IRREDUCIBLE REPRESENTATIONS OF $\bar{G}$

5.1. General considerations

We want to examine which i.r.'s among those classified in Theorem 1 are H.C.-equivalent to unitary ones. Since all of them are integrable the problem can be treated by infinitesimal methods : Let E be the analytic subspace of k-finite vectors ; iff the i.r. is unitary, then one must be able to grant E with a prehilbert structure, that is, define a positive definite sesquilinear form on it, for which $(\varphi|X\psi) + (X\varphi|\psi) = 0$ for any φ,ψ in E and any X in g.

Now, one has :

Lemma 7 : For every i.r. of $\bar{G}$ for which i U has a real spectrum, there is a unique (up to a multiplicative factor) nondegenerate sesquilinear form on E for which $X+X^*=0$ for every X in g.

Proof : First of all, for any sesquilinear form, E(j,n) and E(j',n') must be mutually orthogonal unless j=j' and n=n' , and the same holds for two eigenvectors of J_k corresponding to different eigenvaluess. It follows that if $\varphi \in E(j,n)$ and E(j,n) is 2j+1 dimensional (i.e. of multiplicity one) then $(\varphi|\varphi) \neq 0$, otherwise φ would be orthogonal to all E. This is the case if φ is an LQW vector.

Observe next that $X \to X^*$ can be extended to $\mathcal{U}_g$ by defining $\lambda^*=\bar{\lambda}$ for $\lambda \in \mathbb{C}$, and $(XY)^* = Y^*(-X)$ for $X \in g$ and $Y \in \mathcal{U}_g$. Take now φ in the LQW subspace E(j,n), j denoting the LQW ; we know that $\mathcal{U}_g\varphi = E$, so that $X\varphi = \sum_{\alpha,\beta} X^\alpha_\beta\varphi$ with $X^\alpha_\beta\varphi \in E(j+\alpha,n+\beta)$ for every X in $\mathcal{U}_g$, the sum being finite. Define now the scalar product on E(j,n) as the ordinary scalar product on $\mathbb{C}^{2j+1}$, with respect to which D_j of so(3) is unitary ; and let for $A\varphi, B\varphi \in E$:

$$(99) \qquad (A\varphi|B\varphi) \equiv (\varphi|(A^*B)^0_0\varphi)$$

It is quite easy to show that (99) defines a sesquilinear form. Non-degeneracy follows from the fact that $E^\perp$ is g-invariant and does not contain φ; hence zero because of irreducibility. This form has the desired properties on elements of g, and any other form which coïncide with it on φ must coïncide everywhere, since E is nomogeneous. Hence the lemma is proved.

Since unitarity on $\bar{G}$ implies unitarity on $\bar{K}$, the parameter n will take only real values in what follows ; Applying lemma 7 on sees that an i.r. is unitary iff the sesquilinear form defined by (99) is positive definite, in which case E has the desired prehilbert structure.

Fixing $\varphi \in E(j,n)$, $\psi \in E(j+\alpha,n+\varepsilon)$, one easily sees that $(M_{\varepsilon k}\varphi|\psi) + (\varphi|M_{\varepsilon' k}\psi) = 0$ implies (with $\varepsilon' = -\varepsilon$ as usual)

$$(100) \qquad (F_{\varepsilon})^* = F_{\varepsilon'} \quad ; \quad (M^{\uparrow}_{\varepsilon k})^* = M^{\downarrow}_{\varepsilon' k}$$

For every $E(j,n)$ the space ${}^{\uparrow}_{\varepsilon}E(j,n) = \sum_k M^{\uparrow}_{\varepsilon k} E(j,n)$ (the sum is not direct in general) in an $\underline{so}(3)$-invariant subspace of $E(j+1,n+\varepsilon)$; but not necessarily a $\mathscr{C}$-invariant one.

On the other hand, let $Z^{\downarrow}_{\varepsilon}(j+1,n+\varepsilon) = (\bigcap_k \mathrm{Ker}\, M^{\downarrow}_{\varepsilon k}) \cap E(j+1,n+\varepsilon)$ denote the orthocomplementary subspace of ${}^{\uparrow}_{\varepsilon}E(j,n)$ inside $E(j+1,n+\varepsilon)$. Notice that the two subspace can be defined for either choice of ε, and that one of them may be $\{0\}$ (even both iff $(j+1,n+\varepsilon)$ is not in Range(j,n)). By algebraic direct sum one can also define :

$${}^{\uparrow}_{\varepsilon}E(j) = \bigoplus_n {}^{\uparrow}_{\varepsilon}E(j,n) \ ; \ Z^{\downarrow}_{\varepsilon}(j+1) = \bigoplus_n Z^{\downarrow}_{\varepsilon}(j+1,n)$$

Notice next that $M^{\downarrow}_{\varepsilon' 3}M^{\uparrow}_{\varepsilon 3}\varphi = ((j+1)^2-m^2)(j+1)^{-1}(2j+3)^{-1}M^{\downarrow}_{\varepsilon' k}M^{\uparrow}_{\varepsilon k}\varphi$ if $J_3\varphi = \mathrm{im}\ \varphi$, because of (32), and similar expressions for $M^{\uparrow}_{\varepsilon k}\ M^{\downarrow}_{\varepsilon' k}$ in general, show that $(M^{\uparrow}_{\varepsilon k}\varphi\ |M^{\uparrow}_{\varepsilon k}\varphi)$ conserves the same sign independently of summation or not on the index k, and the same holds with downwards arrows.

Now, a look at (44)

$$(44)\ j^2(j+1)(2j+1)M^{\downarrow}_{\varepsilon' k}M^{\uparrow}_{\varepsilon k} = j(j+1)^2(2j+1)M^{\uparrow}_{\varepsilon k}M^{\downarrow}_{\varepsilon' k} + (j+1)F_{\varepsilon}F_{\varepsilon'} + jF_{\varepsilon'}F_{\varepsilon} + (j^2+j)^2(2j+1+2\varepsilon n)$$

together with what precedes shows that :

<u>Lemma 8</u> : In the sesquilinear form is positive definite on every $E(j')$ with $j_{min} \leqslant j' \leqslant j$ and $j > 0$, and if $2j+1+2\varepsilon n \geqslant 0$ then it is also positive definite on ${}^{\uparrow}_{\varepsilon}E(j,n)$.

<u>Remark</u> : If the form is positive it is also positive definite, since it is not degenerate, by Lemma 7. Another way of proving definiteness is to show that if both numbers of (44) vanish on φ, then $M^{\uparrow}_{\varepsilon k}\ \varphi$ is a LQW vector.

We point out that lemma 8 gives a sufficient condition of positivity. Also, there is always a choice of ε for fixed j and n such that $j + \frac{1}{2}\varepsilon n \geqslant 0$; both choices are good for $(j + \frac{1}{2})^2 \geqslant n^2$.

To see what happens on $Z^{\downarrow}_{\varepsilon}(j,n)$ take φ in this space and use (46), modified by (34a) ; one obtains , for $j > \frac{1}{2}$:

$$(101)\ j(2j+1)^{-1}P_j(\varphi|\varphi) = (\varphi|(F^{\uparrow}_k F^{\downarrow}_k + j(j-1-\varepsilon n)M^{\uparrow}_{\varepsilon k}M^{\downarrow}_{\varepsilon' k} + j(2j-1)^{-1} M^{\uparrow}_{\varepsilon h}M^{\downarrow}_{\varepsilon' k}M^{\uparrow}_{\varepsilon k}M^{\downarrow}_{\varepsilon' h})\varphi)$$

with $P_j = C_4+(C_2-2)(j^2-j)^2$; we recall that $P_j=0$ means that $\pm(2j+1)$ are roots of (59). Using (44) to modify the last term one also obtains for $j \geqslant \frac{3}{2}$.

$$(102)\quad j(2j+1)^{-1}P_j(\varphi|\varphi)=Q+[(j-\varepsilon n)(4j^2-6j+1)+(j-1)]j(2j-1)^{-2}$$

$$(\varphi|M^{\uparrow}_{\varepsilon k}\, M^{\downarrow}_{\varepsilon' k}\varphi),$$

where Q is a linear combination with positive coefficients of squares of vectors in $E(j-1),E(j-2)$.

There is plenty of information in these formulas. One easily sees, at a first glance, that they imply $P_j > 0$, unless $Z^{\downarrow}_{\varepsilon}(j,n) = \{0\}$ whenever $j-\varepsilon n \geqslant 0$ (or $-\varepsilon n \geqslant 0$ for $j=1$), which means $E(j,n) \subset {}^{\uparrow}_{\varepsilon}E(j-1,n+\varepsilon)$. This last relation means of course that the so(3) multiplicity does not increase from $(j-1,n+\varepsilon)$ to (j,n) and it appears when the i.r. in question contains a critical point (j_c,n_c). Indeed one has :

Lemma 9 : If for a given i.r. and any j, $_\beta$Rangen $\subset \nu+\varepsilon\mathbb{N}$ within $E(j)$, for complex ν, then $E(j) \subset \bigoplus_{\alpha,\beta\in\mathbb{N}} \omega^{\beta}_{\varepsilon}\, F^{\beta}_{\varepsilon}\, E(j,\nu)$.

The proof is based on induction upon $\varepsilon(n-\nu)$. One can use (24a) which gives $[\omega_{\varepsilon'},F_{\varepsilon}]$ and $[F_{\varepsilon'},\omega_{\varepsilon}]$ whenever it is possible, that is $\alpha \neq 0$ for $F_{\varepsilon'}$, and $\beta \neq 0$ for $\omega_{\varepsilon'}$; otherwise one can always express $\omega_{\varepsilon'}\omega_{\varepsilon}$ in terms of $F_{\varepsilon}F_{\varepsilon'}$ and $F_{\varepsilon'}F_{\varepsilon}$ and vice versa. Details are left to the reader.

It is easily follows from Lemma 9 that if $E(j)$ contains an $\varepsilon^{\uparrow}$ boundary $E(j,\nu)$, then $E(j+1) \subset {}^{\uparrow}_{\varepsilon'}E(j)$, since this is the case on the boundary, and since operators $M^{\uparrow}_{\varepsilon' k},\omega_{\varepsilon'}$, $F_{\varepsilon'}$ commute. We shall make use of that later, to seek representations which do not fulfill the following condition :

Proposition 5 : (sufficient condition of positivity)
If the sesquilinear for (99) is positive definite on $E(j_{min})$ and, in case $j_{min} = 0$, on ${}^{\uparrow}_{+}E(0)$ and on ${}^{\uparrow}_{-}E(0)$, then, $P_j \geqslant 0$ for every $j \geqslant j_{min}+1$ implies positivity on the whole E.

Proof : Using induction one chooses ε such that $-n\varepsilon \geqslant 0$ for given j,n which yields positivity on $Z^{\downarrow}_{\varepsilon}(j,n)$ or nullity of this space. But (99) is then positive on the orthocomplement ${}^{\uparrow}_{\varepsilon'}E(j-1,n+\varepsilon)$, either by hypothesis if $j=1$, or by Lemma 8 if $j > 1$, since $j-1-\varepsilon(n+\varepsilon)+\frac{1}{2}=j-\ n=\frac{1}{2} \geqslant 0$.
Another condition of positivity concerning i.r.'s with critical subspaces is :

Proposition 6 : An i.r. containing a critical subspace (j_c,n_c) with $E(j+1) \subset {}^{\uparrow}_{\varepsilon}E(j)$ for every $j \geqslant j_c$ is unitary iff (99) is positive on every $E(j)$, $j_{min} \leqslant j \leqslant j_c$.

Proof : For $j > j_c$ one has $j+\varepsilon n \leqslant j_c+\varepsilon n_c$. To apply lemma 8 one needs $j-\varepsilon n+\frac{1}{2} \geqslant 0$. But from (90) one must also have $-\varepsilon n_c-1 \geqslant 0$, so

$$j-\varepsilon n \geqslant 2j-j_c-\varepsilon n_c \geqslant j+1$$

Thus the "if" is proved ; the "only if" is trivial.

We shall now apply these criteria to an exhaustive study of the different cases of j_{min}.

5.2. U.I.R.'S with $j_{min} \geq \frac{1}{2}$

Observe first that $P_j=0$ means that $\pm(2j-1)$ are roots of (59), and this is the case when $j = j_{min}$.

One has

(103) $$16P_j = ((2j-1)^2-(2j_{min}-1)^2)((2j-1)^2-4y^2)$$

where $\pm 2y$ designes the remaining couple of roots of (59). Condition 1) of Proposition 5 implies, in view of (66), that for every n :

(104) $$2j^{-2}F_\varepsilon,\ F_\varepsilon = (2n+\varepsilon)^2 - 4y^2 \geqslant 0$$

and if this quantity vanishes for some n_ε then Rangen $\subset n_\varepsilon - \varepsilon\mathbb{N}$ for $j = j_{min}$. In particular y^2 must be a real number ; we are sure to obtain only representations of $\mathcal{C}_3$ which are also ones of $\underline{sl}(2,\mathbb{R})$, and integrable to u.i.r.'s of $SL(2,\mathbb{R})$, though not all of them.

If Range $n = \nu + \mathbb{Z}$ with $0 \leqslant \nu < 1$, one obtains from (104) .

(105) $$4y^2 < (2\nu-1)^2$$

which is necessary and sufficient to ensure positivity for every n, as well as $P_j > 0$ for every $j > j_{min}$.

If Range $n \subset n_\varepsilon - \varepsilon\mathbb{N}$, then $(2n_\varepsilon+\varepsilon)^2 = 4y^2$; positivity of $F_\varepsilon F_{\varepsilon'}$ for $n=n_\varepsilon$ shows that either $-\varepsilon n_\varepsilon > 0$, which yields i.r.'s of types b,c ; or $n_\varepsilon = 0$. In the last case we have the singleton $(j_{min};[0,0])$, which is unitary, since $4y^2 = 1$ and $P_j > 0$ for $j > j_{min}$.

Now, as long as $4y^2 \leqslant (2j_{min}+1)^2$, that is :

(106) $$0 < -\varepsilon n_\varepsilon \leqslant j_{min} + 1$$

we are sure that $P_j \geqslant 0$ for every j. On the contrary, when $-\varepsilon n_\varepsilon$ becomes bigger than $j_{min} + 1$, we have :

$$P_j = (j-j_{min})(j+j_{min}-1)(j-\varepsilon n_\varepsilon-1)(j+\varepsilon n_\varepsilon)$$

The last factor is the only negative one for $j=j_{min}+1$; looking at the expression of $M^{\downarrow}_{\varepsilon'k}\ M^{\uparrow}_{\varepsilon k}$ in (89), and putting $s=j_{min}-\varepsilon n_\varepsilon-1$, one has to put $\pm d = j_{min}+\varepsilon n_\varepsilon$, so that $4y^2 = (s\mp d)^2$. One has on the $\varepsilon\downarrow$ boundary :

$$(107)\qquad M^{\downarrow}_{\varepsilon'k}\ M^{\uparrow}_{\varepsilon k} = (j-j_{min}+1)(j+\varepsilon n_\varepsilon+1)$$

which is strictly negative for $j = j_{min}$ iff (106) does not hold. So the only u.i.r.'s found here are those which satisfy the sufficient condition of Proposition 5. Notice also that $\varepsilon n_\varepsilon+j_{min}+1=0$. falls also inside Proposition 6: indeed the only unitary i.r.'s among those containing critical subspaces for $j_{min} > \frac{1}{2}$ are these ones, for both choices of ε.

5.3. U.I.R.'s with $j_{min} = \frac{1}{2}$

Naming $\pm x_\varepsilon$, $\pm x_{\varepsilon'}$ the roots of (53), we may write for $n \in \nu + 2\mathbb{Z}$ in Range n :

$$(108)\qquad F_\varepsilon\ F_{\varepsilon'}\ \varphi_n = [(\varepsilon n - \tfrac{1}{2})^2 - x^2_{\varepsilon'}]\varphi_n.$$

so that both x^2_ε, $x^2_{\varepsilon'}$ must be real numbers. The quantity P_j for $j = \frac{3}{2}$ is

$$(109)\qquad 16\ P_{3/2} = (4-(x_{\varepsilon'}+x_\varepsilon)^2)(4-(x_\varepsilon-x_{\varepsilon'})^2)$$

Suppose first that there is an $\varepsilon^{\downarrow}$-boundary $(\frac{1}{2},n_\varepsilon)$ on which $F_\varepsilon=0$. We shall put $x_\varepsilon = s = -\varepsilon n_\varepsilon - \frac{1}{2}$ and $x_{\varepsilon'}=d$. Applying $F_\varepsilon\ F_{\varepsilon'}$ on $D(\frac{1}{2},n_\varepsilon)$ one obtains

$$(110)\qquad d^2 \leqslant (s+1)^2 = (\varepsilon n_\varepsilon - \tfrac{1}{2})^2$$

the equality $d^2 = (s+1)^2$ happening when $\{n_\varepsilon\}$ = Range n. If it is not the case, one gets from $n = n_\varepsilon - 2\varepsilon$:

$$(111)\qquad -\varepsilon n_\varepsilon + \tfrac{1}{2} \geqslant 0 \quad \text{or } s + 1 \geqslant 0$$

Besides the singleton case, (111) is sufficient to assure positivity all over $E(\frac{1}{2})$; indeed, one gets for $n = n_\varepsilon - 2\varepsilon k$ with $k \geqslant 1$:

$$(112a)\qquad F_\varepsilon F_{\varepsilon'} = (2k+1+s)^2-d^2 > (2k+1+s)^2(s+1)^2=4k(k+1+s) > 0$$

$$(112b)\qquad F_{\varepsilon'}\ F_\varepsilon = (2k+s)^2-s^2 = 4k(k+s) \geqslant 0$$

It follows that one has either singletons, or the doubleton $\{-\frac{1}{2},\frac{1}{2}\}$ with $s = -1$ and $d^2 < 0$, or the string $n_\varepsilon - \varepsilon\mathbb{N}$ with $\varepsilon n_\varepsilon < \frac{1}{2}$, $s+1 > 0$, $d^2 < (s+1)^2$. Notice that the singleton $\{n\}$ must satisfy

$n^2 \leqslant 1$ to get a positive form on $\underset{\pm}{\uparrow}E(\frac{1}{2},n)$; by (44).

With these notations, P_j takes the form :

$$(113) \qquad 16P_j = ((2j-1-s)^2-d^2)((2j-1+s)^2-d^2)$$

while on the $\varepsilon^{\downarrow}$ boundary one has

$$(114) \qquad 4M^{\uparrow}_{\varepsilon k}\, M^{\downarrow}_{\varepsilon' k} = c((2j-1-s)^2-d^2)$$

where $c = (2j-1)(2j+1)^{-1}$ which is positive. Now, the second factor of P_j is always positive for $j \geqslant \frac{3}{2}$; indeed, $2j-1+s=2(j-1)+(s+1)$ is positive, so it is minimal for $2j=3$, and $(s+2)^2 > (s+1)^2 \geqslant d^2$. Thus we have to study the first factor.

Now, as long as P_j stays positive when j increases, the sesquilinear form is positive on E(j) because the induction of the proof of Proposition 5 works well ; if P_j becomes strictly negative without vanishing (114) leaves no doubt : the i.r. is not unitary ; but if it becomes zero (without having been negative before) for some j, then j-1 is a critical value and one may apply Proposition 6.

Thus, one has

$$(115) \quad d^2 < \mathrm{Inf}[(s+1)^2,(s-2)^2,(s-4)^2,\ldots,(s-2k)^2,\ldots], \quad k \in \mathbb{N}+1$$

or, in terms of $\varepsilon n_\varepsilon \leqslant \frac{1}{2}$:

$$(116) \quad d^2 < \mathrm{Inf}[(-\tfrac{1}{2}+\varepsilon n_\varepsilon)^2,(2+\tfrac{1}{2}+\varepsilon n_\varepsilon)^2,\ldots,(2k+\tfrac{1}{2}+\varepsilon n_\varepsilon)^2,\ldots], \ k \in 1+\mathbb{N}$$

for representations which are neither singletons nor contain a critical subspace. For singletons one has

$$(117) \quad d^2 = (s+1)^2 = (-\tfrac{1}{2}+\varepsilon n_\varepsilon)^2 \text{ with } (s+\tfrac{1}{2})^2 = n_\varepsilon \leqslant 1.$$

U.I.R.'s with critical subspaces exist for every choice of $\varepsilon' n_\varepsilon \geqslant 1$ (the equality gives the Majorana-Dirac representations which are singletons for every j, satisfying $\varepsilon' n = j + \frac{1}{2}$) ; the values of j_c must then obey

$$(118a) \quad j_c = \tfrac{1}{2} : 1 \leqslant \varepsilon' n_\varepsilon, \quad \pm d = 2 + \tfrac{1}{2} + \varepsilon n_\varepsilon$$

$$(118b) \quad j_c > \tfrac{1}{2} : 2j_c + \tfrac{1}{2} < \varepsilon' n_\varepsilon, \quad \pm d = 2j_c + \tfrac{3}{2} + \varepsilon n_\varepsilon$$

We point ont that in u.i.r.'s for which :

$$(119) \qquad -\tfrac{1}{2} \leqslant -\varepsilon n_\varepsilon \leqslant 1$$

the Inf of (115) is obtained for the first values, which means that the factor $(2j-1-s)^2-d^2$ of P_j takes its minimum within the Range of j already at $j = \frac{3}{2}$. This is in particular the situation for the doubletons $\{-\frac{1}{2}, \frac{1}{2}\}$.

Coming now the Range $n = \nu + \mathbb{Z}$ with $\nu \in]-\frac{1}{2},\frac{1}{2}]$ on which (108) holds for $n = \nu$ it is easy to find the positivity conditions for $E(\frac{1}{2})$; The two parameters x_+, x_- may be chosen independently of each other, but they depend on ν through :

$$(120) \qquad x_\varepsilon^2 < (\varepsilon\nu + \tfrac{1}{2})^2$$

These conditions are necessary ; they are also sufficient by Proposition 5 since it is easily to check $P_j > 0$ for every j.

Notice that (116),(120) are satisfied for pure imaginary numbers $d, x_\pm$.

5.4. U.I.R.'s with $j_{min} = 0$

Let us write down the necessary conditions of positivity on $\overset{\uparrow}{\varepsilon}E(0)$, E(0) and $F_k^{\uparrow}E(0)$, in terms both of C_2, C_4 and of the roots $\pm x_1, \pm x_2$ of (53). For every n such that $E(0,n) \neq \{0\}$ one must have respectively:

$$(121) \quad \lambda_\varepsilon(n) = 2M_{\varepsilon'k}M_{\varepsilon k} = C_2 + n^2 + 3\varepsilon n = -\tfrac{1}{2}(x_1^2 + x_2^2 - 5) + n^2 + 3\varepsilon n \geqslant 0$$

$$(122) \quad 4\omega_{\varepsilon'}\omega_\varepsilon(n) = (n+\varepsilon)^4 + (2C_2-5)(n+\varepsilon)^2 + (C_2-2)^2 - 4C_4 = ((n+\varepsilon)^2 - x_1^2)((n+\varepsilon)^2 - x_2^2) \geqslant 0$$

$$(123) \quad 16P_1 = 16C_4 = ((x_1+x_2)^2-1)((x_1-x_2)^2-1) \geqslant 0$$

Notice that these inequalities imply that the parameters C_2 and C_4 are real, so that the parameters x_i^2 may be either both real or imaginary conjugate. We shall see that this last possibility is not excluded, contrarily to the other choices of j_{min}.

We begin the study with two remarks : since one has

$$(124) \qquad P_{j+1} - P_j = j(C_2 + 2j^2 - 2)$$

whenever we obtain $C_2 \geqslant 0$ inside the above necessary conditions the corresponding i.r. will be unitary by Proposition 5, since P_j will be increasing and P_1 positive.

Next, (122) can be written

$$(125) \qquad 4\omega_{\varepsilon'}\omega_\varepsilon(n) = \lambda_\varepsilon(n).\lambda_{\varepsilon'}(n+2\varepsilon) - 4C_4 = \omega_\varepsilon\omega_{\varepsilon'}(n+2\varepsilon)$$

so that if $\lambda_\varepsilon(n) = 0$ one must have $C_4=\omega_\varepsilon,\omega_\varepsilon(n)= 0$ and $E(0,n)$ will be an $\varepsilon^\uparrow$-boundary space ; by Proposition 6, the i.r. will be unitary iff $\lambda_\pm(n-2\varepsilon k)$, $\omega_\pm\omega_\mp(n-2\varepsilon k)$ are positive for $k > 0$, and $n-2\varepsilon k \in$ Range n.

We shall immediately consider this case : let $\lambda_\varepsilon(n_\varepsilon)=\omega_{\varepsilon'}\omega_\varepsilon(n_\varepsilon)=C_4=0$ One obtains , choosing arbitrarily x_1,x_2 :

(126) $x_1 = \varepsilon n_\varepsilon+1,\ x_2 = \varepsilon n_\varepsilon+2\ ;\ C_2 = -n_\varepsilon^2-3\varepsilon n_\varepsilon\ ,\ C_4 = 0$

One immediatly sees that $\lambda_{\varepsilon'}(n_\varepsilon) = -6\varepsilon n_\varepsilon$ must be positive.

If $n_\varepsilon = 0$ we have the trivial representation of g. If not one has

(127) $4\omega_{\varepsilon'}\omega_\varepsilon(n)=(n-n_\varepsilon)(n-n_\varepsilon-\varepsilon)(n+n_\varepsilon+2\varepsilon)(n+n_\varepsilon+3\varepsilon) \geqslant 0$

with $n \in n_\varepsilon-\varepsilon 2\mathbb{N}$. The first three factors are positive since $-\varepsilon n > 0$ and the last one yields

(128) $$-\varepsilon n_\varepsilon \geqslant \tfrac{1}{2}$$

One easily checks that this condition is sufficient for $\lambda_\pm(n) \geqslant 0$. These u.i.r.'s have a critical subspace at zero and they also fall in these case of Theorem 2, i.e. $F_\mu E = 0$. The limiting cases are the Majorana-Dirac representations, which are n-singletons for every value of j ; they satisfy :

(129) $-\varepsilon n = \tfrac{1}{2} + j$ for every $(j,n) \in$ Range(j,n)

Coming now to the general case, let $\nu \in]-1,1]$ such that Range n $\subset \nu+2\mathbb{Z}$. We shall distinguish the case where n is unbounded towards $\pm\infty$, from the one where it has an upper or lower bound, assuming $\lambda_\varepsilon(n) \neq 0$ all over Range n any how.

Suppose first there is an $\varepsilon^\downarrow$ boundary at $n = n_\varepsilon$.
One must then have, say, $x_1^2 = (\varepsilon n_\varepsilon+1)^2$. One can write down conditions of positivity on the whole $\varepsilon^\downarrow$ boundary string.

Writing $\mu(j)$ for the eigenvalue of $4M^\downarrow_{\varepsilon'k}\ M^\uparrow_{\varepsilon k}$ on $E(j,n)$, one obtains from (87), (89) :

(130) $\mu(j) = (2j+2+\varepsilon n_\varepsilon)^2 - x_2^2 \geqslant 0\ ;\ \ \mu(0) > 0$

(131) $4\Psi = j.\mu(-1) \geqslant 0\ ;\ 16P_j = \mu(j-1).\mu(-j) \geqslant 0$

We shall see that $-\varepsilon n_\varepsilon \geqslant 0$. Indeed one has

(132) $\omega_\varepsilon\omega_{\varepsilon'}(n_\varepsilon) = -\varepsilon n_\varepsilon\ (2C_2+2n_\varepsilon^2-3) \geqslant 0$

(133) $\lambda_\varepsilon(n_\varepsilon)=C_2+n_\varepsilon^2+3\varepsilon n_\varepsilon > 0\ ;\ \lambda_\varepsilon(n_\varepsilon-2\varepsilon) = C_2+n_\varepsilon^2-\varepsilon n_\varepsilon-2 \geqslant 0$

the last relation been justified by (125) even if $n_\varepsilon - 2\varepsilon \notin$ Range n. It is easy to see that the hypothesis $-\varepsilon n_\varepsilon < 0$ leads to $C_2 + n_\varepsilon^2 - \frac{3}{2} \leqslant 0$, which combined to (133) yields successively $\varepsilon n_\varepsilon + \frac{1}{2} > 0$ and $\varepsilon n_\varepsilon + \frac{1}{2} \leqslant 0$. Hence $\varepsilon n_\varepsilon \leqslant 0$. The case $n_\varepsilon = \nu = 0$ yields a singleton $E(0)=E(0,0)$, which is unitary iff $x_2^2 \leqslant 0$; this is necessary because of (131) and sufficient since $\mu(\pm j) > 0$ for every $j \neq -1$, hence Proposition 5 can be applied.

Now, if $-\varepsilon n_\varepsilon > 0$, one always has $\lambda_{\varepsilon'}(n) > \lambda_\varepsilon(n)$ and the minimal value of $\lambda_\varepsilon(n)$ is $\lambda_\varepsilon(n_\varepsilon) = \mu(0)$, except if $0 < -\varepsilon n_\varepsilon < \frac{1}{2}$, in which case the minimum is obtained for

$$(134) \qquad \lambda_\varepsilon(n-2\varepsilon) = \tfrac{1}{2}(n_\varepsilon^2 - 4\varepsilon n_\varepsilon - x_2^2) > \tfrac{1}{2}\,\mu(-1)$$

On the other hand, both factors of $\omega_\varepsilon \omega_{\varepsilon'}(n)$, equal to $(\varepsilon' n+1)^2 - x_i^2$ are increasing functions of $\varepsilon' n \in \varepsilon' n_\varepsilon + 2\,\mathbb{N}$.
For $i = 1$ the corresponding factor is always strictly positive : for $i = 2$ one has always $(n-\varepsilon)^2 - x_2^2 > \mu(-1) \geqslant 0$, so that $\mu(-1) \geqslant 0$, $\mu(0) > 0$ assure positivity all over $E(1)$ and $E(0)$. The range of n in $E(0)$ is the half-string $n_\varepsilon - \varepsilon 2\,\mathbb{N}$.

Looking for the sign of P_j one sees it is the same as that of $\mu(j-1)$, since $\mu(-j) > \mu(-1)$ for $j > 0$.

There are two possibilities : either $\mu(j_c) = 0$ for some $j_c > 0$, with $\mu(j) > 0$ for $0 \leqslant j < j_c$, which yields a u.i.r. having a critical subspace at $j = j_c$ by Proposition 6 (indeed, positivity over every $E(j)$, $j \leqslant j_c$ is guaranteed by the induction procedure) ; or $\mu(j) > 0$ for every $j \geqslant 0$, and the $\varepsilon^\downarrow$ string is unbounded.

In the critical subspace case one has :

$$(135) \qquad x_2 = 2j_c + 2 + \varepsilon n_\varepsilon = j_c + 2 + \varepsilon n_c \quad , \qquad \text{with}$$

$$(136) \qquad -\varepsilon n_\varepsilon > 2j_c + 1 \geqslant 3$$

In the unbounded string case one has

$$(137) \qquad \nu^2 - x_2^2 > 0 \quad \text{with} \quad \nu^2 \leqslant 1,\ n_\varepsilon \in \nu - 2\,\varepsilon\mathbb{N}$$

since this is the minimum for $\mu(j)$. The inequality must be strict, otherwise a critical subspace appears, with the following exception : if the minimum is reached at $\mu(-1) = 0$, which supposes $n_\varepsilon^2 < (n_\varepsilon + 2\varepsilon)^2$, we may have :

$$(138) \qquad \nu^2 - x_2^2 = 0 \qquad\qquad 0 < -\varepsilon\nu = -\varepsilon n_\varepsilon < 1$$

These are u.i.r.'s studied in Theorem 2, for which $F_\mu E = 0$ and the $\varepsilon^\downarrow$ string remains unbounded. Remark that if $\nu = 0$ we obtain the singleton already studied.

We finally come to Range (n) = $\nu+2\mathbb{Z}$. One easily checks in (121) that the minimum of $\lambda_\varepsilon(n)$ is :

$$(139)\qquad C_2-2-|\nu|+\nu^2 = \tfrac{1}{2}(\nu^2+(1-|\nu|)^2-x_1^2-x_2^2) > 0$$

and since $C_2 > 2$ anyhow, the remark following (124) holds.

As for $\omega_\varepsilon, \omega_\varepsilon$, remarking that it is a trinomial with positive derivative as soon as $(n+\varepsilon)^2 = -C_2+\frac{5}{2} < \frac{1}{2}(\nu^2+(1-|\nu|)^2)$, the minimum is reached at $n = \nu$ with $\varepsilon n = -|\nu|$, so that :

$$(140)\quad (1-|\nu|)^4+(2C-5)(1-|\nu|)^2+(C_2-2)^2-4C_4=((1-|\nu|)^2-x_1^2)((1-|\nu|)^2-x_2^2)>0$$

Inequalities (139), (140) determine completely which i.r.'s are unitary, together with (123) ; in the (C_2,C_4) plane they determine a region bounded by a parabola and two straight lines. However, they are not very explicit about the x_i's , so we shall distinguish several cases.

a) Both x_1,x_2 are imaginary numbers

b) x_1 imaginary , x_2 real, $|x_1| < 1-|\nu|$

c) Both x_1,x_2 reals, $|x_1| + |x_2| \leqslant 1$ and
either $|x_i^2| < 1-|\nu|$, i=1,2 ; or $|x_i| > 1-|\nu|$,i = 1,2

d) $x_1 = \bar{x}_2 \in \mathbb{C}$ and $|\text{Re}\ x_1| \leqslant \frac{1}{2}$.

Only conditions c) and d) need a small commentary. In c) the two factors of $\omega_\varepsilon, \omega_\varepsilon$ are either both positive or both negative, (139) confines the x_i's within a circle in the (x_1,x_2) plane and (123) confines them within a square ; notice that $|x_i^2| > 1-|\nu|_2$ cannot occur if $|\nu| \leqslant \frac{1}{2}$. In d) the condition $0 \leqslant 16C_4=[1+(2\text{Im}x_1)^2][1-(2\text{Re}x_1)^2]$ gives the above result and it proves to satisfy (139).

All this description is summarized in the following table.
Some cases are also illustrated in figures 16 to 18.

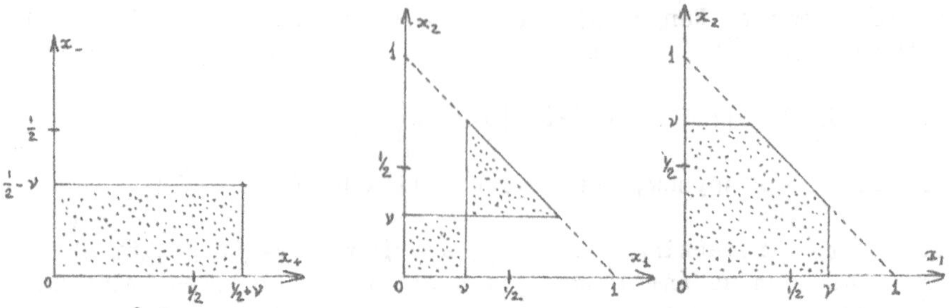

16a: $(\frac{1}{2};\nu;x_+^2,x_-^2)$ with $0<\nu<\frac{1}{2}$ 16b: $(0;\nu;x_1^2,x_2^2)$, $0<\nu<\frac{1}{2}$ 16c: $(0;\nu;x_1^2,x_2^2)$, $\frac{1}{2}<\nu<1$

Fig 16: UIR's of types Ba, Ca for real valued roots of (53).

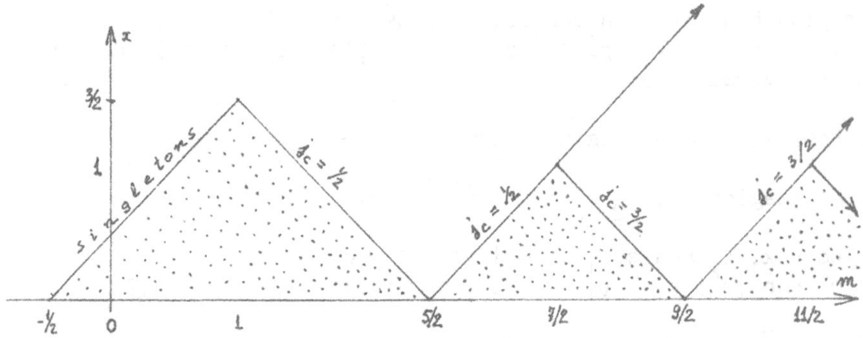

Fig 17: UIR's $(\frac{1}{2};[m,\dots[;x^2)$ for real valued x.

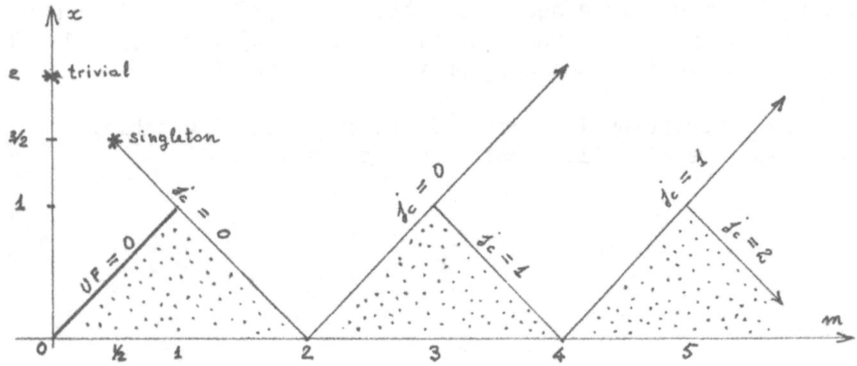

Fig 18: UIR's $(0;[m,\dots[;x^2)$ for real valued x.

6. CONCLUSIONS

6.1. Weight representations and physical interest

It is more or less evident from the discussion of section 4 that i.r.'s having a critical subspace (j_c, n_c) coïncide with the so-called weight representations : the Cartan subalgebra spanned by J_3 and U contains a weight vector in the critical subspace, which is an eigenvector of iJ_3 for the eigenvalue j_c. It is generally believed that weight representations are physically interesting, in the following sense : the spectrum of i U has an upper or lower bound and there is a hope that such i.r.'s are closely related to "physical" representations of the Poincaré group, i.e. ones for which the spectrum of the energy operator P_o is positive or negative.

This is in particular the case when $0 \neq j_c = j_{min} = n_c - 1$ [2], in which case there are 2 straighforward Wigner-Inönü contractions to u.i.r.'s of Poincaré with mass zero and helicity $\pm j_{min}$. However, it is not a priori impossible that such contractions exist (being less straighforward) from representations for which there is no critical subspace but for which εn is bounded for every fixed $j \geqslant j_{min}$, that is i.r.'s with an $\varepsilon^{\downarrow}$ boundary.

6.2. Commentary about the method used

This method contains two distinct features : LQW constraints for the i.r.'s and positivity constraints for the u.i.r.'s. The algebraic constraints issued from LQW considerations (as well as ε-boundaries considerations) are closely related to the reduction of the representations of $\underline{k}$ on $\underline{p}_\varepsilon \otimes E(j,n)$, where $\underline{p}_\varepsilon$ is the $\underline{k}$-module spanned by the $M_{\varepsilon k}$. One could as well start by saying that the analytic subspace of $\underline{k}$-finite vectors is $\oplus E(j,n) = \oplus(\underline{p}_\varepsilon \otimes E(j,n))$ and derive all algebraic constraints from this equality. This feature is typical for all real semi-simple Lie groups ; in fact it is used, more or less explicitely , in most works of exhaustive classification of i.r.'s (or u.i.r.'s) of a given Lie group : Bargman [4] for SL(2,R) and Miller [1] for $SO_o(2,1)$, Naimark [5] for $SO_o(3,1)$, Šijački [6]and the author [7] for $SL(3,\mathbb{R})$, Duflo [3] for $SL(3,\mathbb{C})$, $Sp(2,\mathbb{C})$ and $G_2(\mathbb{C})$.
It is quite natural that, when the subgroup K is easier to handle, this feature is less explicit.

The positivity of the sesquilinear form (99) is a quite distinct feature from the previous one, though it depends closely on the results of LQW constraints. A very interesting observation, which leads to a conjecture for every semi simple Lie group is the following : <u>if the unique sesquilinear form for which generators of g are skew adjoint is positive on every irreducible k-module with</u>

multiplicity one, then it is positive every where. This is of course true when all k-modules have multiplicity one, like $SO_o(p,1)$ moreover, in this case explicite formulas are available, and it is easy to check that, say, n^2-n-C_2 is minimal when $n \in [0,1]$ (with step $\Delta n=1$) . What is more surprising is that in the $SO_o(3,2)$ case there are two formulas inside the enveloping algebra, that is (44) and (46) , which lead to the result above. This means in particular that one can prove unitarity independently of the choice of a basis in D(j,n), which could be a set of eigenvectors of, say, the operator $\Psi \in \mathcal{C}_3$. Similar formulas are available for $SO_o(4,2)$, which lead to the exhaustive classification of u.i.r.'s of this group too [8] , and there are reasons to think that this is not so accidental.

Acknowledgements

I wish to thank Chris Fronsdal, Moshé Flato and Daniel Sternheimer for interesting discussion about this subject.

Bibliography

1. W. MILLER : Lie theory and special functions, Academic Press New York, 1968.
2. E. ANGELOPOULOS, M. FLATO, C. FRONSDAL, D. STERNHEIMER : Phys. Rev. D, 23, 6 (1981), 1278.
3. M. DUFLO : Bull. Soc. Math. France, 107 (1979), 55.
4. V. BARGMAN : Ann. Math. 59, 1 (1954).
5. M.A. NAIMARK : Les représentations linéaires du groupe de Lorentz, Dunod, Paris, 1962.
6. D. SIJACKI : J. Math. Phys. 16, (2) (1975), 298.
7. E. ANGELOPOULOS : J. Math. Phys., 19 (1978), 2108.
8. E. ANGELOPOULOS : to be published.

SUPERGRAVITY [1/]

J. Niederle

International Centre for Theoretical Physics, Trieste, Italy; permanent address: Institute of Physics, Czech.Acad.Sci., 180 40 Prague 8, Czechoslovakia

The present status and some recent developments of supergravity are briefly reviewed.

1. INTRODUCTION

Successful attempts to unify the gravitation theory and the quantum field theory lay undiscovered for long. The turning point appeared in 1975 when new tools of theoretical physics were applied which hopefully not only bring together theory of gravity with the principles of quantum mechanics but also clarify the connection of the gravitation interaction with the other fundamental interactions. There are even hopes that the gravitation, electromagnetic, weak and strong interactions are just four aspects of one, more fundamental interaction.

So far as the energy scales are concerned such a full unification may go through a number of intermediate scales - through the unification scale of the electromagnetic and weak interactions around 100 GeV, the unification scale of the electroweak and strong interactions as low as (10^4-10^6) GeV (for the Pati-Salam model) or higher than 10^{15} GeV to the unification of all interactions at 10^{19} GeV. In this connection let us only note that one expects to reach the energy 300 GeV (c.m.) in 1981 and that one observes events in cosmic rays with the energy of particles 10^{10} GeV.

What are the main tools which allow us to attack the problem of the unification of the basic interactions? First, it is the gauge principle and second, it is the structure of Lie

A. O. Barut (ed.), Quantum Theory, Groups, Fields and Particles, 149–159.

superalgebras.

1/ As is well known, the gauge principle represents a systematic procedure for constructing theories invariant under local (gauge) transformations (i.e. transformations characterized by their spacetime dependence) starting from the theories invariant under global transformations. It is typical for this procedure that it introduces not only gauge potentials A^r_μ (which mediate the considered interactions of matter fields) but also their interactions with themselves (via the kinetic term of the type $-\frac{1}{4} F^r_{\mu\nu} F_r^{\mu\nu}$) and with the matter fields as well (via covariant derivatives D_μ which replace the corresponding derivatives in the initial Lagrangian invariant only under global transformations).

It is important to note three things:
i) all basic interactions can be constructed by using the gauge principle, i.e. within the framework of gauge theories (for more details see [1]);
ii) the gauge theory has to be renormalizable or finite in order to be a physical theory,
iii) its invariance under gauge transformations has to be spontaneously broken (via the Higgs-Kibble mechanism) whenever some gauge fields are massive.

2/ The Lie superalgebra (or the Z_2-graded Lie algebra) is a non-associative algebra the generators of which satisfy commutation or anticommutation relations. More precisely the Lie superalgebra is a graded vector space $\mathfrak{A}$ over $\mathbb{R}$ or $\mathbb{C}$ (i.e. a vector space $\mathfrak{A}$ over $\mathbb{R}$ or $\mathbb{C}$ which can be written as $\mathfrak{A} = \mathfrak{A}_0 + \mathfrak{A}_1$; elements $x_i \in \mathfrak{A}_i$, i=0 or 1, are homogeneous with grading $g(x_i) = i$) with a binary operation $[\![\cdot , \cdot]\!] : \mathfrak{A} \times \mathfrak{A} \to \mathfrak{A}$ which is bilinear and satisfies:
i) $g([\![x_i, x_j]\!]) = g(x_i) + g(x_j) \pmod 2$ (i.e. $[\![\mathfrak{A}_i, \mathfrak{A}_j]\!] \subset \mathfrak{A}_{i+j \pmod 2}$)
ii) $[\![x_i, x_j]\!] = (-1)^{g(x_i)g(x_j)+1} [\![x_j, x_i]\!]$
iii) and the generalized Jacobi identities:

$$(-1)^{g(x_i)g(x_k)} [\![x_i, [\![x_j, x_k]\!]]\!] + \text{cycl.} = 0.$$

Thus $\mathfrak{A}_0$ forms a Lie algebra and $\mathfrak{A}_1$ its representation.

In physics, the Lie superalgebras were introduced by Gol'fand and Likhtman in 1971, by Volkov and Akulov in 1973 and, in particular, by Wess and Zumino in 1974. However, the origin of the Lie superalgebras can be found in the earlier works by Martin, Lipkin, Berezin, Stavraki, Flato, Hillion, Schwinger and others (for references and more details about Lie algebras, their representations etc. see [2]).

The most interesting Lie superalgebras are those which

contain the Lie algebras of spacetime transformations (such as the Poincaré algebra, the conformal algebra etc.) as their subalgebras. They are called supersymmetry algebras. Those which can be symmetry algebras of S-matrix were classified by Haag, Lopuszanski, Sohnius [3] and by Nahm [4] (however, there are also supersymmetries with spinorial generators of higher spin than $\frac{1}{2}$).

As an illustration let us consider the simple supersymmetry consisting of the Poincaré algebra (generated by $M_{\mu\nu}$ and P_μ)

$$[M_{\mu\nu}, M_{\rho\sigma}] = i\eta_{\nu\rho} M_{\mu\sigma} + i\eta_{\mu\sigma} M_{\nu\rho} - i\eta_{\mu\rho} M_{\nu\sigma} - i\eta_{\nu\sigma} M_{\mu\rho}, \quad (1.1)$$

$$[P_\mu, P_\nu] = 0, \quad (1.2)$$

$$[M_{\mu\nu}, P_\rho] = i\eta_{\nu\rho} P_\mu - i\eta_{\mu\rho} P_\nu, \quad (1.3)$$

and the generators Q_α fulfilling the relations

$$[M_{\mu\nu}, Q_\alpha] = -\frac{i}{2} (\sigma_{\mu\nu})_\alpha{}^\beta Q_\beta, \quad (1.4)$$

$$[P_\mu, Q_\alpha] = 0, \quad (1.5)$$

$$\{Q_\alpha, \bar{Q}_\beta\} = -2 (\gamma^\mu)_{\alpha\beta} P_\mu \quad (1.6)$$

where $\bar{Q} = Q^T \gamma_0$, $\sigma_{\nu\mu} = \frac{1}{4}[\gamma_\nu, \gamma_\mu]$, $\mu, \nu = 0,1,2,3$ and $\mathrm{diag}(\eta_{\mu\nu}) = (-1,+1,+1,+1)$. One easily sees that generators Q_α have a spinorial character (due to (1.4)) and that in the irreducible representation of the simple supersymmetry all particles have the same mass (due to (1.5)) and are grouped into spin pairs $(J, J+\frac{1}{2})$ or $(J, J-\frac{1}{2})$ (by decomposing it with respect to the Poincaré subgroup).

Since the simple supersymmetry (as well as other supersymmetries) can be realized by means of interacting fields within the field theory which is local hopefully renormalizable or finite and invariant under the global simple supersymmetry transformations, a theory invariant with respect to local supersymmetry transformations can be constructed by appying the gauge principle. One obtains the theory called simple supergravity. Since the local supersymmetries contain local spacetime symmetries, i.e. the gauge transformations of gravity, in the supergravity the gravitation field emerges as a gauge field. However, one always gets yet another field - the gauge field with respect to pure local supersymmetry transformations generated by Q_α. It is a spinorial Rarita-Schwinger field with helicity $\pm 3/2$. The other possible companion of the gravitation field, a spinorial field with helicity $\pm 5/2$, is usually rejected for technical difficulties to construct a consistent higher spin quantum field theory (cf. [6]). The situation might be changed due to [7]. Since the Rarita-Schwinger field accompanies the gravitation field its quantum is called gravitino.

2. SIMPLE SUPERGRAVITY

In 1976 two formulations of the supergravity based on the local simple supersymmetry appeared: The formulation by Deser and Zumino [7] and the formulation by Freedman, Nieuwenhuizen and Ferrara [8].

In the simple supergravity, the vierbein e^{μ}_{ρ}, the connections $\omega^{\mu\nu}_{\rho}$ and the Rarita-Schwinger field ψ^{α}_{ρ} appeared to be the gauge fields with respect to the transformations generated by P_{μ}, $M_{\mu\nu}$ and Q_{α} respectively. The Lagrangian in the Deser-Zumino formulation is of the form

$$L = L_E + L_{RS} , \tag{2.1}$$

where L_E is the Lagrangian of the Einstein gravitation theory

$$L_E = - \frac{1}{2\varkappa^2} R\, e \tag{2.2}$$

and L_{RS} is the Lagrangian of the Rarita-Schwinger field in a curved spacetime

$$L_{RS} = - \frac{i}{2} \varepsilon^{\rho\sigma\tau\omega} \bar{\psi}_{\rho} \gamma_5 \gamma_{\sigma} D_{\tau} \psi_{\omega} \tag{2.3}$$

Here,

$$\begin{aligned} R &= R^{\mu\nu}_{\rho\sigma}\, e^{\rho}_{\mu}\, e^{\sigma}_{\nu} , \\ R^{\mu\nu}_{\rho\sigma} &= \partial_{\rho} \omega^{\mu\nu}_{\sigma} - \partial_{\sigma} \omega^{\mu\nu}_{\rho} - \omega^{\mu}_{\rho\tau} \omega^{\tau\nu}_{\sigma} + \omega^{\mu}_{\sigma\tau} \omega^{\tau\nu}_{\rho} , \\ e &= \det e^{\mu}_{\rho} , \end{aligned} \tag{2.4}$$

$$g^{\rho\sigma}(x)\, e^{\mu}_{\rho}(x)\, e^{\nu}_{\sigma}(x) = \eta^{\mu\nu} = (-1,+1,+1,+1) \tag{2.5}$$

and covariant derivative D_{τ} is of the form

$$D_{\tau} = \partial_{\tau} - \frac{1}{2} \omega_{\tau\alpha\beta}\, \sigma^{\alpha\beta} \tag{2.6}$$

Lagrangian (2.1) is invariant with respect to gauge transformations generated by $M_{\mu\nu}$, P_{μ}, Q_{α} (see [7], [8]).

From the point of view of quantum field theory the resulting simple supergravity has several advantages:

i) It is the first consistent field theory describing the Rarita-Schwinger field interacting with another field namely

the gravitation field (cf. earlier accausality problems discussed e.g. in [5]).
ii) Its S-matrix elements in one- and two-loop orders are finite [9].
iii) It has off-shell formulations [11].

On the other hand, at any loop order starting with the three-loop one there exist counter terms which do not vanish on-shell. They would spoil on-shell renormalizability of the simple supergravity unless their coefficients vanish.

At present, the role of the simple supergravity is analogous to that of the H-atom in quantum mechanics - it probes various techniques. In any case, the simple supergravity is too small to be considered as a scheme for a unification of physical interactions. However, many extensions of the simple supergravity were suggested. They are based on supersymmetry algebras classified in [3], [4].

3. EXTENDED SUPERGRAVITY

The Lie superalgebra of the <u>extended supergravity</u> consists of the Poincaré algebra, the internal symmetry algebra SO(n) and the spinorial generators Q^i_α, $i=1,2,\ldots,n$, $\alpha=1,2$, which transform according to vector representations of SO(n). If the spin 2 gravitation field is the field with the maximal spin in the theory then the internal symmetry SO(n) is limited to $n=1,\ldots,8$.

The situation in extended supergravity can be summarized as follows:
i) Complete theories are claimed for n=2 [11], n=3 [12] and n=4 [13].
ii) For n>4, the theories have very complicated non-polynomial structure (mainly due to the presence of spinless fields). It turns out, however, that the next most important case is n=8, since n=5 and 6 are expected to be subcases of the SO(8)-extended supergravity and n=7 to be equivalent to the n=8 case.
iii) The case n=8 is extensively studied by Cremmer, Julia and Scherk [14] (some results were obtained also in [15]) by using a special dimensional reduction technique analogous to that of the Kaluza-Klein theory (i.e. they start with the simple (n=1) supergravity but in 11-dimensional spacetime, then they compactify 7 spatial dimensions and reach the SO(7)-extended supergravity in 4 dimensions which is related to the SO(8)-extended supergravity.

What are the advantages of the extended supergravities?
i) The extended supergravity theories unify fields with different helicities.

ii) Their S-matrix elements up to the two-loop order are finite [9].
On the other hand the extended supergravities have also several drawbacks:
i) Their internal symmetries SO(n) are considered as global symmetries. If one considers SO(n) as a local symmetry (which can be done due to the right number of fields in the irreducible representations of the extended supergravity), one gets a cosmological term $\sim eg^2/\varkappa^4$ (for n=4 it is not a constant but a scalar field potential) which is more than 100 orders of magnitude larger than the astronomical upper limit. Let us remark in this connection that it is claimed that one might be able to overcome this problem by using the Higgs mechanism or the Hawking approach [16].
ii) Even the biggest local symmetry SO(8) is small to include the minimal gauge group of strong and electromagnetic interactions SU(3)⊗ SU(2)⊗ U(1). Consequently several well established particles (e.g. μ, τ, $W^{\pm}$) are missing in the representations of the corresponding theories though their field structures are pretty rich (e.g. the irreducible massless representation of the SO(8)-supergravity contains 1 field of helicity 2, 8 fields of helicity 3/2, 28 fields of helicity 1, 56 fields of helicity 1/2 and 70 fields of helicity 0).

There are several suggestions how to treat these difficulties. For instance, one may construct field theory based on several representations with spin 2 of the extended supergravity or one may use the SO(9) or SO(10) extended supergravity. One may also say that the fields in the supermultiplet of the considered supergravity do not correspond to known elementary particles at present energy and that the supermultiplet contains preons or prequarks, and so on. Finally, one may try to construct other approaches of the supergravity based e.g. on superconformal or super-de Sitter algebras.

4. CONFORMAL SUPERGRAVITY

The conformal supergravity is built on the Lie superalgebra generated by the generators of the conformal algebra, the generators of the internal symmetry algebra U(n) (for n=4 also SU(4)), by two sets of spinorial generators Q^i_α and S^i_α, i=1,...,n, α=1,2, which transform according to vector representations of U(n) and by one more generator.

From the point of view of gravity alone, the conformal supergravity is closely related to the Weyl gravitation theory. The present status of the conformal supergravity can be summarized as follows:
i) For n=1 the theory is known [17].

ii) For $n>1$ only partial results are available because the corresponding Lagrangians are pretty complicated (they contain terms with higher order derivatives like $\bar{\psi}_\mu \not\partial \Box g_{\mu\nu} \psi_\nu$).
iii) The internal symmetry $U(n)$ for $n \geq 5$ is big enough to include the gauge group $SU(3) \otimes SU(2) \otimes U(1)$. However, as argued by de Wit and Ferrara [18], this perhaps does not mean anything since the supergravity theory beyond $n=4$ probably does not exist.
iv) It is stated that the conformal supergravity (with the Lagrangian quadratic in R) is renormalizable analogously to the Yang-Mills theory.
v) As shown by Fradkin the conformal supergravity is asymptotically free in $\varkappa^2$.
vi) On the other hand, since $L \sim R^2$, the conformal supergravity does not agree with the macroscopic predictions of general relativity. (However, there is hope that by applying the Higgs mechanism the Lagrangian changes and a new term linear in R appears which will reproduce the usual predictions of general relativity.
vii) Since the Lagrangian consists of higher derivative terms, negative metric ghosts appear in the conformal supergravity. It is not clear how to remedy this difficulty (see, however [19]).

5. GEOMETRICAL FORMULATIONS OF SUPERGRAVITY

Up to now, the supergravity theories have been discussed in the framework of quantum field theory. One may ask whether one cannot construct the supergravity theories by applying geometrical methods. The answer is yes and as a matter of fact the geometrical approaches seem to be natural, systematic and quite comprehensive. One can make steps as Einstein did but on a fibre bundless the bases of which are rather superspaces (or L- and R-chiral superspaces) than spacetime and the structure groups of which are Lie supergroups. Such constructions use various notions which were intuitevely introduced by Salam, Strathdee, Okulov, Volkov, Zumino etc. (see [20]). It is advantageous that we are getting now the precise definitions of these notions by Berezin, Leites, Kac, Kostant, Hochschild and others (for references see e.g. [2]).

There are at least four geometrical approaches treated in the literature. Their primary objects as well as the number of their components are summarized in Table I.

In geometrical approaches (except the Ogievetsky-Sokatchev one) there are technical difficuties to eliminate a large number of unwanted high spin components. Usually it is done by imposing various constraints. Consequently, a serious problem of their physical consistences arises. Many physicists are, therefore, trying to construct a "tensor calculus" which hopefully will simplify not only the construction of Lagrangians but also of invariants which can play a role of counter terms and thus make a systematic discussion of renormalizability possible.

Supergravity approach by:	Primary objects:	Number of components :
Zumino, Wess [21]	supervierbein $E(x,\Theta)$ superconnection $\Gamma(x,\Theta)$	1024 1792
Gell-mann, Brink, Schwarz, Ramond; Mac Dowell, Mansouri; Ne´eman, Regge [22]	$E(x,\Theta)$ $\Gamma(x,\Theta)$	112 indep.
Arnowitt, Nath [23]	supermetric $g_{MN}(x,\Theta)$	1024
Ogievetsky, Sokatchev [24]	axial superfield $H^m(x,\Theta,\bar{\Theta})$	64

Table I.

6. CONCLUSIONS

i) The supergravity is a supersymmetric extension of the Einstein (Einstein-Cartan or Weyl) theory of gravity in which gravitation is unified with very restrictive combinations of lower spin particles in one irreducible representation of the Lie superalgebra. For example, the gravitation field is always accompanied by the spin 3/2 Rarita-Schwinger field.

ii) It is crucial for supergravity to find out gravitinos, the quanta of the Rarita-Schwinger field, in experiments. It is not an easy task since the properties of gravitinos are not well established and probably yield subtle effects.

iii) In any case the supergravity is the first consistent quantum field theory for spin 3/2 field interacting with the spin 2 gravitation field and the spin 1, 1/2 and 0 fields.

iv) Two technical triumphs were remarked in the last two years- the off-shell component approach to simple supergavity (via auxiliary fields) and the superspace formulation of the simple supergravity.

v) The status of divergences may be summarized as follows: The conformal supergravity is believed to be renormalizable. In the simple and extended supergravities (in contradistiction to the Einstein gravitation theory interacting with matter) the S-matrix elements in one- and two-loop orders are finite. However, starting with three-loop order the invariants which can play a role of counter terms do not vanish in the simple supergravity. For the extended supergravities there is no technique available to calculate these invariants for higher loop orders.

vi) The supergravity can be considered as a possible unification scheme of gravitation, electromagnetic, weak and strong interactions. For a realistic scheme it will be necessary to solve the problem of renormalizability. In realizing that, it is perhaps worthwhile to keep in mind the Einstein idea: "Conviction is a good main spring, but a bad regulator".

ACKNOWLEDGMENTS

The author would like to thank Professor Abdus Salam, the IAEA and UNESCO for hospitality at the ICTP, Trieste.

1/ Invited talk at the Colloquium on Mathematical Physics, Bogaziçi University, Istanbul, August 1979

References:

[1] Niederle, J., in the Proc. of the Workshop on theor. problems in elementary particle physics and quantum field theory, Protvino, July 1979; see also
Ivanov, E., Niederle, J.: Phys. Rev. D25 (1982), 976; 988.

[2] Niederle, J.: 1980, Czech. J. Phys. B30, p. 1;
Kac, V., G.: 1977, Commun. Math. Phys. 53, p. 31.

[3] Haag, R., Lopuszanski, J., T., Sohnius, M.: 1975, Nucl. Phys. B88, p. 257.

[4] Nahm, W.: 1978, Nucl. Phys. B135, p. 149; see also
Hlavatý, L., Niederle, J.: 1980, Lett. Math. Phys. 4, p. 301.

[5] Proc. Int. School of Mathematical Physics, Erice 1977; Lecture Notes in Physics Vol. 73 (Ed. by G. Velo, A. S. Wightman), Springer Verlag, Berlin 1978.

[6] Fronsdal, C.: 1978, Phys.Rev. D18,p.3624;
Fang,J., Fronsdal, C.: 1978, Phys.Rev. D18,p.3630.

[7] Deser, S., Zumino, B.: 1976, Phys.Letters 62B, p.335.

[8] Freedman, D.,Z., van Nieuwenhuizen, P., Ferrara, S.: 1976, Phys.Rev. D13, p.3214.

[9] Grisaru, M., Vermaseren, J., van Nieuwenhuizen, P.: 1976, Phys.Rev.Letters 37, p.2662;
Grisaru, M.: 1977, Phys.Letters 66B, p.75;
Tomboulis, E.: 1977, Phys.Letters 67B, p.417;
Deser, S., Kay, J.: Phys.Letters 76B (1978), p.400.

[10] Stelle, K.,S., West, P.,C.: 1978, Phys.Letters 74B, p.330; 1978, Nucl.Phys. B145, p.175;
Fradkin, E.,S., Vasiliev, M.,A.: 1979, Lett.Nuovo Cimento 25, p.79;
Ferrara, S., van Nieuwenhuizen, P.: 1978, Phys.Letters 74B,p. 333;76B, p.404; 78B, p.573.

[11] Ferrara, S., van Nieuwenhuizen, P.: 1976 , Phys.Rev.Letters 37, p.1662.

[12] Freedman, D.: 1977, Phys.Rev.Letters 38, p.105;
Ferrara, S., Scherk,J., Zumino,B.: 1977, Phys.Letters 66B, p.35.

[13] Cremmer,E., Scherk,J.: 1977, Nucl.Phys. B127, p.259;
Cremmer,E., Scherk,J., Ferrara,S.: 1978, Phys.Letters 74B, p.61.

[14] Cremmer,E., et al.: 1978, Phys.Letters 74B, p.333; 1979, Nucl.Phys. B147, p.105;
Cremmer,J., Julia,B.: 1978, Phys.Letters 80B, p.48; 1979, Nucl.Phys. B158, p.357.

[15] de Wit,B., Freedman,D.: 1977, Nucl.Phys. B130, p.105;
Cremmer,E., Julia,B., Scherk,J.: 1978, Phys.Letters 76B, p.409.

[16] Avis,S., Isham,C., Storey,D.: 1978, Phys.Rev. D18, p.3565.

[17] Kaku,M., Townsend,P., van Nieuwenhuizen,P.: 1978, Phys.Rev. D17, p.3179;
Ferrara,S., et al.: 1977, Nucl.Phys. B129, p.125.

[18] de Wit,B., Ferrara,S.: 1979, Phys.Letters 81B, 317.

[19] Ferrara,S., Zumino,B.: 1978, Nucl.Phys.B134, p.301.

[20] Salam,A., Strathdee,J.: 1978, Fortschr.Phys. 26, p.57; Fayet,P., Ferrara,S.: 1977, Phys.Rep. 32C, p.249.

[21] Zumino,B. in the Proc.19th Int.Conf.High Energy Physics, Tokyo 1978, p.555.

[22] Brink,L., et al.: 1978, Phys.Letters 74B, p.336; 76B, p.417; Ne'eman,Y. in the Proc. 19th Int.Conf.High Energy Physics, Tokyo 1978, p.552.

[23] Arnowitt,R., Nath,P., Zumino,B.: 1975, Phys.Letters 56B, p.81;
Arnowitt,R., Nath,P.; 1975, Phys.Letters 56B, p.117; 1976, Phys.Letters 65B, p.73;
Nath,P. in the Proc.19th Int.Conf.High energy Physics, Tokyo 1978, p.563.

[24] Ogievetsky,V., Sokatchev,E.: 1977, Nucl.Phys. B124, p.309; 1978, Phys.Letters B79, p.219; 1979, Czech.J.Phys. B29, p.68.

NOTE ADDED

The lecture was done in 1979. From that time several new developments and results in supergravity have appeared. E.g.: It was proved by Mandelstam that the S-matrix elements of the so called n=4 Yang-Mills supersymmetric theory are finite in all loop orders and thus the first finite quantum field theory discovered. New off-shell formulations of n=1,2 extended supergravities were found (however, for $n > 2$ they are still unknown). The extensions of various geometrical approaches from simple to extended supergravity were developped (the Wess-Zumino approach for n=2, the Gell-mann et al. approach to n=2,3). The constraints on superfields used for eliminating redundant superfield components in geometrical approaches turned out to be superanalogons of the Cauchy-Riemann analyticity conditions. Most of these developments can be found in a recent review on supergravity by P.van Nieuwenhuizen in Phys.Rep. 68 (1981), p.189.

SELF DUAL GAUGE FIELD EQUATIONS ON N DIMENSIONAL MANIFOLDS

D. H. Tchrakian

St. Patrick's College, Maynooth, Co. Kildare, Ireland
& Dublin Institute for Advanced Studies,
10 Burlington Road, Dublin 4, Ireland

ABSTRACT. A framework is described, within which a self dual gauge field system on an N dimensional manifold, possessing finite action associated with a topological invariant can be formulated. This generalizes the well known monopoles and instantons on N=3 and 4. A discussion of possible applications to physical models by means of dimensional reduction techniques is given, and some new results using N=6 are reviewed.

The main purpose of these lectures is to treat instantons and monopoles, as far as possible, as the same type of solutions of the Yang-Mills-Higgs field equations. For this reason, the only type of monopole solutions we consider are those for systems of fields where the self interaction potential of the Higgs fields vanishes, thus allowing the existence of self dual monopole solutions, in concert with the self dual instanton solutions, there being no known non self dual instantons.

Within our context, instantons and monopoles are special types of solutions to the classical equations of motion for gauge fields defined on four and three dimensional Euclidean base manifolds respectively. The special features of these solutions are that they have finite action, and that the value of this action is equal to a topological invariant via the condition of self duality.

The manner in which the qualitative unification of instantons and monopoles is tackled here is by way of developing a framework for the description of such solutions on N dimensional base manifolds. As Euclidean field equations these would be analogous to the static solutions of the Maxwell equations, namely the Laplace equation on N dimensional manifolds. Here we shall find

A. O. Barut (ed.), Quantum Theory, Groups, Fields and Particles, 161–177.

that it is possible to extend the definition of instantons to any even dimensional manifold, and monopoles to any odd N.

We shall first give a brief introduction to classical gauge field theories by way of introducing our notation. Then the case of instantons and of monopoles will be treated separately. The last part will be devoted to the consideration of some explicit solutions as examples of the foregoing framework.

We stress again that we are only concerned with self dual solutions here, thereby not covering the well known non self dual monopole solutions of 't Hooft[1] and Polyakov[2].

GAUGE FIELDS

There are two types of fields in such theories. The gauge covariant fields, like the Higgs field, Dirac fields etc., and the curvature or the second rank antisymmetric tensor fields, and the non gauge covariant fields called connections or vector potentials.

The elements of the gauge group are functions of the base manifold via the space dependence of the group parameter. The groups under consideration will be SU(n). The reason we do not take G the gauge group to be SU(2) for simplicity is, that on N dimensional manifolds with N>4 it will turn out that our new definitions of self duality will only allow non trivial solutions for G=SU(n) for n>2+k with k>1. In what follows the index of the base manifold will be labelled by a Greek index for even N, and a Latin index for odd N.

Also, because of our insistence on self dual solutions of the field equations, we shall consider only gauge covariant Higgs fields Φ that take their values in the algebra $\mathcal{G}$ of G. The Yang-Mills fields and $F_{\mu\nu}$ and potentials A_μ can take their values only in $\mathcal{G}$.

Under the action of an element g of G

$$g(x) = \exp \langle T, \theta(x) \rangle$$

where T are the (antihermitian) generators in $\mathcal{G}$ and $\theta(x)$ the local parameters, gauge covariant fields transform as

$$\Phi \xrightarrow{g} g \quad g^{-1} \tag{1a}$$

$$F_{\mu\nu} \xrightarrow{g} gF_{\mu\nu}g^{-1} \tag{1b}$$

while the Yang-Mills potential or connection is defined by the following transformation rule

$$A_\mu \xrightarrow{g} g\, A_\mu\, g^{-1} + g\partial_\mu\, g^{-1} \qquad (1c)$$

The transformation rule (1c) confirms our statement that A_μ must take its values in $\mathcal{G}$ since the inhomogeneous term $g\partial_\mu g^{-1}$ does take its values in $\mathcal{G}$ only.

It is clear from (1a) that $\partial_\mu\Phi$ is not gauge covariant, but that the covariant derivative

$$D_\mu\Phi = \partial_\mu\Phi + [A_\mu,\Phi] \qquad (2)$$

is covariant, by virtue of (1a,b,c), with

$$(D_\mu\Phi) \xrightarrow{g} g(D_\mu\Phi)g^{-1} \qquad (1d)$$

The Yang-Mills field, or the curvature tensor referred to above is in fact defined by the following relation

$$[D_\mu,D_\nu]\ f = F_{\mu\nu}f \qquad (3a)$$

$$F_{\mu\nu} = \partial_\mu A_\nu - \partial_\nu A_\mu + [A_\mu,A_\nu] \qquad (3b)$$

where f is a function taking its values in some representation of the group G.

An important consequence of the non commuting nature of the covariant derivative is the series of differential identities satisfied by the curvature field, called the Bianchi identities.

In three and four dimensions, the Jacobi identity for the cyclic product of D_μ acting on some function f implies the following identities

$$\varepsilon_{ijk}\ [D_i,[D_j,D_k]] = 0 \qquad (4a)$$

$$\varepsilon_{\mu\nu\rho\sigma}[D_\nu,[D_\rho,D_\sigma]] = 0 \qquad (4b)$$

which lead, because of (3a), to the Bianchi identities

$$D_i{}^*F_i = D_\nu{}^*F_{\mu\nu} = 0 \qquad (5a,b)$$

In (5) we have used the definition of the dual of the curvature

$$^{*}F_i = \frac{1}{2}\,\varepsilon_{ijk}F_{jk} \tag{6a}$$

$$^{*}F_{\mu\nu} = \frac{1}{2}\,\varepsilon_{\mu\nu\rho\sigma}F_{\rho\sigma} \tag{6b}$$

It is clear that in higher dimensional manifolds the curvature will satisfy further Bianchi identities in addition to (5), arising from the extended analogues of (4), for example

$$\varepsilon_{\mu_N\mu_{N-1}\cdot\cdot\mu_5\mu_4\mu_3\mu_2\mu_1}[D_{\mu_5},\{[D_{\mu_4},D_{\mu_3}],[D_{\mu_2},D_{\mu_1}]\}] = 0 \quad (4b')$$

$$\varepsilon_{\mu_N\mu_{N-1}\cdot\cdot\mu_7\cdot\cdot\mu_1}[D_{\mu_7},\{[D_{\mu_6},D_{\mu_5}],\{[D_{\mu_4},D_{\mu_3}],[D_{\mu_2},D_{\mu_1}]\}\}] = 0 \quad (4b'')$$

and so forth, giving rise to

$$\varepsilon_{\mu_N\mu_{N-1}\cdot\cdot\mu_5\mu_4\mu_3\mu_2\mu_1}D_{\mu_5}\{F_{\mu_4\mu_3},F_{\mu_2\mu_1}\} = 0 \quad (5b')$$

$$\varepsilon_{\mu_N\mu_{N-1}\cdot\cdot\mu_7\cdot\cdot\mu_1}D_{\mu_7}\{F_{\mu_6\mu_5},\{F_{\mu_4\mu_3},F_{\mu_2\mu_1}\}\} = 0 \quad (5b'')$$

and so forth.

It is useful at this stage to extend the duality operation (6b) to arbitrary dimensions. To this end we define the following curvature 2k forms

$$F(4)=F_{\mu\nu\rho\sigma}=\{F_{\mu\nu},F_{\rho\sigma}\} - \{F_{\mu\rho},F_{\nu\sigma}\} - \{F_{\mu\sigma},F_{\rho\nu}\} \tag{7}$$

$$F(6)=F_{\mu\nu\rho\sigma\tau\lambda}=\{F_{\mu\nu},F_{\rho\sigma\tau\lambda}\} - \{F_{\mu\rho},F_{\nu\sigma\tau\lambda}\} - \{F_{\mu\sigma},F_{\rho\nu\tau\lambda}\}$$

$$- \{F_{\mu\tau},F_{\rho\sigma\nu\lambda}\} - \{F_{\mu\lambda},F_{\rho\sigma\tau\nu}\} \quad (7')$$

and so forth, noting that these 2k-forms have the same symmetries as the totally antisymmetric symbol.

Using these we extend the N=4 duality operation (6b) to the next two higher dimensions N=6 and 8

$$N=6 \qquad {}^{(+)}F_{\mu\nu} = \frac{1}{4!}\varepsilon_{\mu\nu\rho\sigma\tau\lambda}\, F_{\rho\sigma\tau\lambda} \qquad (8^{+})$$

$$ {}^{(-)}F_{\mu\nu\rho\sigma} = -\frac{1}{2!}\varepsilon_{\mu\nu\rho\sigma\tau\lambda}\, F_{\tau\lambda} \qquad (8^{-})$$

$$N=8 \qquad {}^{(+)}F_{\mu\nu} = \frac{1}{6!}\varepsilon_{\mu\nu\rho\sigma\tau\lambda\kappa\eta}F_{\rho\sigma\tau\lambda\kappa\eta} \qquad (9^{+})$$

$$ {}^{(o)}F_{\mu\nu\rho\sigma} = \frac{1}{4!}\varepsilon_{\mu\nu\rho\sigma\tau\lambda\kappa\eta}F_{\tau\lambda\kappa\eta} \qquad (9^{o})$$

$$ {}^{(-)}F_{\mu\nu\rho\sigma\tau\lambda} = \frac{1}{2!}\varepsilon_{\mu\nu\rho\sigma\tau\lambda\kappa\eta}F_{\kappa\eta} \qquad (9^{-})$$

Note that for N=2p dimensional manifolds with odd p we have even number of duality operations manifolds with even p we have odd number of duality operations.

These extended duality operations possess the following properties[4]

$$ {}^{(-)}[{}^{(+)}F(2)]=F(4), \quad {}^{(+)}[{}^{(-)}F(4)]=F(2) \qquad (10)$$

$$ {}^{(-)}[{}^{(+)}F(2)]=F(6), \quad {}^{(o)}[{}^{(o)}F(4)]=F(4), \quad {}^{(+)}[{}^{(-)}F(6)]=F(2). \qquad (11)$$

Finally, using the notation of (7), (8) and (9) the Bianchi identities for N=6 and 8 condense to the respective forms

$$D_\nu {}^{(+)}F_{\mu\nu} = 0, \quad D_\sigma {}^{(-)}F_{\mu\nu\rho\sigma} = 0 \qquad (12)$$

$$D_\nu {}^{(+)}F_{\mu\nu} = 0, \quad D_\sigma {}^{(o)}F_{\mu\nu\rho\sigma} = 0, \quad D_\lambda {}^{(-)}F_{\mu\nu\rho\sigma\tau\lambda} = 0. \qquad (12')$$

This completes the introduction of our notation for gauge fields on N dimensional manifolds. So far we have not considered the Euler-Lagrange equations or equivalently the action density of these fields. This will be the main item of interest in the next two Sections.

INSTANTONS: even N

Self dual solutions of the Yang-Mills equations for N=4 are the so called instantons(3). In this case there is only one dual $^*F = {}^{(o)}F$ in our notations. The conventional action density

$$S_4 = \frac{1}{4} \operatorname{tr} F_{\mu\nu}F_{\mu\nu} \tag{13}$$

leads to the equations of motion

$$D_\nu F_{\mu\nu} = 0 \tag{14}$$

which would be identically satisfied if the solutions in question were to be required to be self dual

$$F_{\mu\nu} = {}^{(o)}F_{\mu\nu} \tag{15}$$

by virtue of the Bianchi identity (5b).

That such solutions should have finite actions, proportional to a topological invariant with integer values (the Pontryagin number) can best be seen following the demonstration of BPST(3).

Consider the inequality

$$\operatorname{tr} \int (F_{\mu\nu} - {}^{(o)}F_{\mu\nu})^2 \, d_4x \geqslant 0, \tag{16}$$

which yields

$$\int S_4 d_4x \geqslant \operatorname{tr} \int F_{\mu\nu} \, {}^{(o)}F_{\mu\nu} d_4x, \tag{16'}$$

the right side of which is equal to $4\pi^2$ times a normalisation factor times an integer, provided that the curvature vanishes at infinity in the following manner

$$A_\mu \xrightarrow[x \to \infty]{} g \, \partial_\mu g^{-1} \tag{17}$$

where $g(x)$ is some element of G.

It is then clear that the self duality condition (15) causes the inequality to become an equality, and then (16') expresses the fact that the total action corresponding to such a solution is equal to the finite quantity on the right.

The question now is whether the above argument can be modified to apply to gauge fields defined on N=2p dimensional manifolds[4] (p>2). As in the previous Section, we shall demonstrate our point by explicit consideration of the cases p=3 and 4.

From the above argument of BPST for N=4, it is clear that the central agency which controls the behaviour of such solutions is the integer topological invariant, with (17) satisfied. So we start by writing down the topological invariants respectively

N=6,p=3: Chern Class[5]

$$q_6 = \frac{1}{DV_6}\,\varepsilon_{\mu\nu\rho\sigma\tau\lambda}\ \mathrm{tr} \int F_{\mu\nu}F_{\rho\sigma}F_{\tau\lambda}\, d_6x \tag{18}$$

N=8,p=4: Pontryagin Class[4]

$$q_8 = \frac{1}{DV_8}\,\varepsilon_{\mu\nu\rho\sigma\tau\lambda\kappa\eta} \int (\mathrm{tr}\ F_{\mu\nu}F_{\rho\sigma}F_{\tau\lambda}F_{\kappa\eta} - \frac{1}{2}\ \mathrm{tr}\ F_{\mu\nu}F_{\rho\sigma}.\mathrm{tr}F_{\tau\lambda}F_{\kappa\eta})d_8x \tag{19}$$

where D is a normalisation factor depending on the representation matrices of G and V_N is the volume of the N dimensional sphere.

It is now clear that if we are to be able to use an argument similar to the BPST one, where the action integral becomes equal to the topological invariant, then we must expect the action densities to be of progressively higher order in the curvature as N increases.

Thus for N=6 let us consider the following inequality

$$\mathrm{tr} \int (\ F(2) - {}^{(+)}F(2)\ )^2 d_6x \geqslant 0. \tag{20}$$

It is also possible to use the inequality

$$\mathrm{tr} \int (\ F(4) - {}^{(-)}F(4)\ )^2 d_6x \geqslant 0$$

but this simply reduces to (20).

Expanding the integrand of (20), it is readily seen that the csoss term $\mathrm{tr}F(2)\ {}^{(+)}F(2)$ leads to the integral formula for the topological invariant given by (18). It is therefore obvious that if we were to consider the (unconventional) gauge theory with action density

$$S_6 = \text{tr}\ [F(2)^2 + {}^{(+)}F(2)^2], \tag{21}$$

and rewrite (20) as

$$\int S_6 d_6 x \geqslant 2DV_6 q_6 \ , \tag{20'}$$

then, if we imposed the following extended self duality conditions

$$F(2) = {}^{(+)}F(2) \quad \text{or} \quad F(4) = {}^{(-)}F(4) \tag{22}$$

then (20') would become an equality, and the total action corresponding to these solutions will be finite.

Similarly for N=8, we consider the following inequality

$$\text{tr} \int [F(2) - {}^{(+)}F(2)]^2 d_8 x + \frac{1}{2} \int [\text{tr}F(4) + {}^{(o)}F(4)]^2 d_8 x \geqslant 0 \tag{23}$$

or the equivalent inequality obtained by replacing F(2) and ${}^{(+)}F(2)$ by F(6) and ${}^{(-)}F(6)$ respectively.

Then expanding the integrands in (23) we find

$$\int S_8 d_8 x \geqslant 2DV_8 q_8 \tag{24}$$

where the action density for the eight dimensional gauge theory is taken to be

$$S_8 = \text{tr}\ [F(2)^2 + {}^{(+)}F(2)^2] + 2[\text{tr}F(4)]^2. \tag{25}$$

The total action is then equal to the (finite) quantity 2DV q when (24) becomes an equality subject to the extended self duality conditions

$$F(2) = {}^{(+)}F(2) \quad \text{or} \quad F(6) = {}^{(-)}F(6) \tag{26}$$

and

$$F(4) = -{}^{(o)}F(4). \tag{26'}$$

Two remarks are now in order. The first is that the self duality condition (22) shows clearly why we had anticipated a factor of i in the definition of the duality (8), for otherwise (22) would lead to a trivial solution F(2)=0. This situation will always occur for odd p (N=2p), in contrast with the cases of even p, where for example the duality (9) do not involve factors of i.

The second remark is that for N>4, the gauge group G=SU(n) will allow non trivial solutions of (22) and (26) (26') only if n>2. It is easy to check for example that F(2)=0 for n=2 and N=6, by taking the trace of the self duality equation.

Finally it is clear that the extended self duality conditions solve the Euler-Lagrange equations arising from the corresponding action densities. For example, for N=6 the field equations are

$$D_\mu F_{\mu\nu} + \frac{1}{2} D_\mu \{ F_{\mu\nu\rho\sigma}, F_{\rho\sigma} \} = 0 \tag{27}$$

which is identically satisfied if (22) hold by virtue of the Bianchi identities (12).

MONOPOLES: odd N

Our considerations[4] here are restricted to self dual monopoles. In three dimensions these are the solutions[6] found by Prasad and Sommerfield and by Bogomolnyi. The Lagrangian density of these gauge theories include no self interaction term of the Higgs fields, the solutions however possess suitable boundary conditions which otherwise would have been consistent with and implied by the peculiar form of the Higgs potential[1].

We review the N=3 case briefly before going on to the N=5 case as an example of the extension of the notion of self duality for monpoles.

The Lagrange density for N=3

$$\mathcal{L}_3 = \frac{1}{4} \operatorname{tr} F_{ij}^2 + \frac{1}{2} \operatorname{tr} (D_i \Phi)^2 \tag{28}$$

leads to the field equations

$$D_j F_{ij} + [\Phi, D_i \Phi] = 0 \tag{29}$$

$$D_i D_i \Phi = 0 \tag{30}$$

It is easy to verify that the 'self duality' condition

$$F_{ij} = \epsilon_{ijk} D_k \Phi \tag{31}$$

solves (29) immediately, and reduces (30) to the Bianchi identity (5a). Therefore the problem of solving (29) and (30) is reduced to finding solutions of (31).

Now solutions of (31) that satisfy the following boundary conditions

$$D_i \Phi \xrightarrow[r \to \infty]{} 0 \qquad \mathrm{tr}\,\Phi^2 \xrightarrow[r \to \infty]{} \text{const.} \tag{32}$$

and

$$\Phi \xrightarrow[r \to \infty]{} \text{const.}\,\mathbb{1} \tag{33}$$

have finite action (energy), equal to 4πtimes an integer. This is seen by considering the following inequality

$$\mathrm{tr} \int (F_{ij} - \epsilon_{ijk} D_k \Phi)^2 d_3x \geqslant 0. \tag{34}$$

Expanding the integrand, and using Stokes' theorem and the Bianchi identify (5a), this reduces to

$$\int \mathcal{L}_3 d_3x \geqslant \epsilon_{ijk} \; \mathrm{tr} \int \Phi F_{jk} \, dS_i , \tag{34'}$$

and this inequality becomes an equality if (31) is imposed.

It now remains to show that

$$\frac{1}{4\pi} \epsilon_{ijk} \; \mathrm{tr} \int \Phi F_{jk} dS_i = \frac{1}{4\pi} \epsilon_{ijk} \; \mathrm{tr} \int_{\Sigma} \Phi [\partial_j \hat{\Phi}, \partial_k \hat{\Phi}] dS_i , \tag{35}$$

where Σ surface of the sphere at infinity, $\hat{\Phi}$ is the direction of the Higgs field $\Phi/\mathrm{tr}\Phi^2$, and where the right side is a Kronecker integral subject to (32), and so takes on integer values only.

To arrive at (35) we introduce the so called 'electromagnetic' field

$$\mathcal{F}_{ij} = \mathrm{tr}\, \Phi (F_{ij} + \tfrac{1}{4} [D_i \hat{\Phi}, D_j \hat{\Phi}]) \tag{36}$$

which on the $r \longrightarrow \infty$ surface over which (34') is integrated, is equal to the integrand of the right side of (34'), because of (32). But by using (33), we can show that (36) reduces to

$$\mathcal{F}_{ij} = \partial_i B_j - \partial_j B_i + \mathrm{tr}\, \Phi[\partial_i \hat{\Phi}, \partial_j \hat{\Phi}] \qquad (36')$$

with

$$B_i = \mathrm{tr}\, \Phi A_i.$$

This proves (35), since the surface integral of the pure curl term $\partial_i B_j - \partial_j B_i$ vanishes.

Therefore the energy integral is finite and equal to an integer modulo 4π, provided that (31), (32) and (33) hold.

We observe from the above that the integer bounding the action integral is a Kronecker integral. As there exist no global topological invariants on odd dimensional manifolds, we expect that the bounding integer should also be a Kronecker integral for higher dimensional Yang-Mills-Higgs theories.

We proceed to define the extended self duality conditions, that generalise (31), by considering the N=5 case as an example.

The Kronecker integral for N=5 is given by

$$\frac{1}{4\pi^2} \epsilon_{ijklm}\, \mathrm{tr} \int \Phi\{[\partial_i \hat{\Phi}, \partial_j \hat{\Phi}], [\partial_k \hat{\Phi}, \partial_l \hat{\Phi}]\} dS_m \qquad (37)$$

in terms of the Higgs field satisfying (32).

As for the N=3 case, we define an 'electromagnetic' field

$$\mathcal{F}_{ijkl} = \mathrm{tr}\, \Phi\{(F_{ij} + \tfrac{1}{4}[D_i \hat{\Phi}, D_j \hat{\Phi}]), (F_{kl} + \tfrac{1}{4}[D_k \hat{\Phi}, D_l \hat{\Phi}])\} \qquad (38)$$

which at large distances behaves, because of (32), like

$$\mathcal{F}_{ijkl} \xrightarrow[r \to \infty]{} \mathrm{tr}\, \Phi F_{ij} F_{kl}. \qquad (38')$$

On the other hand, using (33) it can be seen that

$$\mathcal{F}_{ijkl} = \frac{1}{2}\, \mathrm{tr}\, \Phi\{[\partial_i \hat{\Phi}, \partial_j \hat{\Phi}], [\partial_k \hat{\Phi}, \partial_l \hat{\Phi}]\} + \text{total divergence}. \qquad (38'')$$

We are now in a position to find a Lagrangian density in five dimensions[4], whose action integral is controlled by the integer (modulo $4\pi^2$) given by (37). To this end we consider the following inequality

$$\mathrm{tr} \int \left(\frac{1}{4!} \varepsilon_{ijklm} F_{jklm} - D_i \Phi\right)^2 d_5x \geqslant 0 \tag{39}$$

Choosing the Lagrangian density

$$\mathcal{L}_5 = \mathrm{tr} \left(\frac{1}{8} F_{ijkl}{}^2 + (D_i \Phi)^2\right) \tag{40}$$

and expanding (39) we get

$$\int \mathcal{L}_5 d_5x > \frac{1}{4!} \varepsilon_{ijklm} \, \mathrm{tr} \int \Phi F_{jklm} dS_i \tag{39'}$$

where in the last step we have used the Stokes' theorem and the Bianchi identity

$$\varepsilon_{ijklm} D_i F_{jklm} = 0.$$

The right side of (39') is now obviously equal to $4\pi^2$ times (37), by virtue of (38), (38') and (38''), and the action integral of this five dimensional Yang-Mills-Higgs theory will be finite and proportional to an integer provided that the following self duality holds

$$\frac{1}{4!} \varepsilon_{ijklm} F_{jklm} = D_i \Phi, \tag{41}$$

which turns (39') into an equality. This is a direct generalisation of (31), for N=5.

Here we remark that a guage theory with G=SU(n), N>3 has non trivial self dual solutions only if n>2. For example with N=5 and n=2 $D_i\Phi$ vanishes everywhere, as can be checked easily by taking the trace of (41).

That such solutions for N=3 were called monopole is because the integral in (35) can be written as

$$\int \frac{1}{2} \varepsilon_{ijk} F_{jk} dS_i = \int \vec{B} \cdot d\vec{S},$$

which is the magnetic flux of a monopole.

Applications:

The application of the above described higher dimensional theories would have to involve some kind of dimensional reduction procedure, so that the end theory is one on a three or four dimensional Euclidean manifold.

Thus, it is envisaged that a hierarchy of gauge theories with finite action associated with a topological invariant can be constructed, by reducing the dimension from N (even), down to four or three Euclidean dimensions. The ensuing theories are not any more Yang-Mills, or Yang-Mills-Higgs theories, and they possess action densities involving the curvature with higher powers than quadratic. In fact the higher the dimension N of the manifold before being subjected to the reduction procedure, the higher will be the non linearity of the action density in the three or four dimensional manifold.

As far as dimensional reduction is concerned, there has been considerable interest(7) in the literature on the reduction of the dimensions of a non-abelian gauge theory. These methods are relevant to our case, or at least can be adapted to the reduction of our non-Yang-Mills non-abelian theories. At present however, we shall illustrate the possibilities by considering the elementary prescription used(8) in the reduction of the four dimensional pure Yang-Mills system to the three dimensional Yang-Mills-Higgs system. This method can be applied also to the higher dimensional theories described above, and, by way of illustration, we present the case N=4 reduced to N=3.

One starts with the pure Yang-Mills action density

$$S_4 = \frac{1}{4} \operatorname{tr} F_{\mu\nu}{}^2 \qquad (13)$$

where $\mu,\nu=1,2,3,4$, and invokes the independence of x_4 of all field quantities, whence, we have for the $\mu=i$ $\nu=4$ component of the curvature

$$F_{i4} = \partial_i A_4 + [A_i, A_4] = D_i A_4, \qquad (43)$$

where $i=1,2,3$ and $D_i A_4$ is the covariant derivation of the field A_4 in the adjoint representation of G. Identifying then A_4 as a Higgs field Φ, which is now gauge covariant under a gauge group G whose parameters now depend only on x_i and not on x_4. The action density (42) then reduces to

$$L_3 = \frac{1}{4} \operatorname{tr} F_{ij}{}^2 + \frac{1}{2} (D_i \Phi)^2 \qquad (28)$$

which is the corresponding quantity for the Yang-Mills-Higgs system. Similarly, it is easy to check that the Euler Lagrange equations (14) and self duality conditions (15) for the pure Yang-Mills system on four dimensions reduce to the corresponding equations for the Yang-Mills-Higgs system, namely equations (29), (30) and (31) respectively.

There remains the identification of the topological charge, which in four dimensions is the Pontryagin index given by the right hand side of (16') defined in terms of the pure Yang-Mills curvature, while in three dimensions it is defined as the magnetic flux (35). To see that this arises also from the above prescribed dimensional reduction, we reconsider the integral inequality (16) which then reduces to

$$\mathrm{tr} \int \mathcal{L}_3 d_3x \, dx_4 \geqslant \epsilon_{ijk} \, \mathrm{tr} \int D_i\Phi \, F_{jk} \, d_3x \, dx_4 \qquad (44)$$

where the x_4 integrations on both sides of the inequality (44) cancel, thus reducing (44) to the inequality (34) with the attendent consequences, namely, that the requirement of self-duality (31) further leads to the condition[8] that

$$\mathrm{tr} \int \mathcal{L}_3 \, d_3x = \frac{1}{2} \int \Delta_3 \, \mathrm{tr} \, \Phi^2 \, d_3x \qquad (45)$$

where Δ_3 is the Laplace operator in three dimensions.
The right hand side of (45) will then take on integer spectrum, n, provided that the solutions in question satisfy the desired boundary conditions, namely that $\mathrm{tr}\Phi^2$ has the following asymptotic form

$$\mathrm{tr} \, \Phi^2 = \mathrm{const.} \frac{2n}{r} + O(r^{-2}), \quad r \longrightarrow \infty \qquad (46)$$

This is consistent with the boundary conditions (32) listed above.

Having described the dimensional reduction technique, we now indicate how this can be applied to higher dimensions. As was the case above, our prescription is to descend always from an <u>even</u> dimension, because it is on even dimensions that the topological invariants, namely the Chern and Pontryagin classes are defined[4], and expressed as gauge invariant polynomials of the curvature field.

Considering the two examples of N=6 and 8 we have for:

N=6: i) reducing from six to four dimensions, we invoke the independence of all field quantities, of the variables x_5 and x_6, and, we identify the following components of the connection field

$$A_5 = \Phi_1 \quad \text{and} \quad A_6 = \Phi_2$$

as Higgs fields, which are gauge covariant under gauge transformations whose parameters depend only on x_μ, $\mu=1,2,3,4$. Then we have, in addition to the curvature $F_{\mu\nu}$, the following components of the original curvature,

$$F_{\mu 5} = D_\mu \Phi_1 , F_{\mu 6} = D_\mu \Phi_2 , \tag{47a,b}$$

$$F_{56} = [\Phi_1 , \Phi_2] . \tag{48}$$

ii) reducing from six to three dimensions, similarly will give rise to three Higgs fields $A_4 = \Phi_1$ $A_5 = \Phi_2$, $A_6 = \Phi_3$, with

$$F_{\mu 4} = D_\mu \Phi_1 , F_{\mu 5} = D_\mu \Phi_2 , F_{\mu 6} = D_\mu \Phi_3 \tag{49a,bc}$$

$$F_{45} = [\Phi_1 , \Phi_2], \; F_{56} = [\Phi_2 , \Phi_3], \; F_{64} = [\Phi_3 , \Phi_1] \tag{50a,b,c}$$

It is clear that for N=8, the same procedure will lead to the following field structures

i) reducing to four dimensions, there will be four Higgs fields, and
ii) reducing to three dimensions, five Higgs fields.

The topological invariants in both above cases are the third and fourth Chern classes, respectively, subjected to this dimensional reduction procedure, just as the Pontryagin index reduced to the right hand side of (45) in the four-to-three reduction[8].

It is interesting to speculate that these theories, on four and three dimensional manifolds, are physically relevant just as the Yang-Mills-Higgs systems are. Their main qualitative difference is that they are much more non-linear in the curvature and the covariant derivatives of the Higgs fields than the Yang-Mills theories that are quadratic. Indeed, our self-duality conditions (22) and (26) (26') for example, are strong enough that this high degree of non-linearity is compensated for and there is the possibility of actually finding finite action (energy) solutions.

In four dimensions, these theories would differ from those with instantons in several respects. Firstly they would be much more non linear, namely, their action densities would involve the 2(N-2)-nd power of the curvature fields, where N is the dimension of the theory before reduction. Then, the topological invariant associated with the finite action solutions would be the $\frac{1}{2}$Nth Chern class. Finally, they would not be pure gauge-field theories, but would involve (N-4) different gauge covariant Higgs fields, each in the algebra of G which, as remarked in Section 1, must be a group larger than SU(2).

Recently, such a model with N=6 has been considered[9] in detail, and it turns out that, the self duality equations (22) result in the following action

$$\int S_{6\to 4} d_4x = \int \Delta_4 \, \mathrm{tr}\, \Phi_1^{\,2} \quad d_4x, \tag{47}$$

where $S_{6\to 4}$ is the reduction form N to four dimensions obtained from S_6 as given by (21), and Δ_4 is the Laplace operator in four dimensions.

In three dimensions, correspondingly, one would obtain models with gauge group larger than SU(2), that are different from the 't Hooft-Polyakov type self-dual monopoles. They are highly non linear, and as above involve the curvature field and the covariant derivatives of the Higgs fields to the power of 2(N-2), where N is the original dimension. Again, the monopole (topological) charge in these cases will be related to the $\frac{1}{2}$Nth Chern classes via the above described prescription for dimensional reduction.

In such theories, topological charge may be related to an electromagnetic field density, which is the direct generalisation of the quantity

$$\mathcal{B}_i = \varepsilon_{ijk} \, \mathrm{tr}\, \Phi F_{jk} \tag{48}$$

that occurs in (35) and (36), and is the flux density giving rise to the topological charge. The flux density $\mathcal{B}_i$ in a three dimensional theory reduced from the N dimensional theory can be defined easily in principle by noting that the Chern classes can always be expressed as surface integrals, for example, the Pontryagin index for four dimensions can be written as

$$\mathrm{tr} \int F_{\mu\nu} {}^*F_{\mu\nu} d_4x = \varepsilon_{\mu\nu\rho\sigma} \int \partial_\mu \, \mathrm{tr}\, A_\nu \, (F_{\rho\sigma} - \tfrac{2}{3} A_\rho A_\sigma) \, d_4x \tag{49}$$

Then, the simple dimensional reduction we use implies that the right hand side of (49) is a three dimensional surface integral, since ε_{ijk4} tr A_i $(F_{jk} - \frac{2}{3}A_jA_k)$ is taken to be independent of x_4. It is then obvious that our definition of $\mathcal{B}_i$ is

$$\mathcal{B}_i = \varepsilon_{i\nu\rho\sigma} \, \mathrm{tr}\, A_\nu \, (F_{\rho\sigma} - \tfrac{2}{3} A_\rho A_\sigma) \tag{50}$$

Since the property (49) is true for all Chern densities, that is they are always expressible as total divergences, it is then a straightforward matter to define the generalisations of (50), when the topological flux density is derived via a reduction from an N-dimensional theory. Though straightforward, these in general involve quite complicated computations.

Recently, a theory of this type, with N=6 was considered in detail, and the following results were obtained

$$\int S_{6\to 3} d_3x = \int \Delta_3 \ \mathrm{tr}\ (\Phi_1^2 + \Phi_2^2 + \Phi_3^2) d_3 x \qquad (51)$$

with the topological flux density

$$\mathcal{B}_i = \varepsilon_{ijk}\ \mathrm{tr}\ [\tfrac{1}{2}F_{jk}\{\Phi_1,[\Phi_2,\Phi_3]\}] - \Phi_1\{D_j\Phi_2, D_k\Phi_3\}]\ \text{cycl.}(1,2,3) \qquad (52)$$

The results (47) and (51) are interesting in that they would lead to finite action integrals provided the solution in question would satisfy the requisite boundary conditions, namely, in the case of (51), that $\mathrm{tr}\ (\Phi_1^2 + \Phi_2^2 + \Phi_3^2)$ should have the same asymptotic form as (46), while in the case of (47), $\mathrm{tr}\Phi_1^2$ should have the following asymptotic form

$$\mathrm{tr}\ \Phi_1^2 = \text{const.} - \frac{1}{3}\frac{n}{r^2} + O(r^{-3}) + \text{---..} \qquad (53)$$

This completes the discussion of the possibilities of deriving 3 and 4-dimensional models from the N-dimensional gauge field theories discussed above. Here, only an elementary prescription of dimensional reduction has been discussed(8), but clearly this does not exclude the possibility that the more sophisticated dimensional reduction techniques developed in references (7) may be just as or even more useful for this purpose.

Finally, some remarks are in order concerning exact solutions to self-duality equations of types (22) and (26). So far no such solutions have been constructed, mainly due to the difficulty arising from the high degree of non linearity. One important feature of such solutions would be that they would involve a dimensional constant, which, has been set equal to unity in all the preceding self-duality equations.

REFERENCES

(1) G.'t Hooft, Nucl. Phys. B 79, 276 (1974)
(2) A.M. Polyakov, JETP Lett. 20, 1940 (1974)
(3) A.A. Belavin, A.M. Polyakov, A.S. Schwarz and Yu. S. Tyupkin, Phys. Lett. 59B, 85 (1975)
(4) D.H. Tchrakian, J. Math. Phys. 21, 166 (1980)
(5) S. Kobayashi and K. Nomizu, Interscience (1969)
(6) M. Prasad and C. Sommerfield, Phys. Rev. Lett. 35, 760 (1975)
E.B. Bogomolnyi, Sov. J. Nucl. Phys. 24, 449 (1976)
(7) R. Kerner, Ann. Inst. H. Poincaré 9, 143 (1968)
A. Trautman, Rep. Math. Phys. 1, 29 (1970)
Y.M. Cho and P.G.O. Freund, Phys. Rev. D12, 1711 (1975)
A.S. Schwarz and Yu. S. Tyupkin, Nucl. Phys. B 187,321 (1981)
(8) R.S. Ward, Commun. Math. Phys. 79, 317 (1981)
(9) T.N. Sherry and D.H. Tchrakian, Dimensional Reduction of a 6-dimensional Self-Dual Gauge Field Theory, DIAS-STP-82-07.

EXPOSITORY LECTURES ON TOPOLOGY, GEOMETRY, AND GAUGE THEORIES

Yilmaz Akyildiz*

Mathematical Sciences, University of Petroleum and Minerals, Dhahran, Saudi Arabia

ABSTRACT. Since the introduction of gauge fields in physics, it has been realized that they have a rich geometric and topological interpretation. Classical gauge field theory turns out to be a theory of connections on a principal fibre bundle, and one needs to know the modern differential geometric tools to penetrate this new era of physics. These lectures were prepared with physicists in mind and an attempt has been made to keep the level of mathematical sophistication to a minimum. The main purpose of this article is to provide an insight into the modern differential geometric ideas which have recently found their way into physics.

> *The most powerful method of advance that can be suggested at present is to employ all the resources of pure mathematics in attempts to perfect and generalize the mathematical formalism that forms the existing basis of theoretical physics, and after each success in this direction, to try to interpret the new mathematical features in terms of physical entities.*
>
> *P.A. Dirac (1931)*

1. TOPOLOGY

Topology studies the properties of geometrical figures that persist even when the figures are subjected to deformations so drastic that all their metric and projective properties are lost. Topology is a very intuitive subject and one will find that by too much insistence on a rigorous form of presentation one may easily lose sight of the essential geometrical content in a mass

* On leave from the Middle East Technical University, Ankara, Turkey

A. O. Barut (ed.), Quantum Theory, Groups, Fields and Particles, 179–234.

of formal detail. Still, it is a great merit of recent work to have brought topology within the framework of rigorous mathematics, where intuition remains the source but not the final validation of truth.

Every transformation of a geometric figure in which the relations of adjacency of various parts of the figure are not destroyed is called continuous; if the adjacencies are not only not destroyed, but also no new ones arise, then the transformation is called topological. Therefore a topological transformation of any geometric figure, considered as the set of points forming it, is not only a continuous, but also a one-to-one transformation: any two distinct points of the figure are transformed into two distinct points. Thus, topological transformations are single-valued and continuous both ways.

Intuitively, a topological transformation of an arbitrary geometric figure (a curve, a surface, etc.) can be represented in the following way: let us imagine that our figure is made of some flexible and stretchable material, for example of rubber. When a closed rubber ring is given in the form of a circle then we can stretch it into the shape of an ellipse, we can give it the form of a regular or an irregular polygon and even of a very complicated closed curve, some of which are illustrated below:

But we cannot, by a topological transformation, turn a circle into a figure of eight. The circle is the simplest closed curve, since it forms only one loop in contrast to the figure of eight, which forms two loops. The property of a circle of being a simple closed curve is a property preserved under an arbitrary topological transformation or, as one says, it is a topological property.

If we take a spherical surface, which we can imagine in the form of a thin rubber sheet, then we can make great changes in its shape by means of topological transformations as shown in the next figure. But we cannot, by a topological transformation, turn our spherical surface into a square or a ring-shaped surface (the surface of a tyre tube), which is called a torus. For the surface of a sphere has the following two properties which are both preserved under an arbitrary topological transformation. First, our surface is closed: there are no edges on it (but the square has edges); second, if we make a cut along a given closed curve traced out on our rubber sheet, then the surface splits into two disconnected parts. The torus does not have this property: if a torus is cut along a meridian, then it is not split into parts but is turned into a surface having the form of a bent tube which we can then easily straighten into a cylinder by a topological transformation. Therefore, the spherical surface cannot be turned into a torus by a topological transformation. We say that the sphere and

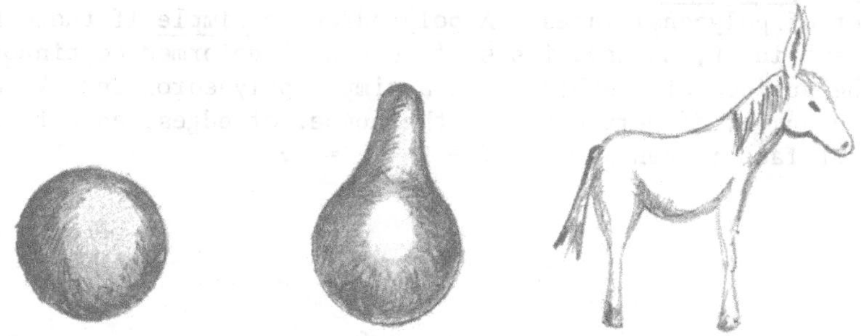

Topologically equivalent surfaces

the torus are topologically distinct surfaces and they are <u>not homeomorphic</u> to each other. Conversely, a sphere and an ellipsoid and, in general, any bounded convex surface belong to one and the same topological type; i.e., they are <u>homeomorphic</u>. This means that they can be carried into one another by a topological transformation.

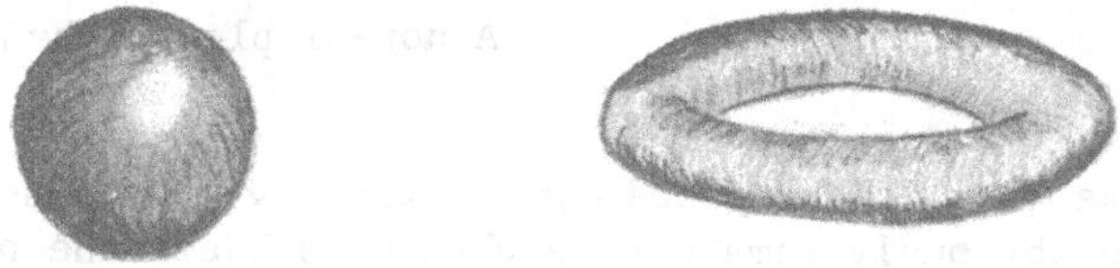

Sphere Torus
Topologically non-equivalent surfaces

As mentioned earlier, every property of a geometrical figure that is preserved under an arbitrary topological transformation of it is called a topological property. Topology studies topological properties of figures; furthermore, it studies topological transformations and also arbitrary continuous transformations of geometrical figures.

While topology is definitely a creation of the last hundred years, there were a few isolated discoveries that later found their place in the modern systematic development. The most important of these is a formula, relating the numbers of vertices, edges, and faces of a simple polyhedron, observed by Euler in 1752. The typical character of this relation as a topological theorem became apparent much later, after Poincaré had recognized "<u>Euler's formula</u>" and its generalizations as one of the central theorems of topology. So, for reasons both historical and intrinsic, we shall begin our discussion of topology with Euler's formula.

By a polyhedron is meant a solid whose surface consists of a number of polygonal faces. A polyhedron is simple if there are no "holes" in it, so that its surface can be deformed continuously into the surface of a sphere. In a simple polyhedron let V denote the number of vertices, E the number of edges, and F the number of faces; then always $V - E + F = 2$.

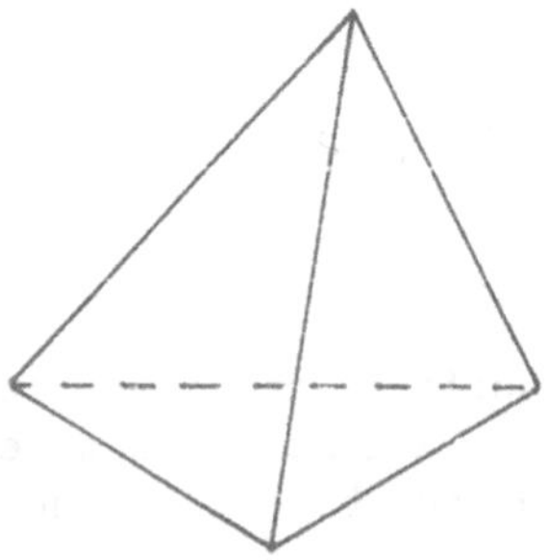

$V - E + F = 2$
A simple polyhedron

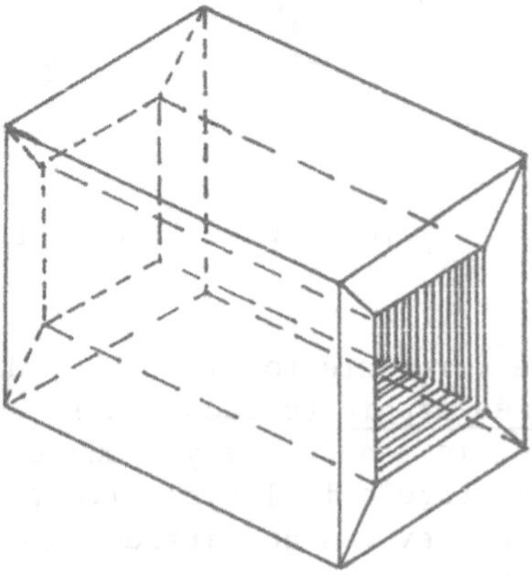

$V - E + F = 16 - 32 + 16 = 0$
A non-simple polyhedron

This geometrical theorem belongs to topology, because our formulae above obviously remain true when we subject the polyhedra in question to arbitrary topological transformations. Under such a transformation the edges will, in general, change from being rectilinear, the faces from being plane and the surface of the polyhedron goes over into a curved surface. But Euler's relation between the number of vertices and the numbers of edges and faces, now curved, remains valid. The most important case is when all the faces are triangles and then we have a so-called triangulation (a division of our surface into triangles, rectilinear or curvilinear). Thus, we can restrict our attention to the case of a triangulation. The combinatorial method in the topology of surfaces consists in replacing the study of such a surface by the study of one of its triangulations, and, of course, we are only interested in the properties of the triangulation which are independent of the accidental choice of one triangulation or another and so, being common to all triangulations of the given surface, express some property of the surface itself.

Euler's formula leads us to one of such properties, and the left hand side of Euler's formula is called the Euler characteristic of this triangulation. Euler's theorem states that for all triangulations of a surface deformable to a sphere the Euler characteristic is equal to two. Now it turns out that for every surface (and not only for a surface homeomorphic to a sphere) all triangulations of the surface have one and the same Euler characteristic.

It is easy to figure out the value of the Euler characteristic for various surfaces. First of all, for the cylindrical surface it is equal to zero. For when we remove from an arbitrary triangulation of the sphere two nonadjacent triangles but preserve the boundaries of these triangles, then we obviously obtain a triangulation of a surface homeomorphic to the curved surface of a cylinder. Here the number of vertices and the edges remains as before, but the number of triangles is decreased by two, therefore the Euler characteristic of the triangulation so obtained is zero. Now let us take the surface resulting from a triangulation of a sphere after removal of 2p triangles of this triangulation that are pairwise not adjacent. Here the Euler characteristic is decreased by 2p units. It is easy to see that the Euler characteristic does not change when cylindrical tubes, "handles", are attached to each pair of holes made in the surface of spheres. This comes from the fact that the characteristic of the tube to be pasted in is, as we have seen, zero and on the rim of the tube the number of vertices is equal to the number of edges. Thus, a closed "two-sided" surface with p "handles", or as it is called,

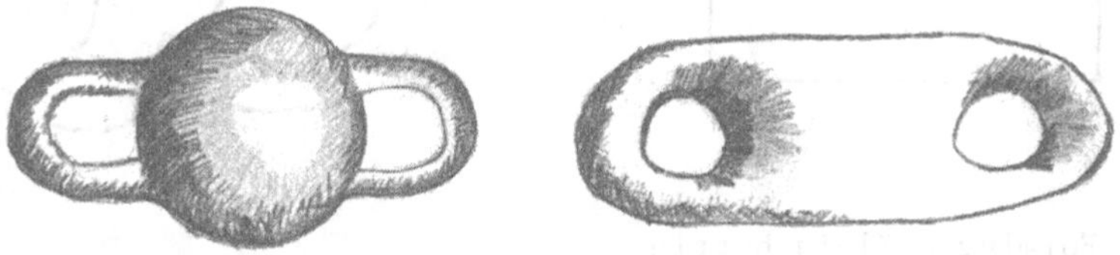

Surfaces of genus 2

a surface of genus p, has the Euler characteristic $2-2p$. Hence, the Euler characteristic of the torus is zero. (Topologically, a torus is the product of two circles, that is $S^1 \times S^1$).

An ordinary surface has two sides. The two sides of such a surface could be painted with different colours to distinguish them. Moebius (1790-1868) made the surprising discovery that there are surfaces with one side only: taking a rectangular piece of paper and bending it round so that one pair of opposite sides meet gives a cylinder. But, first twisting the paper 180° and then making a pair of opposite sides meet results in a Moebius strip.

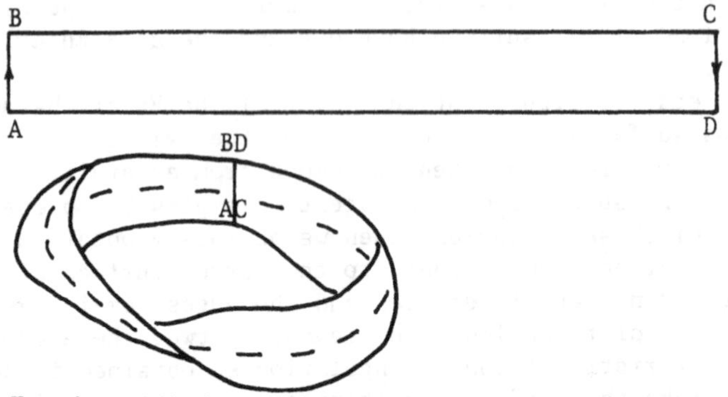

Forming a Moebius strip

Another interesting one-sided surface is the "Klein bottle". This surface is closed, but it has no inside or outside.

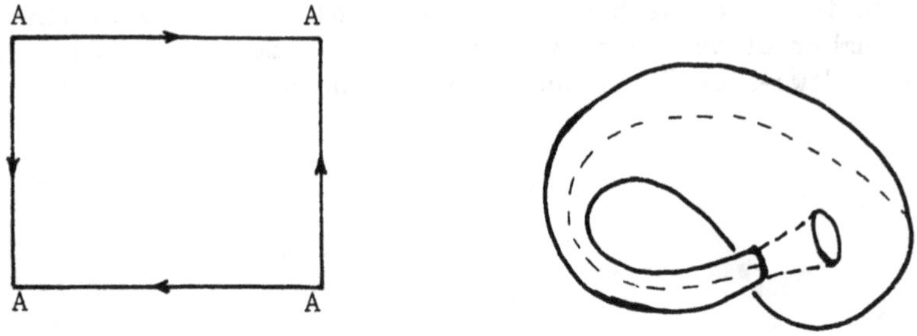

Forming a Klein bottle

It turns out that such one-sided surfaces, no matter how we arrange them in three-dimensional space, always have self-intersections.

Besides the Euler characteristic there is another important topological invariance of geometrical figures. This is the property of orientability. For the time being we will just mention the fact that the two-sided surfaces are orientable whereas one-sided surfaces are non-orientable.

Euler's characteristic and the property of orientability or non-orientability of closed surfaces give us, so to speak, a complete system of topological invariants of closed surfaces. The meaning of this statement is that two surfaces are homeomorphic if and only if, first, their triangulations have one and the same Euler characteristic, and second, both are orientable or non-orientable. But, at the present time we do not have the knowledge of a complete system of invariants for higher dimensional manifolds.

1.1. Homology Theory

The combinatorial method is applicable not only to the study of surfaces (two-dimensional manifolds) but also to manifolds of an arbitrary number of dimensions. But in the case of three-dimensional manifolds, the role of the ordinary triangulations is now taken by decompositions into tetrahedra. The Euler characteristic of a three-dimensional triangulation is defined as the number $\alpha_0 - \alpha_1 + \alpha_2 - \alpha_3$ where α_i, $i = 0, 1, 2, 3$, is the number of i-dimensional elements of this triangulation. For dimensions $n > 3$ the manifolds are divided into n-dimensional simplices. As before, we can speak of an Euler characteristic, interpreting it as the sum

$$\sum_{i=0}^{n} (-1)^i \alpha_i ,$$

where α_i is the number of i-dimensional elements occurring in this triangulation $(i = 0, 1, 2, \ldots, n)$, and as before the Euler characteristic has one and the same value for all triangulations of a given n-dimensional manifold (and for all manifolds homeomorphic to it); i.e., it is a topological invariant. But in the present state of our knowledge, we cannot dream of a complete system of invariants even for three-dimensional manifolds (in the sense in which it is given for surfaces by the Euler characteristic and orientability).

Let us observe that every triangle can be oriented, i.e. that a definite direction of traversing its boundary can be furnished. Note that an oriented 1-simplex has a sense of direction attached to it, an oriented 2-simplex has a sense of rotation attached to it, and so on. Now let us consider all the triangles T_2^i, $i = 1, 2, \ldots, \alpha_2$, that occur in a given triangulation of a surface.

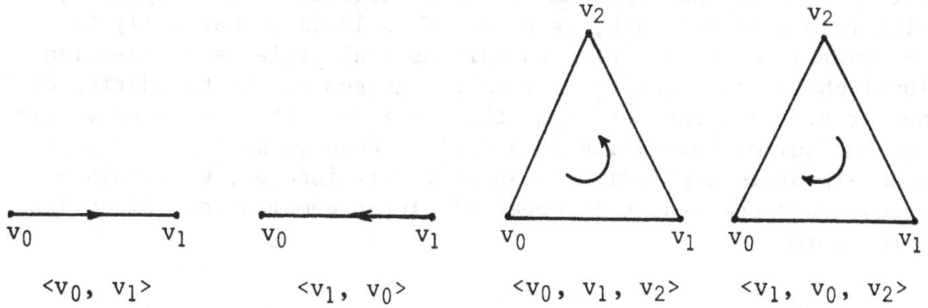

1-simplices

2-simplices

To each of them we can give two orientations; let us denote the triangle T_2^i with one of the two possible orientations by t_2^i and the same triangle with the other (opposite) orientation by $-t_2^i$. Similarly, we denote the one-dimensional segment T_1^k with one of these orientations by t_1^k, and with the other by $-t_1^k$. Now, if the sides of the triangle T_2^i are T_1^1, T_1^2, T_1^3, then the boundary of the oriented triangle t_2^i is the set of the same sides, but taken with a definite direction, i.e., the boundary consists of the directed segments $\varepsilon_1 t_1^1$, $\varepsilon_2 t_1^2$, $\varepsilon_3 t_1^3$; here $\varepsilon_i = 1$ if this direction for the edge T_1^i coincides with its appropriate direction t_1^i, and $\varepsilon_i = -1$ in the opposite case. The boundary of t_2^i is denoted by ∂t_2^i. As we have already seen, $\partial t_2^i = \Sigma \varepsilon_k t_1^k$ and this sum can be imagined as extending over all edges of our triangulation when we consider the coefficients ε_k for segments not in the boundary of t_2^i as being equal to zero.

It now becomes natural to consider more general sums of the form $c_1 = \Sigma a_k t_1^k$ extended over all edges of the given triangulations, where a_k are assumed to be integers. The geometric meaning of such sums is very simple: every summand of the sum is a certain segment that occurs in our triangulation, taken with a definite direction and a definite coefficient ("multiplicity"). The whole algebraic sum so described, expresses a path composed of segments in which every segment is assumed to occur as often as its coefficient indicates.

Sums of the form $c_1 = \Sigma a_k t_1^k$ are called <u>one-dimensional chains</u> of the given triangulation. From the algebraic point of view they represent linear forms (homogeneous polynomials of the first degree); they can be added and subtracted and also multiplied by an arbitrary integer according to the usual rules of algebra. Of particular importance among the one-dimensional chains are the so-called one-dimensional cycles. Geometrically, they correspond to closed paths. For a purely algebraic definition of a cycle, let us make the convention that the boundary of the oriented 1-simplex, $\langle v_0, v_1 \rangle$ in the last figure has the boundary $\partial \langle v_0, v_1 \rangle = v_1 - v_0$. When we accept this convention, we immediately observe that the sum of the boundaries of the segments, which form a closed path, is zero. This leads us naturally to the general definition of a one-dimensional <u>cycle</u> as a one-dimensional chain $z_1 = \Sigma_k a_k t_1^k$ for which the sum of the boundaries of the terms, i.e., the sum $\Sigma_k a_k \partial t_1^k$, is zero. It is easy to verify that the sum of two cycles is a cycle. When we multiply a cycle as an algebraic expression by an arbitrary integer, we obtain a cycle. This enables us to speak of linear combinations of cycles of the form

$$z_1 = \sum_1^s c_\nu z_1^\nu, \quad c_\nu \in Z.$$

In analogy to the concept of a one-dimensional chain of a given triangulation, we can also speak of two-dimensional chains of this triangulation, i.e., of expressions of the form

$$c_2 = \sum_i a_i t_2^i$$

where the t_2^i are oriented triangles of the given triangulation. Since the boundary of each oriented triangle is a one-dimensional cycle the chain $\Sigma a_i \partial t_3^i$ is also a cycle, (i.e., $\partial \circ \partial c_2 = \partial^2 c_2 = 0$). This cycle is taken to be the boundary ∂c_2 of the chain $c_2 = \Sigma a_i t_2^i$.

Let us finally introduce the extremely important concept of <u>homological independence</u> of cycles. The cycles $z_1, \ldots, z_s$ are called homologically independent in the given triangulation if no linear combination $\Sigma c_i z_i$ of them (in which at least one coefficient c_i is different from zero) is <u>homologous to zero</u>, i.e., bounds a piece of surface in this triangulation. As examples of homologically independent cycles on the torus, we can take an arbitrary meridian and equator considered as cycles of a triangulation of the torus (Figure below). If we cut the torus along a meridian γ and an equator δ and spread the outside of it out on the plane we get, after deformation, a rectangle with opposite sides identified. Then one sees clearly that an arbitrary cycle, β, is homologous (i.e., deformable) to an integral linear combination of the cycles, γ and δ.

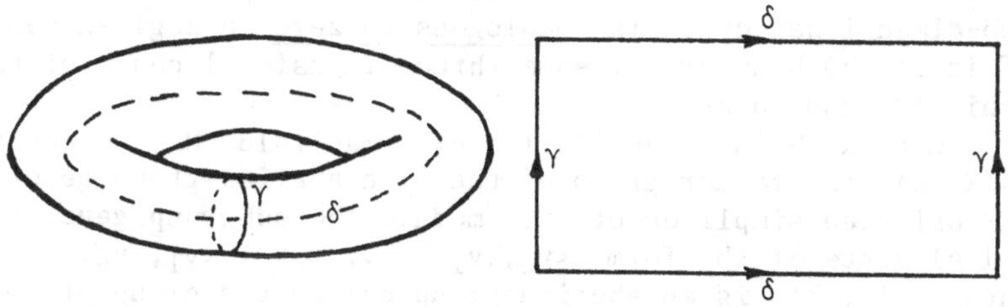

Homology basis on the torus

Indeed, the cycle β can be pushed to the walls in the rectangle. Thus, we can deform β so that the deformed $\beta = m\gamma + n\delta$ where m and n are integers, possibly zero. For this reason we call γ and δ a <u>homology basis</u> on the torus.

These fundamental concepts were defined by us for one-dimensional formations, but they can be extended verbatim to an arbitrary number of dimensions. For example, a two-dimensional chain

$c_2 = \Sigma a_i t_2^i$ is called a cycle of its boundary $\partial c = \Sigma a_i t_2^i$ is equal to zero. A three-dimensional chain is an expression of the form $c_3 = \Sigma a_h t_3^h$, where the t_3^h are oriented three-dimensional simplices (tetrahedra).

As in the case of the triangle, the orientation of the three-dimensional simplex is given by a definite order of its vertices, where two orders of the vertices that can be carried into one another by an even permutation determine one and the same orientation. The boundary of a three-dimensional oriented simplex $t^3 = \langle v_0 v_1 v_2 v_3 \rangle$ is the two-dimensional chain (cycle) $\partial t^3 = \langle v_1 v_2 v_3 \rangle - \langle v_0 v_2 v_3 \rangle + \langle v_0 v_1 v_3 \rangle - \langle v_0 v_1 v_3 \rangle$.

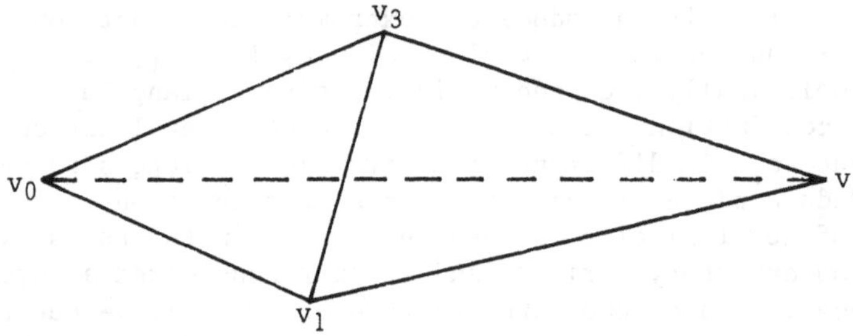

A three-dimensional simplex

The boundary of a three-dimensional chain is defined as the sum of the boundaries of its simplices taken with the same coefficients with which these simplices occur in the given chain. One can easily verify that the boundary of an arbitrary three-dimensional chain is a two-dimensional cycle $(\partial \circ \partial = \partial^2 = 0)$. We say that a two-dimensional cycle is homologous to zero in a given manifold if it is the boundary of some three-dimensional chain of this manifold, and so on.

Let K be a triangulation of a manifold M. We denote by $C_\ell(K, Z)$ the factor group of the free abelian group generated by all oriented simplices of K, modulo the subgroup generated by all elements of the form $\langle v_0, v_1, \ldots, v_\ell \rangle + \langle v_1, v_0, \ldots, v_\ell \rangle$. Thus $C_\ell(K, Z)$ is an abelian group called the group of ℓ-chains of K with coefficients in the integers Z. A typical element of this group is of the form

$$\sum_s n_s \langle s \rangle, \quad (n_s \in Z,\ s = \text{an } \ell\text{-simplex})$$

and we have the boundary map

$$C_\ell(K, Z) \xleftarrow{\partial} C_{\ell+1}(K, Z)$$

which is a group homomorphism defined by

$$\partial(\sum_s n_s \langle s\rangle) = \sum_s n_s \, \partial\langle s\rangle.$$

The maps

$$\cdots \longleftarrow C_{\ell-1}(K,Z) \xleftarrow{\partial} C_\ell(K,Z) \longleftarrow C_{\ell+1}(K,Z) \longleftarrow \cdots$$

satisfy $\partial\circ\partial = 0$.

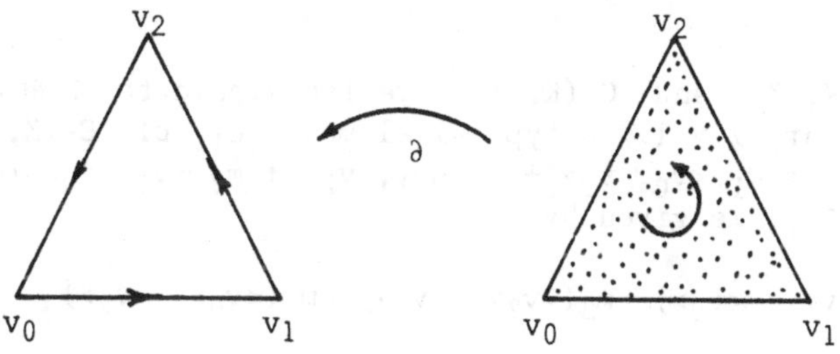

The boundary map

Given a triangulation, K, we form the following subgroups of $C_\ell(K, Z)$:

$$Z_\ell(K, Z) = \{c \in C_\ell(K, Z)\colon \ \partial c = 0\} \quad \text{(cycles)}$$

$$B_\ell(K, Z) = \{\partial c\colon \ c \in C_{\ell+1}(K, Z)\} \qquad \text{(boundaries)}.$$

Since $\partial\circ\partial = 0$ we have $B_\ell(K, Z) \subset Z_\ell(K, Z)$ and the factor group $Z_\ell(K, Z)/B_\ell(K, Z)$ is denoted by $H_\ell(K, Z)$ and is called the ℓ-th homology group of K with integer coefficients. (Instead of integers we can as well use an arbitrary abelian group A, and in the same way form $H_\ell(K, A)$, the ℓ-th homology group of K with coefficients in A. The most interesting cases are where A constitutes the integers Z, the reals $\mathbb{R}$, the complexes C, or the integers Z modulo 2, that is, the group Z_2 of order 2).

It turns out that the groups $H_\ell(K, A)$ depend only on the topology of the triangulation. This means if $f\colon K \to L$ is a homeomorphism between two triangulated spaces K and L, then there is induced an isomorphism

$$f_*\colon H_\ell(K, A) \longrightarrow H_\ell(L, A).$$

In particular, if K_1 and K_2 are two different triangulations of the same space, then they have the same homology groups.

Exercise. Show that the vector space $H_0(K, \mathbb{R})$ has dimension equal to the number of connected components (those who do not split into parts) in K.

Example. Let K be the triangle shown below; so K consists of three vertices v_0, v_1, v_2 and three 1-simplices (v_0, v_1), (v_1, v_2) and (v_2, v_0).

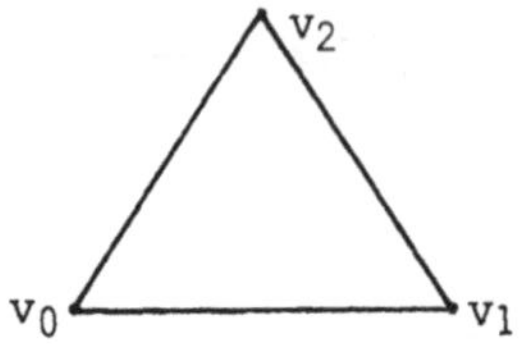

Then both $C_0(K, Z)$ and $C_1(K, Z)$ are isomorphic to $Z \oplus Z \oplus Z$. $C_\ell(K, Z) = 0$ for $\ell > 1$. A typical element c_1 of $C_1(K, Z)$ is of the form $c_1 = m_1 \langle v_0, v_1\rangle + m_2 \langle v_1, v_2\rangle + m_3 \langle v_2, v_0\rangle$ $(m_i \in Z)$. Its boundary ∂c_1 is given by

$$\partial c_1 = m_1(\langle v_1\rangle - \langle v_0\rangle) + m_2(\langle v_2\rangle - \langle v_1\rangle) + m_3(\langle v_0\rangle - \langle v_2\rangle)$$

$$= (m_3 - m_1) \langle v_0\rangle + (m_1 - m_2) \langle v_1\rangle + (m_2 - m_3) \langle v_2\rangle.$$

Thus $c_1 \in Z_1(K, Z)$ if and only if

$$m_3 - m_1 = 0,\ m_1 - m_2 = 0,\ m_2 - m_3 = 0,$$

that is, if and only if $m_1 = m_2 = m_3$, so

$$Z_1(K, Z) = \{n(\langle v_0, v_1\rangle + \langle v_1, v_2\rangle + \langle v_2, v_0\rangle):\ n \in Z\} \simeq Z.$$

Furthermore, $B_1(K, Z) = 0$ because $C_2(K, Z) = 0$. Hence

$$H_1(K, Z) = Z_1(K, Z)/B_1(K, Z) \simeq Z.$$

To compute $H_0(K, Z)$, note that a typical cycle $c_0 \in Z_0(K, Z)$ is of the form

$$c_0 = n_1 \langle v_0\rangle + n_2 \langle v_1\rangle + n_3 \langle v_2\rangle \ (n_i \in Z).$$

Then $c_0 = \partial c_1$ for some

$$c_1 = m_1 \langle v_0, v_1\rangle + m_2 \langle v_1, v_2\rangle + m_3 \langle v_2, v_0\rangle \in C_1(K, Z)$$

if and only if there exist (integer) solutions to the equations

$$m_3 - m_1 = n_1,\ m_1 - m_2 = n_2,\ m_2 - m_3 = n_3.$$

It is easy to check that such a solution exists if and only if $n_1 + n_2 + n_3 = 0$. Thus

$$B_0(K, Z) = \{n_1 \langle v_0\rangle + n_2 \langle v_1\rangle + n_3 \langle v_2\rangle : n_1 + n_2 + n_3 = 0\}.$$

Let $\phi: Z_0(K, Z) \to Z$ be the homomorphism defined by

$$\phi(n_1 \langle v_0 \rangle + n_2 \langle v_1 \rangle + n_3 \langle v_2 \rangle) = n_1 + n_2 + n_3.$$

Then Kernel $\phi = B_0(K, Z)$; thus

$$H_0(K, Z) = Z_0(K, Z)/B_0(K, Z) \simeq Z.$$

Since the triangle is homeomorphic to the circle S^1 we at the same time have computed the homology groups of S^1:

$$H_0(S^1, Z) = Z,$$

$$H_1(S^1, Z) = Z,$$

and

$$H_\ell(S^1, Z) = 0 \quad \text{for} \quad \ell > 1.$$

Example. Let K be the complex consisting of all the faces of a 2-simplex (v_0, v_1, v_2).

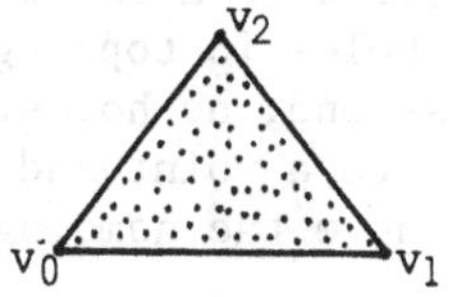

Then, as in the previous example

$$H_0(K, Z) \simeq Z.$$

Moreover as before,

$$Z_1(K, Z) = \{n(\langle v_0, v_1 \rangle + \langle v_1, v_2 \rangle + \langle v_2, v_0 \rangle): \; n \in Z\}.$$

Now, however,

$$C_2(K, Z) = \{n \langle v_0, v_1, v_2 \rangle : \; n \in Z\}.$$

So that

$$\begin{aligned} B_1(K, Z) &= \{\partial(n \langle v_0, v_1, v_2 \rangle): \; n \in Z\} \\ &= \{n(\langle v_1, v_2 \rangle - \langle v_0, v_2 \rangle + \langle v_0, v_1 \rangle): \; n \in Z\} \\ &= \{n(\langle v_0, v_1 \rangle + \langle v_1, v_2 \rangle + \langle v_2, v_0 \rangle): \; n \in Z\} \\ &= Z_1(K, Z). \end{aligned}$$

Hence

$$H_1(K, Z) = Z_1(K, Z)/B_1(K, Z) = 0.$$

Finally, since $\partial(n <v_0, v_1, v_2>) = 0$ if and only if $n = 0$, $Z_2(K, Z) = 0$, and hence

$$H_2(K, Z) \simeq 0.$$

Clearly, these are at the same time homology groups of a 2-dimensional disc:

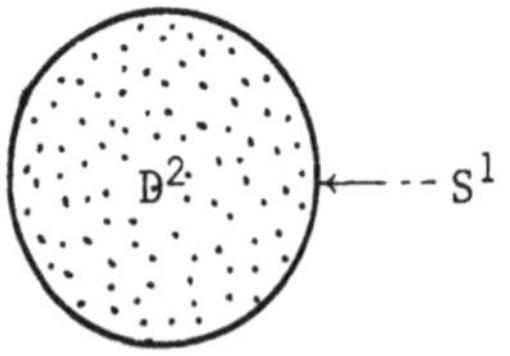

A disc

By comparing these two examples above we can say that the homology groups "measure" the number of holes in topological spaces. The first example has one hole and the second, no holes.

A disc is shrinkable (contractible) to a point and such spaces are called topologically trivial. They have the homology of a point space, i.e.

$$H_0 = Z \quad \text{and} \quad H_r = 0 \quad \text{for} \quad r \geq 1.$$

Exercise. Compute the homology groups of the sphere S^2[11].

We define the ℓ-th Betti number β_ℓ of K as the integer

$$\beta_\ell = \dim H_\ell(K, \mathbb{R}).$$

The Euler-Poincaré characteristic $\chi(K)$ of K is the integer

$$\chi(K) = \sum_{\ell=0}^{\dim K} (-1)^\ell \beta_\ell .$$

We state the following theorem whose proof is just an exercise in linear algebra:

Let K be a triangulation of a space. For each ℓ with $0 \leq \ell \leq \dim K$, let α_ℓ denote the number of ℓ-simplices in K. Then

$$\chi(K) = \sum_{\ell=0}^{\dim K} (-1) \alpha_\ell ,$$

that is, $\chi(K)$ is equal to the number of vertices - the number of edges + the number of 2-faces - ...

Remark. If K is homeomorphic to a connected compact (closed) orientable 2-dimensional manifold, then it turns out that $\beta_0 = 1$ and $\beta_2 = 1$, so that

$$\chi(K) = \beta_0 - \beta_1 + \beta_2 = 2 - \beta_1 \quad \text{or} \quad \beta_1 = 2 - \chi(K).$$

Furthermore, β_1 is always even for such a K. (This is connected with the fact that a 2-dimensional manifold can be given a Kähler structure). It can be shown that any such surface is homeomorphic to a sphere with a certain number of "handles" attached, in fact $\frac{1}{2}\beta_1$ is just the number of handles (see the figures of surfaces of genus 2 in Section 1.).

Thus the homology groups completely determine the homeomorphism classes of connected compact orientable surfaces. However, as we mentioned before, for higher dimensional manifolds, the homology groups contain comparatively little information. In 1905 Poincaré conjectured that the homology groups form a complete system of invariants for S^n, $n > 2$, i.e., a topological space which has the same homology groups as of S^n is in fact homeomorphic to S^n. This conjecture was proved for $n \geq 5$ by S. Smale, J. Stallings and E.C. Zeeman in the 1960's. 3 and 4 are the only dimensions where Poincaré's conjecture is still unresolved, and it is interesting to notice that the physical space is 3 dimensional and the space-time is 4. (Recently, it has been claimed that Poincaré's conjecture in dimension 4 has been solved by Michael H. Freedman).

A manifold is a sophisticated
concept even for a mathematician.
S.S. Chern

2. GEOMETRY

Let us consider the following simple physical apparatus:

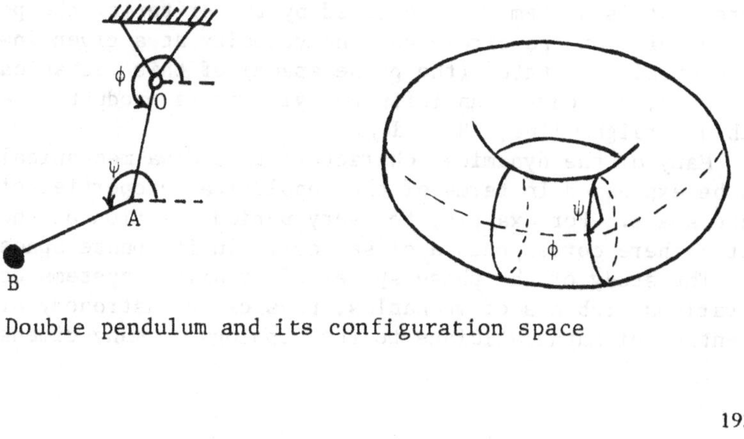

Double pendulum and its configuration space

The rods OA and AB turn freely in a fixed plane around O and A, respectively. Every possible position of our system is completely determined by the magnitude of the angles ϕ and ψ that the rods OA and AB form with an arbitrary fixed direction in the plane. We can regard these two angles, which change from 0 to 2π, as "geographical coordinates" of a point on a torus, counting from the "equator" of the torus and one of its "meridians", respectively. Thus, we can say that the manifold of all possible states of our mechanical system is a manifold of two dimensions, namely a torus, which is the topological product of two circles, $S^1 \times S^1$.

Now we modify our apparatus in the following way: Suppose that the rod OA can turn freely in a definite plane but that AB can turn freely around A in space. Now the position of our system is given by three parameters of which the first is the previous angle ϕ, and the other two determine the direction of the rod AB in space. Thus, the set of all positions of our new system is a certain three-dimensional manifold, and the reader will easily realize that it can be treated as the topological product of a circle and a sphere, $S^1 \times S^2$. This manifold is closed, i.e., it has no edges, and it cannot, therefore, be realized in the form of a figure lying in three-dimensional space.

We can make our apparatus even more complicated if we let the rod OA turn freely in space around O also. The set of possible positions of the system so obtained is then a four-dimensional closed manifold, namely the product of two spheres, $S^2 \times S^2$.

Thus we have seen that even the simplest mechanical considerations lead us to topological manifolds of three and more dimensions. Of even greater value in the practical, more detailed discussion of mechanical problems are certain manifolds (in general, many dimensional), the so-called <u>phase spaces</u> of dynamical systems. It was first the ancient Greeks who realized that <u>physics is determined by positions and velocities</u>. Hence, we take into account not only the configurations that the given mechanical system can have, but also the velocities with which its various constituent points move. Let us confine ourselves to one of the simplest examples. Suppose we have a point that can move freely on a circle with an arbitrary velocity. Every state of this system is determined by two criteria: the position of the point on the circle and the velocity at a given instant. The manifold of states (the phase space) of this mechanical system is, of course, an infinite cylinder (a product of a circle with a straight line, $S^1 \times \mathbb{R}$).

Many of the dynamical characteristics of a mechanical system can be expressed in terms of the topological properties of its phase space. For example, to every periodic motion of the given system there corresponds a closed curve in its phase space.

The study of the phase spaces of dynamical systems occurring in various problems of mechanics, physics and astronomy drew the attention of mathematicians to the topology of many-dimensional manifolds. It was precisely in connection with these problems that the famous French mathematician Poincaré started the systematic construction of the topology of manifolds, by applying the combinatorial method, which, up to the present day, is one of the fundamental methods of topology.

While this is an effective way in deriving topological invariants, there is the following danger: By using a combinatorial approach we may lose sight of the relations of the topological invariants with local geometrical properties.

An equally fundamental and powerful tool for studying the topology of manifolds is the analytical method. This is, in a way, the dual approach to the combinatorial method as will be verified by the famous theorem of de Rham. Since manifolds are locally homeomorphic to Euclidean spaces we can do calculus on them, i.e., define functions on them, take derivatives and integrals. A manifold is described locally by coordinates, but the latter are subject to arbitrary transformations. In other words, it is a space with transient or relative coordinates (principle of relativity). A topological n-manifold is a metrizable topological space M which locally looks like $\mathbb{R}^n$, in the following sense: Each point $p \in M$ has a neighbourhood U homeomorphic to some open set W in $\mathbb{R}^n$. If $x = (x^1, \ldots, x^n): U \to W \subset \mathbb{R}^n$ is such a homeomorphism, the pair (U, x) is called a coordinate chart. Two such charts (U, x) and (V, y) are smoothly related if either (a) $U \cap V = \phi$, or (b) $U \cap V \neq \phi$ and the maps $x \circ y^{-1}: y(U \cap V) \to x(U \cap V)$ and $y \circ x^{-1}: x(U \cap V) \to y(U \cap V)$ (defined on the open subsets $y(U \cap V)$ and $x(U \cap V)$ in $\mathbb{R}^n$) are smooth (i.e., of class C^∞). A differentiable n-manifold is a topological n-manifold M on which a class F of coordinate charts has been singled out. The class F must satisfy (a) every $p \in M$ is in chart of F and (b) if (U, x) and (V, y) are charts in F, then they are smoothly related.

If M is a differentiable n-manifold, we can do calculus on M. For example, we can introduce the notion of differentiable functions on M: a function $f: M \to \mathbb{R}$ is of class C^k if for each chart $(U, x) \in F$ the function $f \circ x^{-1}: x(U) \to \mathbb{R}$ is of class C^k on the open set $x(U)$ in $\mathbb{R}^n$. Similarly, by considering compositions of the form $x \circ \gamma$, one can define a notion of differentiable maps γ from an interval in $\mathbb{R}$ to M.

Consider a point $p \in M$, and all smooth curves on M through p. To any such curve we can associate the operation of taking the directional derivative along this curve at p of any smooth real-valued function f on M. This uniquely characterizes a vector X_p tangent to the curve at p: $X_p f$ is the directional derivative of f at p along the given curve. The tangent vectors at a given point p form an n-dimensional real vector space denoted by T_pM. Let $\{x^1, \ldots, x^n\}$ be a local coordinate system in a neighbourhood U of p. Then $\{\partial/\partial x^1, \ldots, \partial/\partial x^n\}$ is a basis of T_pM. An element $X_p \in T_pM$ can be uniquely written as

$X_p = \Sigma a^i \partial/\partial x^i$, where $a^i \in \mathbb{R}$. Thus, associated with each point p of a differentiable manifold M, there is an n-dimensional real vector space, the tangent space T_pM. The fact that M admits such "linear approximations" is the central feature of the theory of differentiable manifolds; in particular, it explains the constant use of linear and multilinear algebra so characteristic of this theory. $TM = \bigcup_{p \in M} T_pM$ is called the tangent bundle of M.

A vector field X on a manifold M is a smooth assignment of a vector $X_p \in T_pM$ to each point p of M. Locally, that is to say, in a given coordinate chart (U, x), X may be expressed by $X = \Sigma \xi^j \partial/\partial x^j$, where ξ^j are functions defined in U. X is a first-order partial differential operator acting on smooth real valued functions on M. We denote by $X \cdot f$ the result of this action. Let $\mathcal{X}(M)$ denote the set of all smooth vector fields on M. It is a real vector space. Moreover, if f is a function and X is a vector field on M, then $fX \in \mathcal{X}(M)$ and is defined by $(fX)_p = f(p)X_p$, that is, $\mathcal{X}(M)$ is a module over the ring of smooth real-valued functions on M.

Furthermore, we have another important operation defined on $\mathcal{X}(M)$, the Lie bracket:

$$[X, Y] \cdot f = X \cdot (Y \cdot f) - Y \cdot (X \cdot f).$$

With coordinates practically meaningless there is a need for new tools in studying manifolds. The key word is invariance. Invariants are of two kinds: local and global. The former refer to the behaviour under a change of the local coordinates, while the latter are global invariants of the manifold, examples being the topological invariants. Two of the most important local tools are the exterior differential calculus and Ricci's tensor analysis.

The origin of an exterior differential form is the integrand of a multiple integral, such as

$$\iint_D P\, dydz + Q\, dzdx + R\, dxdy,$$

in (x, y, z)-space, where P, Q, R are functions in x, y, z and D is a two-dimensional domain. The following observation is the crucial key to the differential forms: a change of variables in D (supposed to be oriented) will be taken care of automatically if the multiplication of differentials is anti-symmetric, i.e.,

$$dy \wedge dz = -dz \wedge dy, \quad \text{etc.},$$

where the symbol $\wedge$ is used to denote exterior multiplication. For example, suppose we make the following change of variables in $\mathbb{R}^2$: $(x, y) \to (u(x, y), v(x, y))$. Then $du = u_x dx + u_y dy$ and $dv = v_x dx + v_y dy$. Hence $du \wedge dv = (u_x v_y - u_y v_x)\, dx \wedge dy = J\, dx \wedge dy$.

Thus the Jacobian $J = u_x v_y - u_y v_x$ is intrinsically carried along by the d operator. This suggests the introduction of the <u>exterior two-form</u>

$$\omega = P\, dy \wedge dz + Q\, dz \wedge dx + R\, dx \wedge dy$$

and to write the integral above as a pairing (D, ω) of the domain D and the form ω.

If the same is done in n-space, <u>Stokes' theorem</u> can be written

$$(D, d\omega) = (\partial D, \omega), \quad \text{i.e.,} \quad \int_D d\omega = \int_{\partial D} \omega ,$$

where D is an r-dimensional domain and ω is an exterior (r-1)-form; ∂D is the boundary of D; $d\omega$ is the exterior derivative of ω and is an r-form.

<u>Example</u>. Let $n = 1$. Then Stokes' theorem yields the fundamental theorem of integral calculus: For $D = [a, b] \subset R$; $\partial D = \partial([a, b]) = \{b\} - \{a\}$, and $f \in C^\infty(\mathbb{R})$, we have

$$\int_D df = \int_{\partial D} f = f(b) - f(a).$$

Stokes' formula is the fundamental formula of multi-variable calculus, and it shows that ∂ and d are dual operators. The remarkable fact is that, while the boundary operator ∂ on domains is global, the exterior differentiation operator d on forms is local. It is given by the following formula:

$$d(a_{i_1 \ldots i_p} dx^{i_1} \wedge \ldots \wedge dx^{i_p}) = da_{i_1 \ldots i_p} \wedge dx^{i_1} \wedge \ldots \wedge dx^{i_p}$$
$$= \frac{\partial a_{i_1 \ldots i_p}}{dx^j} dx^j \wedge dx^{i_1} \wedge \ldots \wedge dx^{i_p},$$

where $da_{i_1 \ldots i_p}$ is the exact differential (1-form) of the scalar function $a_{i_1 \ldots i_p}$, which is completely anti-symmetric in its indices, and we assume summation over repeated indices. This makes d a powerful tool. When applied to a function (0-form), a 1-form and a 2-form, it gives the gradient, the curl, and the divergence, respectively. We shall illustrate these in the following example.

Example. i) Let $U \subset \mathbb{R}^3$ be an open subset, then for any 0-form f, we have

$$df = \sum \frac{\partial f}{\partial x^i} dx^i .$$

We realize that the coefficients of dx^i in this expression form the gradient vector of the function f.

ii) Consider now a Pfaffian, i.e., a one-form $\omega = \sum_i a_i dx^i$, then

$$d\omega = \sum_i da_i \wedge dx^i = \sum_{i,j} \frac{\partial a_i}{\partial x^j} dx^j \wedge dx^i$$

$$= \sum_{i<j} \frac{\partial a_i}{\partial x^j} dx^j \wedge dx^i + \sum_{i\geq j} \frac{\partial a_i}{\partial x^j} dx^j \wedge dx^i$$

$$= \sum_{i<j} \left(\frac{\partial a_i}{\partial x^j} - \frac{\partial a_j}{\partial x^i}\right) dx^j \wedge dx^i ,$$

where the components $\frac{\partial a_i}{\partial x^j} - \frac{\partial a_j}{\partial x^i} = (\mathrm{Curl}\ \vec{a})_{ij}$ denote the skew-symmetric curl of the vector field $\vec{a} = (a_1, a_2, a_3)$.

iii) For any two-form

$$\omega = P\, dx^2 \wedge dx^3 + Q\, dx^3 \wedge dx^1 + R\, dx^1 \wedge dx^2$$

we have

$$d\omega = \left(\frac{\partial P}{\partial x^1} + \frac{\partial Q}{\partial x^2} + \frac{\partial R}{\partial x^3}\right) dx^1 \wedge dx^2 \wedge dx^3,$$

where the quantity

$$\frac{\partial P}{\partial x^1} + \frac{\partial Q}{\partial x^2} + \frac{\partial R}{\partial x^3}$$

denotes the divergence of the axial-vector (P, Q, R).

Otherwise stated, the operator d extends to $\mathbb{R}^n$, the gradient, the curl and the divergence of vector analysis in $\mathbb{R}^3$.

Example. Cauchy's Integral Theorem may be rephrased in the context of differential forms as follows:

The complex-valued function $f(z)$ *is holomorphic in the open set* $D \subset C$ *if and only if the differential form* $\omega = f(z)dz$ *is closed (i.e.,* $d\omega = 0$*) in* D.

In fact: set $f(z) = u(x, y) + iv(x, y)$ and let

$$\omega = f(z)dz = (u + iv)\,(dx + idy).$$

$$d\omega = (du + idv) \wedge (dx + idy)$$

$$= du \wedge dx - dv \wedge dy + i(dv \wedge dx + du \wedge dy)$$

$$= -\left(\frac{\partial u}{\partial y} + \frac{\partial v}{\partial x}\right) dx \wedge dy + i\left(\frac{\partial u}{\partial x} - \frac{\partial v}{\partial y}\right) dx \wedge dy.$$

$$d\omega = 0 \Leftrightarrow \frac{\partial u}{\partial x} = \frac{\partial v}{\partial y} \quad \text{and} \quad \frac{\partial u}{\partial y} = -\frac{\partial v}{\partial x}$$

which are exactly the Cauchy-Riemann differential equations $\Longleftrightarrow$ $f(z)$ is holomorphic in $D \subset C$. Hence by Stokes' theorem

$$\int_{\partial D} f(z)dz = \int_{D} d(fdz),$$

one arrives at Cauchy's Integral Formula: Let D be a domain in C and $f: D \subset C \to C$ be a holomorphic map in D which is continuous in $\bar{D} = D \cup \partial D$. Then the integral of f is zero along the oriented boundary ∂D:

$$\int_{\partial D} f(z)dz = 0.$$

Remark. The above statement "f is holomorphic iff fdz is closed" is also available as follows:

$$d\omega = df \wedge dz = \left(\frac{\partial f}{\partial z} dz + \frac{\partial f}{\partial \bar{z}} d\bar{z}\right) \wedge dz = \frac{\partial f}{\partial \bar{z}} d\bar{z} \wedge dz = 2i \frac{\partial f}{\partial \bar{z}} dx \wedge dy.$$

$$d\omega = 0 \Longleftrightarrow \frac{\partial f}{\partial \bar{z}} = 0 \Longleftrightarrow f \text{ is holomorphic.}$$

Exercise. By using the definition of d earlier in this section show that for a 1-form ω we have

$$d\omega(X, Y) = X\cdot\omega(Y) - Y\cdot X(\omega) - \omega([X, Y]),\ X, Y \in \mathcal{X}(M).$$

2.1. De Rham Cohomology Theory

All the smooth forms, of all degrees ($\leq$ dim. of manifold), on a differentiable manifold constitute a ring with the exterior differentiation operator d. Elie Cartan used exterior differential calculus most efficiently in local problems of differential geometry and partial differential equations. The global theory was founded by George de Rham after the initial work of Henri Poincaré.

It turns out that, while homology theory depends on the boundary operator ∂, there is a dual cohomology theory based on the exterior differentiation operator d, the latter being a local operator. The resulting de Rham cohomology theory can be summarized as follows: The operator d has the fundamental property that, when applied repeatedly it gives the zero form; that is, for any k-form α, the exterior derivative of the (k+1)-form $d\alpha$ is zero (i.e., $d \circ d = d^2 = 0$). This corresponds to the geometrical fact that the boundary of any chain (or domain) has no boundary (i.e., $\partial \circ \partial = \partial^2 = 0$, see Section 1.1).

A smooth differential form ω on M is called <u>closed</u> if $d\omega = 0$. A form ω is <u>exact</u> if it is the differential of another form on M, that is, ω is exact if $\omega = d\tau$ for some smooth form τ. (Note that every exact form is closed, because $d^2 = 0$. The converse question is fundamental to our subject. It is where the topological information lies.) Let $Z^k(M, d)$ denote the vector space of closed k-forms on M, and let $B^k(M, d)$ be the space of exact k-forms on M. Then $B^k(M, d) \subset Z^k(M, d)$ because $d^2 = 0$. Let $H^k(M, d) = Z^k(M, d)/B^k(M, d)$. $H^k(M, d)$ is called the k-th <u>de Rham cohomology group</u> of M. Cohomology computations answer the fundamental question regarding the conditions for a closed form to be exact.

Although these cohomology groups are defined in terms of the differentiable structure on M, they turn out to be topological invariants; that is, if two manifolds are homeomorphic (by a not necessarily smooth homeomorphism), then they have isomorphic cohomology groups. In fact, these groups can be defined directly by using only the topological structure of M: De Rham's theorem states that the de Rham cohomology groups of M are isomorphic to the algebraic duals of the homology groups with coefficients in $\mathbb{R}$. Hence, Betti numbers can also be defined as $\beta_k = \dim H^k(M,d)$.

Let $\psi: M \to N$ be smooth. Then we have the induced homomorphism $\psi^*: H^k(N, d) \to H^k(M, d)$ which is dual to $\psi_*: H_k(M, \mathbb{R}) \to H_k(N, \mathbb{R})$ (notice the direction of the arrows!). It is an important fact that d and ψ^* commute: $d \circ \psi^* = \psi^* \circ d$.

Thus we have attached to each smooth manifold M new algebraic invariants $H^k(M, d)$ such that given smooth maps between manifolds, there are induced algebraic maps between these algebraic objects (hence the name "Algebraic Topology"). Furthermore, the direct sum

$$\bigoplus_{\ell=0}^{\dim M} H^{\ell}(M, d)$$

can be given the structure of an associative algebra under exterior multiplication. From now on we shall drop d and write $H^{\ell}(M)$ for de Rham's cohomology groups.

Example. Clearly for a manifold M of dimension n, we have $H^p(M) = 0$ for $p < 0$, and also for $p > n$, since there are no non-vanishing $(n+1)$-forms on an n-dimensional manifold due to the alternating property of the exterior differential forms.

Example. Let M be connected. We shall show that $H^0(M) \simeq \mathbb{R}$: To compute $H^0(M)$ we first note that there are no forms of degrees less than 0. Thus $H^0(M) = Z^0(M) = \{f \in C^{\infty}(M, \mathbb{R}): df = 0\}$. If M is connected, $df = 0$ implies that f is constant. In fact: if U is any (connected) coordinate neighbourhood of M with coordinate functions $(x^1, x^2, \ldots, x^n)$, then $df = 0$ on U means:

$$0 = df = \sum \frac{\partial f}{\partial x^i} dx^i \Rightarrow \frac{\partial f}{\partial x^i} = 0, \quad i = 1, 2, \ldots, n.$$

But this means that f is constant on U. Since M is connected, and since f is constant on each connected coordinate neighbourhood in M, then f must be constant on M, that is, $Z^0(M) = \{\text{constant functions on } M\} \simeq R$.

This result may be generalized: Let M consist of k connected components, then $H^0(M) = \mathbb{R}^k$. In fact, $df = 0$ yields $f = \text{const.}$ on each connected component. Thus the dimension of $H^0(M)$ measures the number of connected components of M just as the zero-th Betti number β_0 does.

Example. An n-dimensional manifold M is said to be orientable iff there is a volume-form on M, that is, an n-form ω such that $\omega(x) \neq 0$ for all $x \in M$. Let M be a connected closed (compact) differentiable manifold of dimension n. Then one can prove that

a) $H^n(M) = \mathbb{R}$ if M is orientable

b) $H^n(M) = 0$ if M is non-orientable.

Indeed, let us choose a form ω with $\int_M \omega \neq 0$. Then, for any other n-form ω' we can find a real number, a, such that $\omega' - a\omega = d\alpha$, i.e., ω' and $a\omega$ represent the same cohomology class $[\omega] \in H^n(M)$. To find a:

$$\int_M \omega' - \int_M a\omega = \int_M d\alpha = 0, \quad \text{since } \partial M = \phi. \quad \text{So } a = \int_M \omega' / \int_M \omega.$$

Notice that for an orientable manifold $H^n(M) = \mathbb{R}$ is equivalent to the assertion that:

$$[\omega] \to \int_M \omega$$

is an isomorphism: $H^n \to \mathbb{R}$.

Example. We leave it to our readers' intuition to see that the topologically trivial spaces, which have no holes or missing points in them (such as a disc in $\mathbb{R}^2$ or $\mathbb{R}^n$ in general), have trivial cohomology (and homology) groups except at the zero-th level. This is nothing but the famous Poincaré Lemma which states that closed forms are exact on star-shaped regions:

$$H^p \text{ (Contractible differentiable manifold)} \simeq H^p \text{ (Point)}$$

$$= \begin{cases} 0 & \text{for } p > 0 \\ \mathbb{R} & \text{for } p = 0 \end{cases}$$

In particular $H^0(\mathbb{R}^n) = \mathbb{R}$ and $H^p(\mathbb{R}^n) = 0$ if $p > 0$.

Example. $H^1(S^1) = \mathbb{R}$, where S^1 is the circle. For since there are no non-zero k-forms on S^1 for $k > 1$, $Z^1(S^1, d) =$ one-forms on S^1. Moreover,

$$B^1(S^1, d) = \{df: f \in C^\infty(S^1, \mathbb{R})\}.$$

Now, if θ denotes the polar coordinate on S^1, then $\partial/\partial\theta$ is a non-zero vector field and its dual 1-form, $d\theta$ is a non-zero 1-form on S^1. Furthermore, $d\theta$ is not exact (in spite of the notation!) but given any 1-form $\omega = g(\theta)d\theta$ on S^1, $\omega - (cd\theta)$ is exact for some $c \in \mathbb{R}$. Thus $Z^1(S^1, d)/B^1(S^1, d) = \{cd\theta: c \in \mathbb{R}\} \simeq \mathbb{R}$.

$$\text{(Verify that } c = \frac{1}{2\pi}\int_0^{2\pi} g(\theta)d\theta).$$

The de Rham Theorems. Let $\omega \in Z^p(M)$ be a closed p-form on a differentiable manifold. We seek a (p-1) form α on M such that $\omega = d\alpha$. By Poincaré's Lemma such a form exists locally. But we are asking the global question: is there such a form on all of M? This problem is answered by a fundamental theorem in algebraic topology, de Rham's theorem. First, we have as direct consequence of Stokes' theorem:

The integral of an exact form over a cycle vanishes:

$$\int_c \omega = \int_c d\alpha = \int_{\partial c} \alpha = 0.$$

Similarly, *the integral of a closed form over a boundary* b *vanishes:*

$$\int_b \omega = \int_{\partial c} \omega = \int_c d\omega = 0.$$

Also, *the integral of a closed form over a cycle depends only on the homology class of the cycle:*

$$\int_{c'} \omega = \int_{c''+\partial c} \omega = \int_{c''} \omega + \int_{\partial c} \omega = \int_{c''} \omega .$$

Finally, we have, *the integral of a closed form over a cycle depends only on the cohomology of the form:*

Let $\omega' - \omega'' = d\alpha$, i.e., ω' and ω'' represent the same cohomology class. Then

$$0 = \int_c d\alpha = \int_c \omega' - \int_c \omega'' \quad \text{and} \quad \int_c \omega' = \int_c \omega''.$$

Denote by H_p and H^p the vector spaces of real homology and cohomology, respectively, then we arrive at de Rham's Theorem:

The bilinear map $Z^p(M) \times Z_p(M, \mathbb{R}) \to \mathbb{R}$ given by $(\omega, c) \to \int_c \omega$, induces a bilinear map:

$$H^p(M) \times H_p(M, \mathbb{R}) \to \mathbb{R}, \quad ([\omega], \ [c]) \to \int_c \omega,$$

and this bilinear map is non-degenerate.

By virtue of this theorem, we can relate $H^p(M)$ to the dual space of $H_p(M)$. In fact, $[H_p(M, \mathbb{R})]^*$ denoting the dual of the p-homology group, define the linear map: $\omega \to [\omega]$,

$$H^p(M) \to [H_p(M, \mathbb{R})]^* \ \text{by} \ [\omega]\ ([c]) = \int_c \omega.$$

De Rham's theorem asserts that this is an isomorphism.

Moreover, de Rham's theorem yields a second theorem, commonly referred to as de Rham's existence theorem:

Suppose ω is fixed. Then there exists a homomorphism:

$$h: H_p(M) \to \mathbb{R}, \quad [c] \to \int_c \omega.$$

This homomorphism defines the periods of the closed form ω. The injectiveness of de Rham's isomorphism gives the answer to our global question: when is a closed form exact?. Namely: *if a closed form has all of its periods zero then it is an exact form:*

$$\int_c \omega = 0 \quad \forall c \implies \omega = d\alpha.$$

The surjectivity of de Rham's isomorphism says that if a real number $per(c) \in \mathbb{R}$ is assigned to each cycle $c \in Z_p$ in such a way that $per(ac_1 + bc_2) = a\, per(c_1) + b\, per(c_2)$ and per (boundary) = 0, then there is a closed form ω on M, such that for all cycles c:

$$\int_c \omega = per(c).$$

The story of UPSTAIRS → DOWNSTAIRS

3. FIBRE BUNDLES

On a manifold M it is natural to consider continuous vector fields, i.e., the attachment of a tangent vector at each point, varying in a continuous manner. A theorem of Hopf says that if the Euler-Poincaré characteristic $\chi(M)$ is not zero, there is at least one point of M at which the vector vanishes. More precisely at an isolated zero of a continuous vector field, an integer, called the index, can be defined, which describes to a certain extent the behaviour of the vector field at the zero, i.e., whether it is a source, a sink, or otherwise. For the continuous vector fields with only a finite number of zeros the theorem of Poincaré-Hopf says that the sum of its indices at all the zeros is a topological invariant which is precisely $\chi(M)$.

This is a statement on the tangent bundle of M, i.e., the collection of the tangent spaces of M. More generally, a family of topological spaces (fibres) parametrized by the points of a manifold M and satisfying a local product condition is called a fibre bundle over M.

A fundamental question is whether a fibre bundle is globally a product. The above discussion shows that the tangent bundle is not a product if $\chi(M) \neq 0$; for if it were a product, there would exist a continuous vector field which is nowhere zero. The existence of a space which is locally but not globally a product, such as the tangent bundle of the sphere S^2, is not easy to visualize. Geometry thus enters a more sophisticated phase.

More precisely, let E, M, F be topological spaces (manifolds), $\pi: E \to M$ a surjective mapping, and G a Lie transformation group acting on F freely on the left. If there exists an open covering $\{U_\alpha\}$ of M and a diffeomorphism $\psi_\alpha: U_\alpha \times F \approx \pi^{-1}(U_\alpha)$ for each index α, having the following three properties, then the system $(E, \pi, M, F, G, U_\alpha, \psi_\alpha)$ is called a fibre bundle: (1) $\pi\psi_\alpha(x, y) = x$, $x \in U_\alpha$, $y \in F$. (2) Define $\psi_{\alpha x}: F \approx \pi^{-1}(x)$, $x \in U_\alpha$, by $\psi_{\alpha x}(y) = \psi_\alpha(x, y)$, and let $g_{\alpha\beta}(x) = \psi_{\alpha x}^{-1} \psi_{\beta x} \in G$ for $x \in U_\alpha \cap U_\beta$. (3) $g_{\alpha\beta}: U_\alpha \cap U_\beta \to G$ is differentiable. E is called the total space, π the projection, M the base space, F the fibre, and G the structure group, or (in physics language) gauge group. Also, U_α is called the coordinate neighbourhood, ψ_α the local trivialization, or (in physics language) a choice of gauge, and $g_{\alpha\beta}$ the transition function. For a system $\{g_{\alpha\beta}\}$ of transition functions of a fibre bundle, we have $g_{\alpha\beta}\, g_{\beta\gamma} = g_{\alpha\gamma}$ on $U_\alpha \cap U_\beta \cap U_\gamma$. Conversely, given a system of $g_{\alpha\beta}: U_\alpha \cap U_\beta \to G$ satisfying this condition, there is a unique G-bundle with $\{g_{\alpha\beta}\}$ as a system of its transition functions.

Principal fibre bundles. A fibre bundle (P, p, M, G, G) is called a principal fibre bundle if G operates on G by left translations. It is denoted by $P(M, G)$. This is also defined by the following conditions: G acts freely on P on the right, and there exists an open covering $\{U_\alpha\}$ of M and diffeomorphisms $\psi: U \times G \approx p^{-1}(U)$ with $p\psi(x, g) = x$, $\psi(x, g)\cdot g' = \psi(x, gg')$, $x \in U$, $g, g' \in G$. Clearly, if u is a point of $p^{-1}(x)$, then $p^{-1}(x)$ is the set of points ug, $g \in G$ and is called the fibre through u. Every fibre is diffeomorphic to G, but there is no canonical way to choose an identity element on $p^{-1}(x)$. Principal fibre bundles constitute, in a certain sense, a generalization of the notion of Lie groups (cf. Section 3.2).

Let $\eta = (P, p, M, G)$ be a principal fibre bundle, and let F be a manifold having G as a left transformation group. Then G is a right topological transformation group of the product space $P \times F$ by $(x, y)\cdot g = (x\cdot g, g^{-1}\cdot y)$, $x \in P$, $y \in F$, $g \in G$. Consider the orbit space $P\times_G F = (P\times F)/G$, and define the mapping $\pi: P\times_G F \to M$ by $\pi\{(x, y)\} = p(x)$. Then we have the fibre bundle, $\eta\times_G F = (P\times_G F, \pi, M, F, G)$, which is called the associated fibre bundle of η with fibre F. On the other hand, η is called an associated principal bundle of $\xi = (E,\pi,M,F,G)$ if $\xi = \eta\times_G F$. A principal bundle η having the same coordinate transformations as ξ is an associated principal bundle of ξ, and two fibre bundles are equivalent if and only if their associated principal bundles are equivalent. Therefore, given a fibre bundle ξ, there is a principal bundle η such that $\xi = \eta\times_G F$.

A vector bundle is nothing but a fibre bundle with typical fibre $\approx \mathbb{R}^n$ and structure group $\approx GL(n, \mathbb{R})$. A one-dimensional vector bundle is called a line bundle.

Example. The cartesian product $X \times Y$ of two spaces X and Y is naturally the total space of the fibre bundle $pr_1: X \times Y \to X$ with the typical fibre Y. The group $G = \{Id_Y\}$ can be taken as the structure group. This is called a trivial bundle. Fibre bundles generalize the notion of the cartesian product of spaces. In general, a fibre bundle locally has the product structure $(\pi^{-1}(U) \approx U \times F)$, but globally it may not be written as a cartesian product of two spaces.

Example. The cylinder is a line bundle over the circle S^1 with typical fibre $\approx \mathbb{R}$. There is no twist in its construction, so its structure group $G = \{Id_{\mathbb{R}}\}$. Similarly, the torus $S^1 \times S^1$ is a trivial fibre bundle over S^1 with fibres $\approx S^1$.

Example. The Moebius strip is a fibre bundle over S^1 with structure group $G = \{\pm Id_{\mathbb{R}}\} \approx Z_2$. Thus, it is a non-trivial bundle and it has a global twist in it. This is also clear in its physical construction in Section 1. Similarly, the Klein bottle can be exhibited as a fibre bundle over S^1 with fibre S^1. This bundle is also non-trivial with structure group $\approx Z_2$.

A (global) cross section of a fibre bundle is a map s from the base space to the bundle space such that $\pi(s(x)) = x$ for all $x \in M$. While vector bundles always have global sections (e.g., the zero-section which assigns the zero vector of the fibre to every point on the base), in general a fibre bundle may not admit any global section. In particular, *a principal fibre bundle is trivial if and only if there exists a global section* $s: M \to P$; we can just let $(x, g) \in M \times G$ correspond to $s(x)\cdot g \in P$, thus proving that $P \approx M \times G$.

3.1. Connections on Fibre Bundles

Ordinary functions on a manifold M are sections of the trivial line bundle $M \times \mathbb{R}$, and sections of fibre bundles generalize the notion of functions on manifolds. Since directional derivative operation is intrinsically defined, we can differentiate ordinary functions along tangent vectors on manifolds. Fibre bundles being only locally trivial, there is *a priori* no natural way of differentiating its sections.

On the global setting, the same problem shows itself again when we consider the related problem of parallel translations of vectors. On a Euclidean space we have an intrinsic notion of parallelism: in $\mathbb{R}^n$, two vectors X and Y at two different points x and y, respectively, are parallel if X, Y and the line segment between x and y form three sides of a parallelogram.

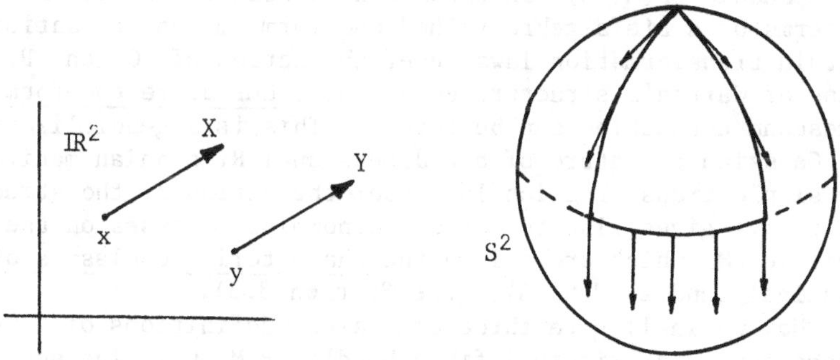

On the other hand, as is clear from investigating the tangent vectors on S^2 along great circles, if X and Y are tangent to a manifold M at distinct points, then direct comparison of X and Y is meaningless (unless M is an open subset of $\mathbb{R}^n$). The development of some sort of connection between distinct tangent spaces on M is the fundamental problem in the definition of the required geometric structure on M to overcome the above difficulties.

For this purpose, various indirect procedures are possible. For example, the ordinary directional derivative operation on $\mathbb{R}^n$ can be generalized to the concept of a connection or covariant derivative on M; the covariant derivative of a vector field (section of the tangent bundle TM) $X \in \mathcal{X}(M)$ with respect to a tangent vector $V \in T_xM$ is a tangent vector $\nabla_V X$ on M at x. The operation ∇ satisfies linearity rules with respect to V and X, furthermore

$$\nabla_{fV} X = f \nabla_V X\,, \quad f \in C^\infty(M),$$

and

$$\nabla_V (fX)(x) = (V \cdot f) X(x) + f(x)\, \nabla_V X, \quad f \in C^\infty(M).$$

A vector field X on M is then defined to be parallel if

$\nabla_V X = 0$ for all $V \in TM$. This is, in fact, Koszul's axiomatized version of the classical affine connection of Ricci and Levi-Civitá. In general, covariant derivative is not commutative. The notion of curvature is a measure of the non-commutativity of covariant differentiation (see the Remark in this section).

A conceptually clarifying and at the same time an invariant definition of a connection, due to Ehresmann, giving rise to the development of modern differential geometry, was formulated in 1950. Ehresmann defined a connection in an arbitrary fibre bundle as a field of horizontal spaces on the total space. On a principal fibre bundle $P(M, G)$ an Ehresmann connection can be characterized in terms of a Lie algebra valued one-form ω on P satisfying certain transformation laws under the action of G on P. By means of Cartan's structure equation, a curvature two-form of an Ehresmann connection can be formed. This is a generalization of the Gaussian curvature of a 2-dimensional Riemannian manifold and its simple transformation law under the action of the structure group G, gives rise to certain cohomology classes on the base manifold M which are called the characteristic classes of the principal bundle $P(M, G)$ (see Section 3.3).

Now we shall give three equivalent definitions of an Ehresmann connection on a principal fibre bundle $P(M, G)$. The equivalence of these definitions can be found in Kobayashi and Nomizu [14].

(1) A connection assigns to each $p \in P$ a subspace $H_p \subset T_p P$ such that for $V_p = \{X \in T_p P\colon dp(X) = 0\}$ we have $T_p P = H_p \oplus V_p$. We require that $dR_g(H_p) = H_{pg}$, where dR_g denotes the differential of the right translation operator R_g on P, induced by $g \in G$. We call V_p the vertical subspace of $T_p P$, while H_p is the horizontal subspace.

(2) Let LG be the Lie algebra of G. A connection is an LG-valued 1-form ω defined on P such that the following properties hold:

(a) $\omega(A^*_p) = A$, where $A \in LG$ and A^* is the fundamental vector field on P defined by

$$A^*_p = \frac{d}{dt} (p \operatorname{Exp} (tA))\Big|_{t=0} .$$

(b) $R^*_g \omega = (\mathrm{Ad}g^{-1})\, \omega$, for all $g \in G$, where Ad stands for the adjoint representation of G on LG.

(3) Chern has defined a connection on a principal fibre bundle as an assignment to each local trivialization $\psi_U\colon U \times G \approx p^{-1}(U)$ (i.e., choice of gauge) an LG-valued 1-form ω_U on U. If ψ_V is another local trivialization and $g_{UV}\colon U \cap V \to G$ is the corresponding transition function, then we

require

$$\omega_V = (\mathrm{Ad}\, g_{UV}^{-1})\, \omega_U + g_{UV}^{-1}\, dg_{UV} \quad \text{on} \quad U \cap V.$$

In case of a matrix group G, we have $(\mathrm{Ad}\, g^{-1})\, \omega = g^{-1} \omega g$.

Clearly, the relation between the first and the second definitions of connection is $H_p = \text{Kernel}\ \omega_p = \{X \in T_pP\colon \omega(X) = 0\}$, that is, ω vanishes on horizontal vectors.

3.2. Curvature of a Connection

Let $p\colon P \to M$ be a principal fibre bundle with structure group G. Consider a k-form α on P, with values in a vector space V. We define a V-valued $(k+1)$-form $\nabla\alpha$, the <u>covariant differential</u> of α, by

$$\nabla\alpha(X_1, \ldots, X_{k+1}) = d\alpha(hX_1, \ldots, hX_{k+1}),$$

where $X_i \in \mathcal{X}(P)$ are vector fields on P, d the exterior derivative and $h\colon T_pP \to H_p$ the projection of the tangent space T_pP onto its horizontal subspace H_p given by the connection on P. In particular, we define the <u>curvature form</u> Ω of ω by $\Omega = \nabla\omega = d\omega \circ h$. Thus, Ω is an LG-valued 2-form on P. Since ω has the property $R_g^*\omega = (\mathrm{Ad}\, g^{-1})\, \omega$, we have

$$R_g^*\, \Omega = R_g^*\, \nabla\omega = R_g^*\, d\omega \circ h = dR_g^*\, \omega \circ h$$

$$= d(\mathrm{Ad}\, g^{-1})\, \omega \circ h = (\mathrm{Ad}\, g^{-1})\, d\omega \circ h,$$

hence

$$R_g^*\, \Omega = (\mathrm{Ad}\, g^{-1})\, \Omega.$$

<u>Remark</u>. In terms of Koszul's connection ∇, the curvature can also be defined in an equivalent way as the law which assigns to each pair $(X, Y) \in \mathcal{X}(M) \times \mathcal{X}(M)$ a mapping $Z \to R(X, Y)\, Z$ of $\mathcal{X}(M)$ into itself, where

$$R(X, Y) = \nabla_X \nabla_Y - \nabla_Y \nabla_X - \nabla_{[X,Y]}.$$

This gives the following interpretation of curvature: We can view ∇ as a mapping between the two Lie algebras $LG_1 = (\mathcal{X}(M), [,] =$ Lie bracket of vector fields) and $LG_2 = (\mathrm{Endom}\ \mathcal{X}(M), [,] =$ operator bracket). The curvature R measures the failure of this mapping to be a Lie algebra homomorphism.

The structure equation of Elie Cartan.

$$d\omega(X, Y) = \Omega(X, Y) - \frac{1}{2}[\omega, \omega](X, Y),$$

for all $X, Y \in \mathcal{X}(P)$, where $[,]$ denotes the bracket operation in LG. Recalling that $d\omega = -\frac{1}{2}[\omega, \omega]$ is the Maurer-Cartan equation for the Lie group G, we realize that in order to obtain the structure equation of a principal G-bundle we just add the curvature term Ω to the right hand side of the Maurer-Cartan equation; that is, principal fibre bundles constitute, in a certain sense, a generalization of the notion of Lie groups.

Bianchi identity. $\nabla\Omega = d\Omega - [\Omega, \omega] = 0$, that is the exterior covariant derivative of the curvature form of a connection in a principal fibre bundle vanishes. Indeed, taking the exterior covariant derivative of Cartan structure equation gives

$$\nabla\Omega(X, Y, Z) = \nabla d\omega(X, Y, Z) - \frac{1}{2}\nabla[\omega, \omega](X, Y, Z)$$

$$= dd\omega(hX, hY, hZ) - \frac{1}{2}d[\omega, \omega](hX, hY, hZ) = 0,$$

since ω vanishes on horizontal vectors.

3.3. Characteristic Classes

Here, at last, connections in principal fibre bundles play their true role via the Bianchi identity. Given a connection on a bundle P over M, we shall construct, in a universal way, certain elements of the de Rham cohomology ring $H^*(M, \mathbb{R})$ of the base manifold M. Even though we employ the connection in their construction, these classes themselves do not depend on the choice of connections used; instead they depend on the bundle P; for this reason they are called characteristic classes of P. It is to be emphasized again that characteristic classes of P are cohomology classes on the base manifold M.

Characteristic classes are the simplest global invariants which measure the deviation of a local product structure of a bundle from a global structure. They are intimately related to the notion of curvature in differential geometry, in fact, a "real" characteristic class is a "total curvature" according to a well-defined scheme.

The simplest characteristic class is the Euler characteristic. As we have seen before in Section 1.1, if K is a finite triangulation of M, its Euler characteristic is defined by

$$\chi(M) = \sum_k (-1)^k \alpha_k = \sum_k (-1)^k \beta_k ,$$

where α_k is the number of k-cells in K and β_k is the k-th Betti number of M. The second equality above is a theorem known as the Euler-Poincaré formula, which is combinatorial in nature.

In two dimensions, the celebrated Gauss-Bonnet theorem states that

$$\chi(M) = \frac{1}{2\pi}\int_M \kappa \, dA,$$

where M is a compact, oriented surface and the Gaussian curvature κ is defined intrinsically in terms of a metric on M. The Gauss-Bonnet formula has many interesting consequences. For example, if M has a metric such that $\kappa > 0$ everywhere, then it is homeomorphic to S^2, and if the metric is such that $\kappa < 0$ everywhere, then it must have genus $p > 1$, that is, it must be homeomorphic to a sphere with two or more handles attached (see Section 1).

The Gauss-Bonnet formula gives a differential geometric meaning to $\chi(M)$ and it has been generalized by Chern (1944) to higher dimensional manifolds. Chern's extraordinary discovery has led to much deeper understanding of modern differential geometry. It is of paramount importance because it expresses the global (topological) invariant $\chi(M)$ as the integral of a local differential invariant (curvature), which is, perhaps, the most desirable relationship between local and global properties of manifolds. The Gauss-Bonnet-Chern theorem is a beautiful example of how integration may be used to obtain relations between the topology of a manifold and some of its local geometric invariants, in this case the curvature.

The main theme of this section is to show that certain quantities which *a priori* depend on the local differential geometry are actually global, topological invariants. Let us reconsider the transformation laws of the connection and curvature forms under change of local trivializations (gauge transformations):

$$\omega_V = (\mathrm{Ad}\, g_{UV}^{-1})\, \omega_U + g_{UV}^{-1}\, dg_{UV}\,,$$

$$\Omega_V = (\mathrm{Ad}\, g_{UV}^{-1})\, \Omega_U\,.$$

These formulae show that $\{\omega_U\}$ or $\{\Omega_U\}$ do <u>not</u> form global differential forms on M. They are, in fact, just the pull-backs of the global connection and curvature forms ω and Ω, respectively, on P, by the local trivializing isomorphisms of the principal bundle P(M, G). However, the very simple transformation law of the curvature forms implies that we can still obtain globally defined objects living downstairs. For example Trace Ω_V = Trace Ω_U in $U \cap V$, that is {Trace Ω_U} constitute an LG-valued 2-form globally defined on M. (We choose a basis for LG and then identify G with the matrix representation via its adjoint action on LG). Taking the trace of the Bianchi identity $d\Omega = [\Omega, \omega]$, gives $d(\mathrm{Tr}\Omega) = 0$, since trace of a commutator vanishes. That is, $\{\mathrm{Tr}\Omega_U\}$ forms a closed 2-form on M; hence it represents a cohomology class in $H^2(M, \mathbb{R})$.

More generally, $\mathrm{Tr}(\Omega \wedge \ldots \wedge \Omega)$ (k-times) is a 2k-form, globally defined on M. Similarly, $\mathrm{Tr}\Omega^k$ can be proved, by the use of the Bianchi identity, to be a closed form, and the cohomology class $\{\mathrm{Tr}\Omega^k\} \in H^{2k}(M, \mathbb{R})$ it represents, in the sense of de Rham's theorem, can be identified with a characteristic class of P.

We now look for more general functions similar to the trace. The transformation law of the curvature form ω suggests that we should introduce functions $F(X_1, \ldots, X_k)$, $X_i \in LG$, which are real or complex valued and satisfy the conditions:

i) F is k-linear and remains unchanged under any permutation of its arguments,

ii) F is Ad-invariant, that is $F(\mathrm{Ad}g(X_1), \ldots, \mathrm{Ad}g(X)_k)) = F(X_1, \ldots, X_k)$, for all $g \in G$.

To this k-linear function $F(X_1, \ldots, X_k)$ there corresponds the polynomial $F(X) = F(X, \ldots, X)$, $X \in LG$, of which $F(X_1, \ldots, X_k)$ is the "complete polarization". We call $F(X)$ an <u>invariant polynomial</u>. All invariant polynomials under G form a ring, to be denoted by $I(G)$.

The invariance condition ii) above implies its infinitesimal version

$$\sum_{1 \le i \le k} F(X_1, \ldots, [Y, X_i], \ldots, X_k) = 0, \text{ for all } X_i, Y \in LG.$$

Now let us consider the case of $X = \Omega$ and $Y = \omega$. If F is an invariant polynomial of degree k, then $F(\Omega)$ is a form of degree $2k$, globally defined on M. Moreover, by the Bianchi identity, the last formula above gives

$$dF(\Omega) = kF([\omega, \Omega], \Omega, \ldots, \Omega) = 0.$$

Thus, $F(\Omega)$ is a closed form and its cohomology class $\{F(\Omega)\}$ is an element of $H^{2k}(M, \mathbb{R})$. Furthermore, one can prove that this class depends only on the polynomial F and is independent of the connection used; that is, if Ω_1 and Ω_2 are two curvature forms of two connections ω_1 and ω_2, respectively, then the forms $F(\Omega_1)$ and $F(\Omega_2)$ are co-homologous, i.e., $F(\Omega_1) - F(\Omega_2) = d\beta$ for a $(2k-1)$-form β. Hence $F(\Omega_1)$ and $F(\Omega_2)$ represent the same cohomology class in $H^{2k}(M, \mathbb{R})$.

By letting $W(F) = \{F(\Omega)\}$, we define a mapping

$$W: I(G) \to H^*(M, \mathbb{R}),$$

which is called the Weil-homomorphism, after the famous contemporary French mathematician André Weil. Thus, the invariant polynomials, via the Weil-homomorphism, constitute a link between the local properties of a connection and its global consequences.

Example. The characteristic classes of a trivial principal fibre bundle all vanish. Indeed, on the product bundle $M \times G$ there is the trivial connection ω with horizontal subspaces tangent to the slices $M \times \{g\}$. Since $\omega | M \times \{g\}$ vanishes, so does $d\omega | M \times \{g\}$, whence $\Omega = d\omega \circ h = 0$, and $F(\Omega) = 0$ for all $F \in I(G)$.

This example explains why characteristic classes are obstructions for bundles to be a product globally.

Chern classes. Let a connection, with a curvature form Ω, be given on a principal $GL(n, C)$-bundle P. One can prove that the set of the coefficients $\{F_1, \ldots, F_n\}$ of the following polynomial in $\lambda \in C$

$$\det(\lambda I + \frac{i}{2\pi} X) = \lambda^n + F_1(X)\, \lambda^{n-1} + \ldots + F_n(X),$$

is linearly independent and generates the whole algebra of invariant polynomials on $GL(n, C)$. Ω is a complex matrix of two-forms and we define

$$c_i(P) = \{F_i(\Omega)\} \in H^{2i}(M, Z),$$

which is called the i-th Chern class of P. The factor $i/2\pi$ above is so chosen that the corresponding classes have integer coefficients. For example,

$$F_1(\Omega) = \frac{i}{2\pi} \operatorname{Tr}\Omega = \frac{i}{2\pi} \sum \Omega^j_j ,$$

$$F_2(\Omega) = \frac{-1}{(2\pi)^2} \sum_{j<k} \det \begin{bmatrix} \Omega^j_j & \Omega^j_k \\ \Omega^k_j & \Omega^k_k \end{bmatrix},$$

$$\vdots$$

$$F_{n-1}(\Omega) = \frac{i}{2\pi} \sum_j \text{Cofactor of } \Omega^j_j ,$$

$$F_n(\Omega) = \left(\frac{i}{2\pi}\right)^n \det(\Omega^i_j) .$$

Example. The tangent bundle of S^2 can be considered as a complex line bundle. The group $U(1)$ acts on a typical fibre C by rotations. Let us consider the associated principal fibre bundle $P(S^2, U(1))$ (see Section 3). Note that there is an isomorphism $U(1) \simeq SO(2)$ given by

$$e^{i\alpha} \to \begin{bmatrix} \cos\alpha & \sin\alpha \\ -\sin\alpha & \cos\alpha \end{bmatrix},$$

with induced Lie algebra isomorphism

$$i\beta \to \begin{bmatrix} 0 & \beta \\ -\beta & 0 \end{bmatrix}.$$

Let

$$F\left(\begin{bmatrix} 0 & \beta \\ -\beta & 0 \end{bmatrix}\right) = \beta,$$

and note that $F \in I(SO(2))$. If Ω is the curvature form of the Levi-Cività connection induced by the embedding $S^2 \subset \mathbb{R}^3$, then one can show that

$$F(\Omega)_x = x^1 dx^2 \wedge dx^3 + x^2 dx^3 \wedge dx^1 + x^3 dx^1 \wedge dx^2,$$

$x = (x^1, x^2, x^3) \in S^2$. Clearly, $F(\Omega)$ is the volume form on S^2 and the Chern class becomes $\{(1/2\pi)F(\Omega)\}$. Integrating this over S^2 we obtain the first Chern number

$$\int_{S^2} c_1(S^2) = \frac{1}{2\pi} \cdot 4\pi = 2,$$

which is just the Euler characteristic of S^2. In fact, the integral equation above is really an example of the Gauss-Bonnet-Chern theorem and, in general, it can be shown that the top dimensional Chern class of a complex bundle is the Euler class of the underlying oriented real bundle.

Pontrjagin classes. Let a connection, with curvature form Ω, be given on a principal $GL(n, \mathbb{R})$-bundle P. We consider

$$\det(\lambda I + \frac{1}{2\pi} X) = \lambda^n + E_1(X)\lambda^{n-1} + \ldots + E_n(X).$$

One can show that $\{E_1, \ldots, E_n\}$ is a basis for the invariant polynomials on $GL(n, \mathbb{R})$. We define

$$p_k(P) = \{E_{2k}(\Omega)\} \in H^{4k}(M, Z)$$

which is called the k-th Pontrjagin class of P. Even though the polynomials E_i for odd i are non-zero, one can prove that $\{E_{2k+1}(\Omega)\} = 0$ for $GL(n, \mathbb{R})$-bundles.

Signature theorem. Let M be a compact connected oriented manifold of dimension 4k. Consider over $H^{2k}(M, \mathbb{R})$ the quadratic form

$$Q(\alpha, \beta) = \int_M \alpha \wedge \beta, \quad \alpha, \beta \in H^{2k}(M, \mathbb{R}).$$

Since Q is a bilinear symmetric form over a real vector space, it can be diagonalised. We define

Signature(M) = no. of positive entries - no. of negative entries

in a diagonalised form of Q. Hirzebruch signature theorem states that Signature (M) can be represented by a "universal linear combination" of Pontrjagin numbers $\int_M p_k$. This is a special case of the famous Atiyah-Singer index theorem. In the simplest case of $k = 1$, the relation is

$$\text{Signature}(M) = \frac{1}{3}\int_M p_1(M), \quad \dim M = 4.$$

In particular, it shows that the integral on the right hand side is divisible by 3.

3.4. What is the Atiyah-Singer Index Theorem About ?

During the 1960's I.M. Gelfand in Moscow prophesied that the index (= dim Kernel - dim Co-kernel) of an elliptic operator on a compact manifold could be expressed in terms of the topological data of the manifold. In later years this has been proved by the British and American mathematicians, Atiyah and Singer, in a much broader context. Here, we shall try to give the flavour of this deep subject of modern mathematics which has recently found its way into physics (see Section 4.3).

A differential operator acting on the space of sections of a vector bundle over a manifold is called elliptic if, in a local coordinate system $\{x^i\}$ on M, the substitution $\partial/\partial x^i \to \xi_i$ transforms the highest order terms in the local expression of this operator into a non-degenerate form (called its symbol) in $\{\xi_i\}$.

Since the property of non-degeneracy is preserved under coordinate changes, this definition of ellipticity is, in fact, intrinsic, that is, independent of the local coordinates used. For example, the Laplacian ∇^2 acting on functions on $\mathbb{R}^n$ is an elliptic partial differential operator of order two whose symbol is $\xi_1^2 + \xi_2^2 + \ldots + \xi_n^2$, which is obviously a non-degenerate quadratic form.

Example. Referring to the Example on Cauchy's Integral Theorem in Section 2, $f\colon C \to C$ is holomorphic if and only if $\partial f/\partial \bar{z} = 0$. $\partial/\partial \bar{z} = 1/2(\partial/\partial x + i\partial/\partial y)$ is an example of an elliptic, first order partial differential operator, and many of the remarkable properties of analytic functions can be related to the fact that they are solutions of an elliptic equation. Conversely, solutions of elliptic equations have many properties in common with analytic functions, such as regularity, the maximum (minimum) principle and the mean value formula.

Suppose that V and W are two inner product spaces, and let $L\colon V \to W$ be a linear map with adjoint $L^*\colon W \to V$ defined by $(v, L^*w) = (Lv, w)$ for all $v \in V$ and $w \in W$. The index of L is defined to be the integer

$$\text{Index } L = \dim(\text{Kernel } L) - \dim(\text{Kernel } L^*)$$

if this is finite. It is a known fact that elliptic operators on compact manifolds belong to the Fredholm class which have finite dimensional kernels [12]. Thus, to an elliptic operator on a compact manifold we have an associated integer, its index.

Now, let us consider the exterior differentiation operator d on the space of differential forms on a manifold M of dimension n (see Section 2.1). We have, what is called, the de Rham complex

$$0 \to \Lambda^0(M) \xrightarrow{d} \ldots\ldots \Lambda^k(M) \xrightarrow{d} \Lambda^{k+1}(M) \to \ldots \Lambda^n(M) \to 0,$$

where $\Lambda^p(M)$ stands for the space of p-forms on M. We define the co-differential operator

$$\delta\colon \Lambda^{k+1}(M) \to \Lambda^k(M)$$

to be the adjoint of d with respect to the following inner product on $\Lambda(M) = \oplus_{k=0}^n \Lambda^k(M)$

$$(\alpha, \beta) = \int_M \langle\alpha(x), \beta(x)\rangle_x \, d\text{Vol}(x), \quad \alpha, \beta \in \Lambda^k(M),$$

$$(\alpha, \beta) = 0 \quad \text{if degree } \alpha \neq \text{degree } \beta,$$

where $\langle , \rangle_x$ stands for an inner product on the fibre at x of the vector bundle $\Lambda^k(M) \to M$ and dVol(x) is the volume form of a Riemannian metric on M, which we assume to be orientable and compact. Since $d^2 = 0$ it follows that $\delta \circ \delta = \delta^2 = 0$. Now we introduce the Laplace-Beltrami operator

$$\Delta = d\circ\delta + \delta\circ d = (d+\delta)^2 \quad \text{on} \quad \Lambda(M) = \bigoplus_{k=0}^{n} \Lambda^k(M),$$

which is of second order, positive, self-adjoint and elliptic.

A differential form $\omega \varepsilon \Lambda(M)$ is called harmonic if it satisfies the equation $\Delta\omega = 0$. Since

$$\Delta\omega = 0 \Leftrightarrow (d+\delta)\,\omega = 0 \Longleftrightarrow d\omega = \delta\omega = 0,$$

this particularly implies that a harmonic form is automatically closed and it therefore represents a de Rham cohomology class. In fact, one of Hodge's theorems states that *each de Rham cohomology class of a compact, orientable Riemannian manifold* M *has a unique harmonic representative, that is,* $H^k(M, \mathbb{R})$ *is isomorphic to the vector space of harmonic* k*-forms on* M. Since the dimension of the kernel of an elliptic operator on a compact manifold is finite, Hodge's theorem, in particular, proves that *the de Rham cohomology groups of a compact, orientable manifold are all finite dimensional.*

The fact that the Laplace-Beltrami operator Δ and its "square root" $d + \delta$ have the same kernel, namely the harmonic forms, has an interesting consequence: Let us calculate the index of the first order, elliptic operator

$$d+\delta\colon \Lambda^{even}(M) \to \Lambda^{odd}(M)$$

where $\Lambda^{even}(M)$ is the space $\oplus\Lambda^{2k}(M)$ of even-degree forms on M, and $\Lambda^{odd}(M)$ is the space of odd-degree forms on M; the adjoint of $d+\delta$ is the map $\delta + d$ in the opposite direction. Since the harmonic forms on M represent the de Rham cohomology of M by the Hodge theorem,

$$\begin{aligned}\operatorname{Index}(d+\delta) &= \dim H^{even}(M, \mathbb{R}) - \dim H^{odd}(M, \mathbb{R}) \\ &= \sum_{k=0}^{n} (-1)^k \dim H^k(M, \mathbb{R}) = \sum_{k=0}^{n} (-1)^k \dim H_k(M, \mathbb{R}).\end{aligned}$$

($\beta_k = \dim H_k(M,\mathbb{R}) = \dim H^k(M,\mathbb{R})$ by de Rham's theorem, see Sections 1.1 and 2.1). But by definition the alternating sum of the Betti numbers of M is the Euler characteristic $\chi(M)$ of M; thus the index of $d+\delta$ (an analytic invariant of M) equals the Euler characteristic of M (a topological invariant). As the differential forms are continuous objects and the index is an integer, the above equality expresses a relation between the continuous and the discrete ("quantization"). We recall that the Gauss-Bonnet-Chern theorem in Section 3.3 gives an integral representation of the same integer in terms of a local invariant (curvature). The expression of the index of an elliptic operator on a manifold as the integral of a local invariant culminates in the famous Atiyah-Singer Index Theorem. It includes as special cases such important theorems as the Hodge Signature Theorem, the Hirzebruch Signature Theorem and the Riemann-Roch Theorem for complex manifolds.

Physics is not so mysterious as many mathematicians seem to consider it. It is rather that physicists have different values and a different viewpoint, and this leads them to explain things in a manner uncongenial to mathematicians. If one works at it, it is possible to translate practically all of physics into well-defined mathematics. Moreover, when one does so, one finds a beautiful coherent scheme, which can be rather briefly summarized.

G.W. Mackey

4. GAUGE THEORIES

"Gauge Theories" of physics is the collective name for a large variety of elementary particle interactions, all of which share one feature: invariance of their physical predictions under a group of transformations of the basic variables of the field theory. Mathematically speaking, a common formal property of all gauge theories is that the natural dynamical variables, such as the potentials A_μ or the metric coefficients $g_{\mu\nu}$, are determined only up to transformations containing functional parameters which happen to have their values in various Lie algebras. Generally speaking, this arbitrariness cannot be eliminated by any natural choice of gauge. Also, if one considers only the gauge-invariant quantities, such as Maxwell's field strength $F_{\mu\nu}$, they do not contain all physically essential information about the field. It is now believed that the ultimate possible unification of natural laws of physics will come from gauge theories.

4.1. Electromagnetism as a Gauge Theory

The first unification in physics occurred between electricity and magnetism in Maxwell's equations, and today electromagnetism is the best-understood example of a gauge theory. If the basic variable is taken to be the one-form ("vector") potential $\alpha = A_\mu dx^\mu$, then the physically measurable quantity is the two-form $F = d\alpha$ corresponding to Maxwell's field tensor $F_{\mu\nu}$. We note that α is <u>not uniquely defined</u>: $\alpha' = \alpha + df$ for an arbitrary function f, also gives the same field strength $F = d\alpha'$. The gradient transformation $\alpha \to \alpha + df$ is the first example of a <u>gauge transformation</u>. On the other hand, if $\psi(x)$ represents the quantum wave function of a particle in an electromagnetic field, then only $\psi\bar{\psi}$ (probability density) has a physical significance; it is, therefore, to be assumed that the laws which govern ψ remain invariant on replacing ψ by $(\exp i\lambda)\,\psi$ (phase transformation), where $\lambda(x)$ is any <u>real</u> function on a space-time manifold. On examining the equation of motion for these two transformations above, we find that it is not invariant under each of them separately, but that there must exist a certain relation between λ and f: As the <u>principal</u> of <u>minimal coupling</u>

suggests, in the equations of motion the potential field $A_\mu(x)$ always appears in the following combination: $(\partial_\mu - i A_\mu)\psi = \nabla_\mu \psi$. If ψ transforms according to the rule $\psi(x) \to \exp(i\lambda(x))\, \psi(x)$, $\bar{\psi}(x) \to \exp(-i\lambda(x))\, \bar{\psi}(x)$, then $\nabla_\mu \psi$ transforms in the same way provided $f = \lambda$.

Indeed:

$$(\partial_\mu - iA_\mu)\, \psi \to (\partial_\mu - i\partial_\mu\lambda - iA_\mu)\, e^{i\lambda}\psi$$

$$= e^{i\lambda}(\partial_\mu - iA_\mu)\, \psi.$$

As a result, the equations of motion are co-variant under the transformations $\alpha \to \alpha + d\lambda$ and $\psi \to (\exp i\lambda)\, \psi$. This means that if the pair (ψ, α) is a solution, the pair $(e^{i\lambda}\psi), \alpha + d\lambda)$ is also a solution describing the same physical phenomenon. In other words, a local change in phase of the field ψ is equivalent to the appearance of an additional electromagnetic field. We see here a complete analogy with the equivalence principle in Einstein's theory of gravity, where a change of the coordinate system leads to the appearance of an additional gravitational field. This observation belongs to Hermann Weyl (1918 [21]).

Let us consider a principal fibre bundle $P(M^4, SO(2))$ over a space-time manifold M^4. In the representation

$$SO(2) = U(1) \simeq S^1 = \{e^{i\theta} : \theta \in \mathbb{R}\}, \quad i = \sqrt{-1},$$

we have the Lie algebra of $SO(2) \simeq i\mathbb{R}$. If a connection in P is to be characterized by an $i\mathbb{R}$-valued connection 1-form ω it must satisfy the following transformation law under the change of the frames (fibre coordinates) (see Section 3.1):

$$\omega \to \omega' = g^{-1}\, \omega g + g^{-1}\, dg, \quad g \in SO(2).$$

$$= e^{-i\theta}\, \omega e^{i\theta} + e^{-i\theta}\, e^{i\theta}\, id\theta$$

$$= \omega + id\theta.$$

Comparing this with the gradient transformation $\alpha \to \alpha' = \alpha + d\lambda$ of electrodynamics, we realize that $i\alpha$ behaves just like a connection 1-form on a principal fibre bundle $P(M^4, SO(2))$ with the transition functions $g(x) = \exp(i\lambda(x))$, $\lambda : U \subset M^4 \to \mathbb{R}$, $x \in U$. Also, the phase transformation $\psi \to e^{i\lambda}\, \psi$, suggests that ψ is a section of the associated complex line bundle over M^4 with the same transition functions. Clearly, the curvature form of the above connection is iF. Thus, we arrive at the following conclusion: Electromagnetic field theory is a connection in a circle bundle.

This mathematical formulation, in particular, predicts an interesting physical reality explained below: It is a well-known result in modern differential geometry that over a non-simply connected manifold we can have non-isomorphic line bundles with the same curvature form. Thus, we can construct a non-trivial bundle with zero curvature over a non-simply connected region in M^4. This means that while the electromagnetic field strength is zero, we can still have a non-trivial phase factor. That is, Maxwell's equations do not describe all the electric and magnetic phenomena. In fact, an experiment proposed by Aharanov and Bohm in 1959 and carried out by Chambers in 1960 gives an electri-magnetic field in a non-simply connected domain whose field strength is zero, but the phase factor can still be observed.

Is it not mysterious that one can know more about things which do not exist than about things do exist?

Socrates

4.2. Yang-Mills Gauge Theory

In the previous section we have shown that electromagnetism is a U(1)-gauge theory, that is, electromagnetism is the theory of connections on a principal U(1)-bundle over the space-time manifold. In 1953 Yang and Mills generalized the principle of gauge invariance of the interaction of a charged particle with an electromagnetic field to the case of weak interactions: interacting isopsins (proton-neutron doublet) [22]. They generalized some ideas of H. Weyl to introduce a new geometric technique into quantum physics; the result was essentially the independent discovery of parallel transport in a fibre bundle. It is interesting to note that the general concept of a connection, equivalent to Y-M field, appeared in mathematical literature only in 1950 (Ehresmann), that is, practically simultaneously with the work of Yang and Mills (cf. this with Heisenberg's rediscovery of matrices in 1926 while working out Quantum Mechanics).

Yang-Mills equations are a non-linear generalization of Maxwell's equations. They can be described in terms of connections and curvatures on principal SU(2)-bundles over the compactified Euclidean four-space, that is, S^4. A far more significant application of this approach has led via the group SU(2) × U(1) to a highly successful unified theory of electromagnetic and weak forces (Weinberg-Salam model). It is hoped that a similar approach with an appropriate group may incorporate the strong force into the theory. The final goal in this programme would be to unify the fourth force-gravitation-with the other three. Recent research in the supersymmetry and supergravity theories seems promising in this direction.

A gauge field theory can be associated with any compact semi-simple Lie group G. It is given by a differential 1-form $A(x)$ on the space-time manifold M^4, taking its values in the Lie algebra, LG, of G, and enjoying a gauge freedom. As we have just seen, in the case of electromagnetism the group is U(1) and the electromagnetic field $A = i\alpha$ is the analogous object, and the gauge freedom is nothing but the well-known gradient transformation $A(x) \to A(x) + id\lambda(x)$. Realizing that this was just like the transformation of a connection form on a principal fibre bundle $P(M^4, U(1))$, Yang and Mills postulated the following direct generalization of this transformation in the case of non-abelian gauge group G:

$$A(x) \to (\mathrm{Ad}g^{-1}(x))\, A(x) + g^{-1}(x)\, dg(x), \quad g \in G.$$

The associated covariant derivative operator is $D_\mu = \partial_\mu + A_\mu$. Thus, the Yang-Mills fields A play the same role as the Christoffel symbols in gravitation theory. Analogously, the Y-M fields describe parallel translation in the associated vector bundle whose typical fibre is a representation space for the gauge group G.

Even though there is this close analogy between the gauge theories and the general relativity, they still have a striking difference between them: We note that in gauge theories we consider only the transformations that change the phase (fibre) coordinates at various points x on the space-time manifold, but leave the points themselves fixed. That is, gauge transformations are defined as base-preserving automorphisms of the principal fibre bundles. Locally, they amount to a variable (space-time dependent) change of "<u>internal</u>" reference frame or gauge. On the other hand, the Einstein field equation of general relativity is based on the principle of invariance under transformations of space-time itself. In this sense, general relativity is an <u>external gauge theory</u>. Here the gauge transformations can be regarded as coordinate changes, which leave the geometry of space-time unchanged.

There is an interesting and unexpected harmony that arises from certain blending of the internal and external gauge theories through the natural imposition of a geometry on the bundle space P over the space-time manifold M, which results in a natural unification of gauge fields and gravitation. Given a geometry (a metric) on M and a gauge potential on P, there is a natural way to define a bundle metric on P. The action density for a variational problem can be defined to be the scalar curvature of this bundle metric. By demanding that the integral of the action density be stationary under variations of the metric on M, one obtains the Einstein field equations, with the energy-momentum source arising from the field strength of the gauge potential. Similarly, variation of the gauge potential leads to the gauge field equations for the field strength of the potential (see the following pages). Thus, the Einstein field equation and Yang-Mills equation arise simultaneously from a single variational principle [5].

Actually, the first model of this type in which $G = U(1)$ goes back to 5-dimensional model of Kaluza-Klein for the unification of the gravity and electromagnetism. In spite of the foregoing, these Kaluza-Klein type models are not regarded as solutions to the unified field theory problem. One reason is that these models do not predict the values of certain universal constants, such as the ratio of strength of the gravitational to the electric forces between the proton and electron.

After this long discussion, we now start the mathematical formulation of gauge theories. Let A be a gauge field (connection 1-form) on a principal fibre bundle $P(M, G)$. The curvature of this bundle is given by Cartan's structure equation

$$F = dA + \frac{1}{2} [A, A],$$

where in local coordinates

$$dA = d(A_\mu \, dx^\mu) = \partial_\mu A_\mu \, dx^\mu \wedge dx^\nu$$

$$= \frac{1}{2} (\partial_\mu A_\nu - \partial_\nu A_\mu) \, dx^\mu \wedge dx^\nu ,$$

$[,]$ stands for the Lie bracket in LG and we have by definition

$$[A, A] = [A_\mu, A_\nu] \, dx^\mu \wedge dx^\nu .$$

Thus,

$$F = \frac{1}{2} F_{\mu\nu}(x) \, dx^\mu \wedge dx^\nu,$$

where

$$F_{\mu\nu} = [D_\mu, D_\nu] = [\partial_\mu + A_\mu, \ \partial_\nu + A_\nu]$$

$$= \partial_\mu A_\nu - \partial_\nu A_\mu + [A_\mu, A_\nu]$$

is the field strength tensor.

Under the gauge transformation, F transforms according to the law $F \to (\mathrm{Ad}g^{-1})F$. In particular if $F_{\mu\nu}$ vanishes one can show that the potential A_μ is necessarily in the form $A_\mu = g^{-1} \partial_\mu g$, for some $g \in G$ and it may therefore be transformed to zero by g^{-1}. Thus, to summarize, in gauge theory, the Yang-Mills fields A_μ correspond to the concept of a connection in a principal fibre bundle. While A describes parallel translation in this bundle, the field strength F measures the curvature.

We consider the gauge-invariant action

$$S(A) = \frac{1}{2} \int d^4x \ \mathrm{Tr}(F_{\mu\nu} F^{\mu\nu}),$$

where we assume that $F_{\mu\nu} \to 0$ as $|x| \to \infty$. Because of the conformal invariance of the above integrant it is not necessary to be precise about the underlying space ($\mathbb{R}^4$ or S^4) over which one is integrating provided, of course, that one uses the Euclidean metric $\delta_{\mu\nu}$ on $\mathbb{R}^4$ and conformal metric $g_{\mu\nu}(x) = 2/(1+|x|^2)$ on S^4 to lower and raise indices. The Y-M field equations follow on from the requirement that the action S(A) be stationary under arbitrary changes of field variable A, i.e., $\delta S(A) = 0$. Using

$$\delta F^{\mu\nu} = \partial^\mu \delta A^\nu + \delta A^\mu \cdot A^\nu + A^\mu \delta A^\nu - \partial^\nu \delta A^\mu - \delta A^\nu \cdot A^\mu - A^\nu \delta A^\mu$$

together with $F_{\mu\nu} = - F_{\nu\mu}$ gives

$$\int d^4x \ \mathrm{Tr}\{F_{\mu\nu}(\partial^\mu \delta A^\nu + \delta A^\mu \cdot A^\nu + A^\mu \delta A^\nu\} = 0.$$

Integrating the first term by parts and throwing away the boundary terms and then using the cyclic properties of trace give

$$\int d^4x \ \mathrm{Tr}\ \{(\partial^\mu F_{\mu\nu} + [A^\mu, F_{\mu\nu}])\ \delta A^\nu\} = 0.$$

Thus, we obtain the Yang-Mills equations:

$$D^\mu F_{\mu\nu} = \partial^\mu F_{\mu\nu} + [A^\mu, F_{\mu\nu}] = 0,$$

where D^μ is the covariant derivative operator acting on 2-forms. We note that this is a system of non-linear partial differential equations in the field variables A_μ.

The Y-M equations have an exceedingly rich structure. They have, in Minkowski space, many solutions. Just as in electrodynamics, there are plane wave solutions. They have infinite energy (but finite energy density). However, unlike in Maxwell's theory, they cannot be superimposed to produce finite energy solutions because of the non-linear nature of these equations.

In a Euclidean (Riemannian) space-time we introduce the following two-form which is dual to $F_{\mu\nu}$

$$^*F_{\mu\nu} = \frac{1}{2}\,\varepsilon_{\mu\nu\rho\sigma}\,F^{\rho\sigma}\,.$$

Here $\varepsilon_{\mu\nu\rho\sigma}$ is skew-symmetric in all its indices with $\varepsilon_{0123} = \det g$, where g is the Riemannian metric on the base manifold. The Bianchi identity now becomes $D^{\mu}{}^*F_{\mu\nu} = 0$, and the Y-M field equations are automatically satisfied if $F_{\mu\nu} = {}^*F_{\mu\nu}$. Such fields are said to be self-dual (the antiself-dual condition $F_{\mu\nu} = -{}^*F_{\mu\nu}$ becoming self-duality when the orientation of space-time is reversed). We would like to emphasize that the Bianchi identity is not an equation of motion since it is automatically satisfied by expressing $F_{\mu\nu}$ in terms of the potentials A_{μ}.

Euclidean space can be regarded as Minkowski space with imaginary time, and in quantum mechanics, process with imaginary time evolution formally corresponds to tunnelling which happens instantly in time. As a result of the Euclidean (more generally, Riemannian) nature of the base space, the primary differential operators in Y-M theory become elliptic and not hyperbolic, which makes possible reduction of the Y-M equations to a specific version of the Cauchy-Riemann equations. Specific conditions at infinity then make it possible to show that the complex-analytic subjects of interest are indeed algebraic (polynomial) by virtue of Serre's general theorem, of which the simplest classical version asserts that an entire function that grows no faster than a polynomial is itself a polynomial. A large literature exists on the algebro-geometric meaning of the self-duality equations, leading to a complete study of the set of their solutions, and even to an explicit construction of these solutions just in terms of linear algebra [3], [10].

In a Euclidean space we have

$$\mathrm{Tr}\{(F_{\mu\nu} - {}^*F_{\mu\nu})\,(F_{\mu\nu} - {}^*F_{\mu\nu})\} \geq 0,$$

since it is the sum of squares. Using $\mathrm{Tr}({}^*F_{\mu\nu}\,{}^*F^{\mu\nu}) = \mathrm{Tr}(F_{\mu\nu}\,F^{\mu\nu})$ it follows that $\mathrm{Tr}(F_{\mu\nu}\,F^{\mu\nu}) \geq \mathrm{Tr}({}^*F_{\mu\nu}\,F^{\mu\nu})$, which establishes, upon integration, a lower bound for the value of the Y-M Euclidean action:

$$S(A) \geq \frac{1}{2}\left|\int d^4x\,\mathrm{Tr}({}^*F_{\mu\nu}\,F^{\mu\nu})\right|.$$

Clearly, the above inequality is saturated if $F_{\mu\nu} = \pm\,{}^*F_{\mu\nu}$, that is, if the field is self-dual or antiself-dual.

If a self-dual field has a finite Euclidean action (such solutions were named "instantons" by the Dutch physicist 't Hooft), its field strength should fall quite rapidly at infinity. A natural condition for this is that as $|x| \to \infty$, the field becomes asymptotically gauge trivial, i.e., $A_\mu \sim g^{-1}\partial_\mu g$ on a sphere of large radius $|x| = R$. This gives a mapping: $S^3 \to G$; $x \to g(x)$, of the 3-sphere into the gauge group G. This mapping is characterized by an integer which is called its winding number. Intuitively, it is the number of times this mapping wraps S^3 onto the gauge group G. The self-dual (or antiself-dual) field action can be shown to be proportional to the winding number which is independent of the detailed features of the field configuration, but which depends only on the general topological properties of the boundary values of the potential A. Since A is a continuous function of its arguments (one of them being the time) we note that its winding number does not change in time. This is a kind of conservation law, but of a different nature from those associated with symmetries by virtue of Noether's theorem. However, winding number is conserved in time for topological reasons, completely independent of the equations of motion. Such laws have come to be called topological conservation laws or topological charges.

Remark. One can easily check that if A_0 and A_1 are two connection forms on a principal fibre bundle P, then $A_t = (1 - t) A_0 + tA_1$ is also a connection form on P; that is, the set of all connection forms on a principal bundle is an affine space and is consequently topologically trivial (shrinkable to a point). However, the requirement of the finiteness of the action leads to a restricted set of connections: Since the action is gauge-invariant, it is actually defined on the quotient space of connection forms by gauge transformations, and this quotient space has a non-trivial topology. Recently, works have appeared about the Morse theory on these infinite-dimensional manifolds [6].

Now we want to elaborate on the special case of $G = SU(2)$: The Pauli spin matrices

$$\sigma^1 = \begin{bmatrix} 0 & 1 \\ 1 & 0 \end{bmatrix}, \quad \sigma^2 = \begin{bmatrix} 0 & -i \\ i & 0 \end{bmatrix}, \quad \sigma^3 = \begin{bmatrix} 1 & 0 \\ 0 & -1 \end{bmatrix}$$

can be taken as a base for the Lie algebra $su(2)$. Hence, the gauge potential A is a 2×2-complex matrix of one-forms with zero trace. The curvature form F may be similarly considered to be a traceless 2×2-complex matrix of two-forms. Let us compute the characteristic classes of the associated principal $SU(2)$-bundle over the Riemannian manifold S^4 (cf. Section 3.3):

$$\det(\lambda I + \frac{i}{2\pi} F) = \lambda^2 + \lambda \frac{i}{2\pi} \operatorname{Tr} F - \frac{1}{4\pi^2} \det F$$

$$= \lambda^2 - \frac{1}{4\pi^2} \det F.$$

Thus, for the case of SU(2)-gauge field theory we have only one topological invariant (this is true for SU(n) also), that is the second Chern number

$$c_2 = -\frac{1}{4\pi^2}\int_M \det F,$$

which is necessarily an integer. In physics literature this number is re-named as the Pontrjagin index and denoted by q. We note that

$$q = c_2 = -\frac{1}{8\pi^2}\int \mathrm{Tr}(F \wedge F)$$

$$= -\frac{1}{16\pi^2}\int \mathrm{Tr}({}^*F_{\mu\nu}\, F^{\mu\nu}),$$

hence the lower bound for the Y-M action becomes

$$S(A) \geq 8\pi^2 |q|.$$

Thus, Euclidean solutions to Y-M field equations with finite action are labelled by their Pontrjagin numbers which give the lower bound for the action, and the self-duality condition becomes the equation for the absolute minimum of the action within a definite Pontrjagin class. The associated bundles $P(S^4, SU(2))$ are characterized by the integer q, and two gauge fields (connections) with different Pontrjagin numbers cannot be defined on the same bundle over S^4.

As an illustration, we give the one-instanton $(q = 1)$ solution of the self-duality equation [4]:

$$A_\mu = \frac{x^2}{x^2 + a^2}\, g\, \partial_\mu\, g^{-1}$$

with

$$g(x) = \frac{x_0 + i\vec{x}\cdot\vec{\sigma}}{\sqrt{x^2}},$$

where $x = (x_0, x_1, x_2, x_3) \in \mathbb{R}^4$, σ_0 is the 2×2 identity matrix and σ_i = Pauli matrices.

Solutions of $F_{\mu\nu} = \pm\, {}^*F_{\mu\nu}$ have been found with an arbitrary value of the topological index (charge) q [19]. But, it is still not known whether all finite action solutions are self-dual. It is also an open problem to decide if there exist non-self-dual solutions of the Yang-Mills equations. Another open question is to clarify what supplementary hypotheses are to be added to the finiteness of the action to lead to the construction of some bundle over $\mathbb{R}^4$ and *a fortiori* ensure that q is an integer (see [17]).

There is no branch of mathematics,
however, abstract, that may not some day
be applied to phenomena of the real world.
Lobacevskii

4.3. Atiyah-Singer Index Theorem in Physics

Yang-Mills equations are non-linear. At present only some particular ("instanton") solutions are known. Still, even without explicitly knowing finite-action solutions to the duality equation, ${}^*F_{\mu\nu} = F_{\mu\nu}$, it is remarkable that we are able to compute the dimension of the manifold of self-dual solutions by using an index theorem which we shall present in this section.

We shall study fermions in a field of instantons and observe the surprising relation between the solutions of Dirac's equation for iso-vector fermions and the infinitesimal deformations of the self-dual gauge potentials. By using an Atiyah-Singer type index theorem we shall prove that the self-dual equations of the Yang-Mills theory has 8q - 3 gauge invariant deformations, where q is the Pontrjagin index of the gauge configuration.

Fermions in the field of instantons. The conformal invariance of the Yang-Mills theory puts us to work on the compactified Euclidean 4-space S^4. We consider the Euclidean Dirac operator D in the gauge field A, where A is a Lie algebra valued connection 1-form on S^4. We assume that the gauge group of our principal fibre bundle is SU(2).

D acts on bispinors ψ. Let us suppose that ψ transforms according to a representation of the group SU(2), and has T isospin indices. We shall study the normalizable zero-modes of D. We consider the Dirac equation

$$i\gamma^{\mu}(\partial_{\mu} + A_{\mu})\,\psi = 0,$$

on the compact manifold S^4 in the following representation of the Dirac matrices

$$\gamma^{\mu} = \begin{bmatrix} 0 & \alpha^{\mu} \\ \bar{\alpha}^{\mu} & 0 \end{bmatrix}, \quad \mu = 1, 2, 3, 4,$$

and

$$\gamma^5 = \gamma^1 \gamma^2 \gamma^3 \gamma^4 = \begin{bmatrix} I & 0 \\ 0 & -I \end{bmatrix},$$

where

$$\alpha^\mu = (-i\vec{\sigma}, I) \quad \text{and} \quad \bar{\alpha}^\mu = (\alpha^\mu)^\dagger = (i\vec{\sigma}, I).$$

In this representation, Dirac's equation decouples into two separate equations for 2-component spinors of definite chirality:

$$\psi = \begin{bmatrix} \psi^+ \\ \psi^- \end{bmatrix}, \quad L\psi^- = i\alpha^\mu(\partial_\mu + A_\mu)\,\psi^- = 0,$$

$$L^+\psi^+ = i\bar{\alpha}^\mu(\partial_\mu + A_\mu)\,\psi^+ = 0,$$

and the full Dirac equation becomes

$$\begin{bmatrix} 0 & L \\ L^+ & 0 \end{bmatrix} \begin{bmatrix} \psi^+ \\ \psi^- \end{bmatrix} = \begin{bmatrix} 0 \\ 0 \end{bmatrix}.$$

We note that L is a first order linear differential operator. Since γ^5 has eigenvalues ± 1 and since $\gamma^5 \gamma^\mu = -\gamma^\mu \gamma^5$, we have $\gamma^5\psi^\pm = \pm\,\psi^\pm$, that is, γ^5 separates the solutions of Dirac's equation into two distinct classes: positive chirality zero eigenvalue solutions, and negative chirality zero eigenvalue solutions. Let

$$n_+ = \dim \text{Kernel } L^+ = \text{no. of positive chirality solutions},$$

$$n_- = \dim \text{Kernel } L = \text{no. of negative chirality solutions}.$$

(All solutions are to be normalizable). We note that L and L^+ are adjoint of each other and we define

$$\text{Index } L = n_- - n_+ .$$

The index theorem evaluates this integer in terms of the topological properties of the gauge fields. To find this expression we need a basis for our function space to carry out generalized expansions, i.e., we need to consider the full eigenvalue problem of the Dirac operator:

$$i\gamma^\mu(\partial_\mu + A_\mu)\,\psi_E = E\psi_E, \quad (\psi_E \text{ normalizable}).$$

Since this is a self-adjoint equation we have a complete set of eigenfunctions (necessarily infinitely many of them). We choose the zero eigenvalue modes to be eigenstates of γ^5, and $\gamma^5 \gamma^\mu = - \gamma^\mu \gamma^5$ implies that $\gamma^5 \psi_E = \psi_{-E}$. Hence we have the following orthogonality relations:

$$\int d^4x \, \psi_E^\dagger(x) \, \gamma^5 \, \psi_E(x) = 0 \quad \text{for} \quad E \neq 0,$$

and

$$\int d^4x \, \psi_0^\dagger(x) \, \gamma^5 \, \psi_0(x) = \pm 1.$$

Now we form the <u>resolvent</u> of the Dirac operator:

$$R(x, y; m) = \sum_E \frac{\psi_E^\dagger(y) \, \psi_E(x)}{E + im},$$

which satisfies

$$[i\gamma^\mu(\partial_\mu + A_\mu) + im] \, R(x, y; m) = \delta^4(x - y).$$

We note that $E = - im$ makes R singular, and soon we will take $x = y$, which also makes R blow up. This is why one uses the Pauli-Villars scheme to make a gauge invariant regularization of R:

$$R_{reg}(x, y; m) = \lim_{M\to\infty} [R(x, y; m) - R(x, y; M)].$$

Now we define the <u>axial-vector-current</u>

$$J_5^\mu(x) = \mathrm{Tr}\{i\gamma^\mu \, \gamma^5 \, R_{reg}(x, x; m)\}$$

as the axial-vector projection of $R_{reg}(x, x; m)$. We would like to compute the divergence of this vector. After a tedious computation one arrives at

$$\partial_\mu \, J_5^\mu(x) = 2im \sum_E \frac{\psi_E^\dagger(x) \, \gamma^5 \, \psi_E(x)}{E + im} - \frac{1}{8\pi^2} \mathrm{Tr}({}^*F_{\mu\nu} \, F^{\mu\nu}),$$

where $F_{\mu\nu} = \partial_\mu A_\nu - \partial_\nu A_\mu + [A_\mu, A_\nu]$ is the field strength tensor. This is, in fact, the <u>local form of the index theorem</u>. We integrate it over the manifold to find the global relation

$$\int d^4x\, \partial_\mu J_5^\mu(x) = 2im \int d^4x \sum_E \frac{\psi_E^\dagger(x)\gamma^5\psi_E(x)}{E + im}$$

$$- \frac{1}{8\pi^2}\int d^4x\, Tr({}^*F_{\mu\nu}\, F^{\mu\nu}).$$

Since we work on a closed manifold we have no boundary terms and the left hand side is automatically zero. Hence we get

$$im \sum_E \frac{1}{E+im}\int d^4x\, \psi_E^\dagger(x)\, \gamma^5\, \psi_E(x) = \frac{1}{16\pi^2}\int d^4x\, Tr({}^*F_{\mu\nu}\, F^{\mu\nu}).$$

Upon using the orthogonality relations between ψ_E we observe many cancellations taking place and finally arrive at

$$im\frac{1}{0 + im}(-n_- + n_+) = \frac{1}{16\pi^2}\int d^4x\, Tr({}^*F_{\mu\nu}\, F^{\mu\nu}),$$

or

$$\text{Index } L = -\frac{1}{16\pi^2}\int d^4x\, Tr({}^*F_{\mu\nu}\, F^{\mu\nu}).$$

(It is interesting to compare the cancellations above with the similar type of cancellations in the heat equation treatment of the index theorem by P. Gilkey [12]).

As we have assumed at the beginning, ψ transforms according to some definite irreducible representation of SU(2), and A_μ is the Yang-Mills potential in an anti-hermitian matrix representation. If $\{T^j\}$ is a Lie algebra basis for this representation we shall write

$$iA^\mu = A_j^\mu\, T^j \quad \text{and} \quad iF^{\mu\nu} = F_j^{\mu\nu}\, T^j,$$

with

$$[T^j, T^k] = i\varepsilon_{jk\ell}\, T^\ell, \quad i = \sqrt{-1},\ j, k, \ell = 1, 2, 3,$$

which is just the structure equation of su(2).

In general, A_μ does not solve the Y-M equation $D_\mu F^{\mu\nu} = 0$, but it has finite action, i.e., its gauge configuration is characterized by an integer-valued Pontrjagin index. For fermions with total isospin T and gauge fields with Pontrjagin index q, we have

$$\text{Index } L = n_- - n_+ = \frac{1}{16\pi^2} \text{Tr}\{T^j T^k \int d^4x ({}^*F_j^{\mu\nu} F_k^{\mu\nu})\}.$$

By using the Killing form of the representation

$$\text{Tr}(T^j T^k) = \frac{2}{3} T(T+1)(2T+1) \delta^{jk},$$

we obtain

$$\text{Index } L = \frac{2}{3} T(T+1)(2T+1) q.$$

Until now we have only assumed that the gauge potential leads to a finite action and that it carries Pontrjagin index q. When A_μ is self (or antiself)-dual then we have the following <u>vanishing theorems</u>:

s.d. ⇒ $n_+ = 0$, (only negative chirality solutions) and Index $L = n_-$.

a.s.d. ⇒ $n_- = 0$, (only positive chirality solutions) and Index $L = -n_+$.

The proof is obtained by applying L and L^+ successively on ψ^+ and ψ^- and by using the fact that $D_\mu^2 = (\partial_\mu + A_\mu)^2$ is a positive definite operator, so it cannot have a nontrivial kernel. (It is also interesting to compare these with the vanishing theorems in algebraic geometry [16]).

At the moment, all the known solutions to the Yang-Mills equations with finite actions are self (or antiself)-dual. Hence, for these potentials there are precisely 2/3 T(T+1) (2T+1) q zero-eigenvalue modes with definite chirality (plus for a.s.d. and minus for s.d.).

Now at last, we shall prove the following: <u>The self-duality equation</u> $F_{\mu\nu} = {}^*F_{\mu\nu}$ <u>has</u> 8q − 3 <u>gauge invariant deformations</u>. (Also for the antiself-dual case).

This is an elliptic system. Therefore, we shall look at the linearized problem and find the number of parameters there (= dimension of tangent space). By the infinite dimensional version of the implicit function theorem we will have the same number of parameters around a self-dual solution.

Using the Pauli spin matrices $\{\sigma^a\}$ we shall represent the potentials and field strengths as anti-hermitian matrices in the space of infinitesimal (Lie algebra) generators of SU(2):

$$A^\mu = A^\mu_j \frac{\sigma^j}{2i}, \quad F^{\mu\nu} = F^{\mu\nu}_j \frac{\sigma^j}{2i}, \quad i = \sqrt{-1},\ j = 1, 2, 3.$$

This is just the isospin $T = 1$ (adjoint) representation. We define

$$\alpha^\mu = (-i\vec{\sigma}, I), \quad \bar{\alpha}^\mu = (\alpha^\mu)^\dagger = (i\vec{\sigma}, I),$$

$$\sigma^{\mu\nu} = {}^*\sigma^{\mu\nu} = \frac{1}{4i}(\bar{\alpha}^\mu \alpha^\nu - \bar{\alpha}^\nu \alpha^\mu),$$

$$\bar{\sigma}^{\mu\nu} = -{}^*\bar{\sigma}^{\mu\nu} = \frac{1}{4i}(\alpha^\mu \bar{\alpha}^\nu - \alpha^\nu \bar{\alpha}^\mu),$$

$$\sigma^{j4} = \frac{\sigma^2}{2}, \quad \sigma^{jk} = \varepsilon^{jk\ell} = \varepsilon^{jk\ell}\frac{\sigma^\ell}{2}$$

$$\bar{\sigma}^{j4} = -\sigma^{j4}, \quad \bar{\sigma}^{jk} = \sigma^{jk}, \quad j, k, \ell = 1, 2, 3.$$

Around a self-dual solution A_μ we perform an infinitesimal variation $A_\mu \to A_\mu + a_\mu$. This generates a variation of $F_{\mu\nu}$

$$\delta F_{\mu\nu} = D_\mu a_\nu - D_\nu a_\mu, \quad \text{modulo lower order terms in } a_\mu,$$

where

$$D_\mu a_\nu = \partial_\mu a_\nu + [A_\mu, a_\nu].$$

The self-duality condition implies that $\delta F_{\mu\nu} = \delta {}^*F_{\mu\nu}$. Since the matrix $\bar{\alpha}^{\mu\nu}$ is anti-symmetric and antiself-dual, the above is completely equivalent to

$$\bar{\alpha}^{\mu\nu} D_\mu a_\nu = 0.$$

Using the identity

$$\bar{\sigma}^{\mu\nu} = \frac{1}{2i}(\alpha^\mu \bar{\alpha}^\nu - \delta^{\mu\nu})$$

and imposing the background gauge condition $D_\mu a^\mu = 0$, which is analogous to the Lorentz gauge condition in electromagnetism, we arrive at

$$\alpha^{\mu} D_{\mu}(\bar{\alpha}^{\nu} a_{\nu}) = 0.$$

We recognize this to be two decoupled Dirac equations for two Dirac 2-component spinors in the adjoint representation (T =1). These two spinors form the two columns of the matrix $\bar{\alpha}^{\nu} a_{\nu}$ and move in the external potential A_{μ}. If ψ_j, $j = 1, 2, 3$ solve $i\alpha^{\mu} (\partial_{\mu} + A_{\mu}) \psi = 0$ with self-dual potential A^{μ}_{j}, then $A^{\mu}_{j} + u^{\dagger} \bar{\alpha}^{\mu} \psi_j$ is self-dual to the first order, where u is an arbitrary two-component constant spinor with defintie chirality (i.e., u has two degreės of freedom). The index formula, Index L = 2/3 T(T + 1) (2T + 1) q, gives 4q for T = 1 and this implies that $i\alpha^{\mu}(\partial_{\mu} + A_{\mu}) \psi = 0$ has 4q solutions. Therefore, we have by construction 2×4q = 8q small deformations. One can show that they can be rearranged into exactly 8q linearly independent real combinations, and one notes, furthermore, that 3(= dim su(2)) of them are infinitesimal gauge transformations. Hence, the maximum number of infinitesimal deformations of the self-duality equation $^{*}F_{\mu\nu} = F_{\mu\nu}$ is 8q - 3 ∎.

ACKNOWLEDGEMENTS

While writing these notes I have made extensive use of the references listed below. I would also like to thank Prof. Asim O. Barut for giving me the opportunity to talk on this subject at a conference held in the Black Sea region where I grew up. Special thanks are due to Ayşe/Ruth for her invaluable help with the manuscript.

REFERENCES

1. Aleksandrov, A.D., Kolmogorov, A.N., and Lavrent'ev, M.A.: 1977, Mathematics, its Content, Methods and Meaning, The MIT Press, Cambridge, Mass.
2. Arnold, V.I.: 1978, Mathematical Methods of Classical Mechanics, Springer-Verlag, New York.
3. Atiyah, M.F., Hitchin, N.J., Drienfel'd, V.G., and Manin, Yu. I.: 1978, *Construction of Instantons*, Phys. Lett. A, 65, pp. 285-287.
4. Belavin, A.A., Polyakov, A.M., Schwarz, A.S., and Tyupkin, Yu. S.: 1975, Phys. Lett. B, 59, pp. 85-87.
5. Bleecker, D.: 1981, Gauge Theory and Variational Principles, Addison-Wesley, Mass.
6. Bott, R.: 1980, *Morse Theory and the Yang-Mills Equations*, in Differential Geometric Methods in Mathematical Physics, Lecture Notes in Mathematics, Vol. 836, Springer-Verlag, New York.

7. Chern, S.S.: 1979, *From Triangles to Manifolds*, Amer. Math. Monthly, 86, pp. 339-349.
8. Chern, S.S.: 1979, Complex Manifolds without Potential Theory, 2nd ed., Springer-Verlag, New York.
9. Courant, R., and Robbins, H.: 1969, What is Mathematics, Oxford Univ. Press, Oxford.
10. Drinfel'd, V.G., and Manin, Yu. I.: 1980, *Gauge Fields and Holomorphic Bundles*, in Methods of Algebraic Geometry in Contemporary Mathematical Physics, Soviet Science Reviews, Physics Reviews, Over. Pub. Ass., Amsterdam.
11. Flanders, H.: 1963, Differential Forms, Acad. Press., New York.
12. Gilkey, P.: 1974, The Index Theorem and the Heat Equation, Publish or Perish, Boston.
13. Jackiv, R., Nohl, C., and Rebbi, C.: 1977, Lectures delivered at the Banf Summer Institute.
14. Kobayashi, S., and Nomizu, K.: 1963, Foundations of Differential Geometry, Vols. I & II, Wiley, New York.
15. Poor, W.A.: 1981, Differential Geometric Structures, McGraw-Hill, New York.
16. Rawnsley, J.H.: 1979, *On the Atiyah-Hitchin-Drinfel'd-Manin Vanishing Theorem for Cohomology Groups of Instanton Bundles*, Math. Ann. 241, pp. 43-56.
17. Schlafly, R: 1982, *A Chern number for gauge fields on* $\mathbb{R}^4$, J. Math. Phys., 23 (7), pp. 1379-1384.
18. Singer, I.M., and Thorpe, J.A.: 1976, Lecture Notes on Elementary Topology and Geometry, Springer-Verlag, New York.
19. 't Hooft, G.: 1976, Phys. Rev. D14, pp. 3432-3450.
20. Westenholz, C. von.: 1981, Differential Forms in Mathematical Physics, 2nd ed., North-Holland, Amsterdam.
21. Weyl, H.: 1952, Space-Time-Matter, 4th ed., Dover, New York.
22. Yang, C.N., and Mills, R.L.: 1954, *Conservation of Isotopic Spin and Isotopic Gauge Invariance*, Phys. Rev. 96, pp.191-195.

Part III

Particle Dynamics

PARTICLES, DYNAMICS AND COVARIANCE

Constantin Piron

Department of Theoretical Physics
University of Geneva
1211 Geneva 4, Switzerland

1. INTRODUCTION

One way to understand and finally to solve the difficulties of the relativistic quantum dynamics is to compare different models and to consider the limits to the classical case and to the non-relativistic case. To realize as completely as possible such a program it is useful, if not necessary, to derive the different possible models of particles corresponding to the classical, quantum, Galilean and relativistic case from the same mathematical framework. This means not only that the mathematical objects must be the same for each of the models but that also the physical interpretation must be the same.

Let us first define the general framework which describes both classical and quantum systems. This is a simple generalization of the usual ones. In classical physics a state of the system is described by a point of a set Ω called the phase space, and an observable is a function defined on Ω with values in $\mathbb{R}^n$. In usual quantum mechanics a state is described by a ray of a separable complex Hilbert space H, and an observable is a self-adjoint operator defined in H. A general framework is obtained by describing the system by a family of Hilbert spaces $\{H_\alpha\}$ where the indices α are in a set Ω. A state is described by an index $\alpha_o \epsilon \Omega$ and a ray ψ_{α_o} in the corresponding Hilbert space H_{α_o}, and an observable by a family of self-adjoint operators $\{A_\alpha\}$ each A_α acting in the corresponding H_α. The usual quantum mechanics then corresponds to the case where Ω reduces to one element, and classical mechanics to the case where each of the

A. O. Barut (ed.), Quantum Theory, Groups, Fields and Particles, 237–249.

H_α has dimension one. An observable is a correspondence between some properties of the apparatus and some properties of the system. To simulate such a correspondence in the formalism and then explain the physical interpretation both in classical and quantum mechanics, let us define the notion of property. A property of a given physical system is always associated to an experiment with a well-defined outcome (more exactly, an equivalence class). If the physical system has been prepared in such a way that it is true that in the event of the corresponding experiment is performed the expected result would be certain, we will say that the property is an actual property of the system. In classical physics a property is represented by a subset of the set Ω. In the usual quantum mechanics a property is represented by an orthogonal projector of the space H. In general a property is represented by a family of orthogonal projectors $\{P_\alpha\}$ each P_α being defined on the corresponding space H_α. The state of the system is by definition the subset of all the properties which are actual. Following this definition the state can be represented by giving an index α_o and a vector $\psi_{\alpha_o} \neq 0$ in the corresponding space H_{α_o}. A property $\{P_\alpha\}$ is then actual if and only if $P_{\alpha_o}\psi_{\alpha_o} = \psi_{\alpha_o}$. This is the physical justification of the usual definition of a pure state. Given a self-adjoint operator A_α in the space H_α there exists a spectral family, i.e. a map E from the Borel sets of $\mathbb{R}$ to the orthogonal projectors of H_α which decomposes this operator. $\mathcal{B}_\alpha$ the set of equivalence classes of Borel sets modulo an element of the kernel of E is a complete Boolean lattice. The quotient of E by this equivalence relation is a map $\underline{A}_\alpha$. Physically such a map defines a correspondence between some properties of the apparatus and some properties of the system. In the same way in the general case, given a family of self-adjoint operators $\{A_\alpha\}$, we define a complete Boolean lattice $\mathcal{B}$ as the set of families of elements taken in the $\mathcal{B}_\alpha$ and a map $\underline{A}$ by the relation $\underline{A}\{\Delta_\alpha\} = \{\underline{A}_\alpha\Delta_\alpha\}$ where $\Delta_\alpha \in \mathcal{B}_\alpha$. In the following we will call such a map $\underline{A}$ an observable since there will be no possible confusion with the associated family of operators $\{A_\alpha\}$. It is possible to characterize abstractly such a map $\underline{A}$ [1].

2. THE NOTION OF SYSTEM OF IMPRIMITIVITY

Let us consider an observable $\underline{A}$ as already defined. It is a map from a complete Boolean lattice $\mathcal{B}$ into $\mathcal{L}$ the lattice of all the families of projectors $\{P_\alpha\}$ defined for the spaces H_α, $\alpha\in\Omega$. It is clear from the interpretation that the elements of $\mathcal{B}$ correspond to some properties of the apparatus, for example the different possible positions of the pointer. Thus the way to distinguish this observable from another is to study the group which acts on the scale of its apparatus. Mathematically this leads one to define an observable as a solution of an imprimitivity system [1]. Given a group G, a complete Boolean lattice $\mathcal{B}$ and a representation σ of the group G by automorphisms of $\mathcal{B}$ we will say that the observable $\underline{A}$ is a solution of an imprimitivity system if the following diagram is commutative for all $g \in G$:

$$\begin{array}{ccccc} & \mathcal{B} & \xrightarrow{\underline{A}} & \mathcal{L} & \\ \sigma(g) & \downarrow & & \downarrow & S(g) \\ & \mathcal{B} & \xrightarrow[\underline{A}]{} & \mathcal{L} & \end{array}$$

Here S is a representation of the group G by automorphisms of $\mathcal{L}$, i.e. by transformations

$$\{P_\alpha\} \to \{Q_\alpha = U_{p^{-1}\alpha}\, P_{p^{-1}\alpha}\, U^{-1}_{p^{-1}\alpha}\}$$

where p is a permutation of the indices of Ω and U_α is a unitary or an antiunitary map of the space H_α onto the space $H_{p\alpha}$.

3. THE GALILEAN PARTICLE

An elementary Galilean particle is by definition a physical system described by a lattice of perperties $\mathcal{L}$ which admits the observables momentum $\underline{\vec{p}}$, position in space-time $\underline{q}^\mu = (\underline{\vec{q}},\underline{t})$, and no other independent observables. The group G corresponds to the different possible choices of the zeros of the scale of the apparatus measuring $\vec{p}, q^\mu$. This group is the semi-direct product of $\mathbb{R}^3 \bullet \mathbb{R}^3 \bullet \mathbb{R}$ by $SO(3)$ defined by :

the momentum translations $\vec{\pi}$

$$\vec{p} \mapsto \vec{p} + \vec{\pi}, \quad \vec{q} \mapsto \vec{q}, \quad t \mapsto t$$

the space translations $\vec{a}$

$$\vec{p} \mapsto \vec{p}, \quad \vec{q} \mapsto \vec{q} + \vec{a}, \quad t \mapsto t$$

the time translations τ

$$\vec{p} \mapsto \vec{p}, \quad \vec{q} \mapsto \vec{q}, \quad t \mapsto t + \tau$$

the rotations R $\qquad \vec{p} \mapsto R\vec{p}, \quad \vec{q} \mapsto R\vec{q}, \quad t \mapsto t$.

The translations $\vec{\pi}, \vec{a}, \tau$ correspond to the possible choices of the origins of the scales of the instruments. Rotations R correspond to possible choices of three orthogonal directions in space. They also act on $\vec{p}$ since the direction of the vector $\vec{p}$

is given in space. Finally, we impose two imprimitivity systems :

$$\begin{array}{ccc} \mathcal{B}_p & \xrightarrow{\;\underline{p}\;} & \mathcal{L} \\ \sigma_p(g)\downarrow & & \downarrow S(g) \\ \mathcal{B}_p & \xrightarrow{\;\underline{p}\;} & \mathcal{L} \end{array} \qquad \begin{array}{ccc} \mathcal{B}_p & \xrightarrow{\;\underline{q}^\mu\;} & \mathcal{L} \\ \sigma_q(g)\downarrow & & \downarrow S(g) \\ \mathcal{B}_p & \xrightarrow{\;\underline{q}^\mu\;} & \mathcal{L} \end{array}$$

The complete Boolean lattice $\mathcal{B}_p$ is defined from $\mathbb{R}^3$ and the corresponding $\mathcal{B}_q$ from $\mathbb{R}^4$. The automorphism $\sigma_p(g)$ is induced by the action of g on the $\vec{p}$ and $\sigma(g)$ by the action of g on the q^μ. There exist two and only two elementary Galilean particles [3] :

a) The classical particle : $\mathcal{L}$ is the lattice $\mathcal{P}(\mathbb{R}^7)$ of the subsets of $\mathbb{R}^7 = \{(\vec{p},q^\mu)\}$, $S(g)$ is the canonical representation, $\mathcal{B}_p = \mathcal{P}(\mathbb{R}^3)$, $\mathcal{B}_q = \mathcal{P}(\mathbb{R}^4)$ and

$$\underline{\vec{p}}(\Delta) = \{(\vec{p},q^\mu) \mid \vec{p}\in\Delta \;,\; (\vec{p},q^\mu)\in\mathbb{R}^7\}$$

$$\underline{q}^\mu(\Delta) = \{(\vec{p},q^\mu) \mid q^\mu\in\Delta,\; (\vec{p},q^\mu)\in\mathbb{R}^7\} .$$

These two observables can also be defined as theinverse images of the following two vector-valued functions :

$$(\vec{p},q^\mu) \to \vec{p} \quad \text{and} \quad (\vec{p},q^\mu) \to q^\mu .$$

b) The quantal particle : $\mathcal{L}$ is defined via a family of complex Hilbert spaces $\{H_t\}$, $t\in\mathbb{R}$, all identical to $L^2(\mathbb{R}^3,dv)$. $S(g)$ is induced by the following unitary transformations :

$$(U(\vec{\pi})\phi)_t(\vec{x}) = \exp(i\hbar^{-1}\vec{\pi}\vec{x})\phi_t(\vec{x})$$

$$(U(\vec{a})\phi)_t(\vec{x}) = \phi_t(\vec{x} - \vec{a})$$

$$(U(\tau)\phi)_t(\vec{x}) = \phi_{t-\tau}(\vec{x})$$

$$(U(R)\phi)_t(\vec{x}) = \phi_t(R^{-1}\vec{x}) .$$

As transformations acting on $L^2(\mathbb{R}^3,dv)$, $U(q)$ is a unitary projective representation since a "phase" appears in the relations

$$U(\vec{\pi})U(\vec{a}) = \exp(i\hbar^{-1}\vec{\pi}\vec{a})U(\vec{a})U(\vec{\pi}) .$$

These are the well-known Weyl commutation relations defining the

Planck constant $\hbar$ [4]. The existence and "unicity" of such a phase is related to $H^2(G,R)$, the 2^d-cohomology group of G, which here is siomorphic to $\mathbb{R}$. Up to a constant there exists only one 2-form on $\mathbb{R}^7$ which is invariant under the action of $SO(3)$ defined according to the semi-direct product. This 2-form also plays a role in analytical mechanics; in our theory the introduction of such a 2-form in classical mechanics is justified by these quentum considerations.

Let us first define $\underline{q}^\mu$ the position in space-time. The complete Boolean lattice $\mathcal{B}_q$ is the lattice of families of elements $\{\Delta_t\}$, $t\in\mathbb{R}$ which are Borel sets in $\mathbb{R}^3$ modulo Borel sets of Lebesgue measure zero. The observable $\underline{q}^\mu$ is then given by the family of projectors $\{\underline{q}_t(\Delta_t)\}$ defined by :

$$\underline{q}_t(\Delta_t)\phi_t(\vec{x}) = \chi_{\Delta_t}(\vec{x})\phi_t(\vec{x})$$

where $\chi_{\Delta_t}(\vec{x})$ is the characteristic function associated to Δ_t.

Let now define $\vec{\underline{p}}$, the momentum. The complete Boolean lattice $\mathcal{B}_p$ is the lattice of Borel sets in $\mathbb{R}^3$ modulo Borel sets of Lebesgue measure zero, and the observable $\vec{\underline{p}}$ is given by the family of projectors $\{\vec{\underline{p}}_t(\Delta)\}$ defined by :

$$p_t(\Delta)\hat{\phi}_t(\vec{k}) = \chi_\Delta(\vec{k})\hat{\phi}_t(\vec{k})$$

where $\hat{\phi}_t(\vec{k})$ is obtained via the Fourier unitary transformation defined on a dense subset by :

$$\hat{\phi}_t(\vec{k}) = (2\pi\hbar)^{-3/2}\int \exp(-ih^{-1}\vec{k}\vec{x})\phi_t(\vec{x})dv \ .$$

Such observables can also be defined by the families of self-adjoint operators :

$$\{q_t^k = x^k\},\ \{t_t = tI\},\ \{p_t^k = -i\hbar\partial_{x^k}\} \ .$$

We see in this model that we have been able to describe the time as an observable. It corresponds to a superselection variable since it is compatible with $\vec{\underline{p}},\vec{\underline{q}}$ and this with all the other observables. Since $\vec{p}_t$ and $\vec{q}_t$ do not explicitly depend on t, the representation given here is the Schrödinger picture. By performing an automorphism on $\mathcal{L}$ we can pass to another representation; in particular, we can pass to an interacting picture by a family of unitary operators $\{U_t\}$ which transforms each H_t onto itself. In the interacting picture we must remember that for different t the q_t (for example) act in different Hilbert spaces and are compatible even if the corresponding operators do not commute in $L^2(\mathbb{R}^3, dv)$.

4. THE LORENTZ PARTICLE

Similarly, we define an elementary Lorentz particle as a physical system described by a lattice of properties $\mathcal{L}$ which admits the observables momentum-energy $\underline{p}^\mu = (\vec{\underline{p}}, \underline{E}/c^2)$ and position in space-time $\underline{q}^\mu = (\vec{\underline{q}}, \underline{t})$, and no other independent observables. The corresponding group G is now the semi-direct product of $\mathbb{R}^4 \bullet \mathbb{R}^4$ by $SO(3,1)$ defined by the momentum-energy translation π^μ

$$p^\mu \to p^\mu + \pi^\mu, \quad q^\mu \to q^\mu$$

the space-time translation a^μ

$$p^\mu \to p^\mu, \quad q^\mu \to q^\mu + a^\mu$$

the Lorentz transformation $\Lambda^\mu\nu$

$$p^\mu \to \Lambda^\mu \nu p^\mu, \quad q^\mu \to \Lambda^\mu \nu q^\mu \ .$$

Here also we will impose two imprimitivity systems :

$$p^{(g)} \begin{array}{ccc} \mathcal{B}_p & \xrightarrow{\underline{p}} & \mathcal{L} \\ \downarrow & & \downarrow \\ \mathcal{B}_p & \xrightarrow{\underline{p}} & \mathcal{L} \end{array} S(g) \qquad q^{(g)} \begin{array}{ccc} \mathcal{B}_q & \xrightarrow{\underline{q}} & \mathcal{L} \\ \downarrow & & \downarrow \\ \mathcal{B}_q & \xrightarrow{\underline{q}} & \mathcal{L} \end{array} S(g)$$

but now $\mathcal{B}_p$ as well as $\mathcal{B}_q$ are defined from $\mathbb{R}^4$. Again, there exist two and only two models in the Lorentz case :

a) The classical particle : $\mathcal{L}$ is the lattice $\mathcal{P}(\mathbb{R}^8)$ of the subsets of $\mathbb{R}^8 = \{(p^\mu, q^\mu)\}$, $S(g)$ the canonical representation $\mathcal{B}_p = \mathcal{P}(\mathbb{R}^4)$, $\mathcal{B}_q = \mathcal{P}(\mathbb{R}^4)$ and

$$\underline{p}^\mu(\Delta) = \{(p^\mu, q^\mu) \mid p^\mu \in \Delta, \ (p^\mu, q^\mu) \in \mathbb{R}^8\}$$

$$\underline{q}^\mu(\Delta) = \{(p^\mu, q^\mu) \mid q^\mu \in \Delta, \ (p^\mu, q^\mu) \in \mathbb{R}^8\} \ .$$

b) The quantal particle : $\mathcal{L}$ is the lattice of orthogonal projectors of $L^2(\mathbb{R}^4, dvdt)$ and $S(g)$ is induced by the following unitary transformations :

$$(U(\pi^\mu)\phi)(x^\mu) = \exp(i\hbar^{-1} g_{\mu\nu} \pi^\mu x^\mu)\phi(x^\mu)$$

$$(U(a^\mu)\phi)(x^\mu) = \phi(x^\mu - a^\mu)$$

$$(U(\Lambda^\mu{}_\nu)\phi)(x^\mu) = \phi(\Lambda^{-1\mu}{}_\nu x^\mu)$$

where $g_{\mu\nu} = (1,1,1,-c^2)$. Here also, as transformations acting on $L^2(\mathbb{R}^4, dvdt)$, $U(g)$ is a unitary projective representation since we have the Weyl commutation relations :

$$(U(\pi^\mu)U(a^\mu) = \exp(i\hbar^{-1} g_{\mu\nu}\pi^\mu a^\nu)U(a^\nu)U(\pi^\mu) \ .$$

The observable $\underline{q}^\mu$ is defined by the projector :

$$(\underline{q}^\mu(\Delta)\phi)(x^\mu) = X_\Delta(x^\mu)\phi(x^\mu)$$

and the observable $\underline{p}^\mu$ by :

$$(\underline{p}^\mu(\Delta)\hat{\phi})(k^\mu) = X_\Delta(k^\mu)\hat{\phi}(k^\mu)$$

where

$$\hat{\phi}(k^\mu) = (2\pi h)^{-2}\int \exp(-i\hbar g_{\mu\nu}k^\mu x^\nu)\phi(x^\nu)dvdt \ .$$

Finally the corresponding self-adjoint operators are :

$$q^\mu = x^\mu \quad \text{and} \quad p^\mu = -i\hbar\, g^{\mu\nu}\partial_\nu \ .$$

5. UNIFICATIONS OF THE TWO QUANTAL PARTICLE MODELS

Manifestly the two quantal particle models, the Galilean and the Lorentz one are very different, the Galilean quantal model cannot be obtained as the limit of the Lorentz model. The essential difference is that in the Lorentz case we have an observable $E = i\hbar\partial_t$, thus t does not appear as a superselection variable as in the Galilean case. The way to avoid such an unpleasant feature is to modify one of the models and there are obviously two possibilities.

a) The modified Galilean particle : We impose the existence of not only the momentum $\underline{\vec{p}}$ and the position in space-time $\underline{q}^\mu$, but also the energy $\underline{E}$. The new group G, which contains the old as a subgroup, is obtained by adding the translation of energy W

$$\vec{p} \to \vec{p}, \quad E \to E + W, \quad \vec{q} \to \vec{q}, \quad t \to t$$

and the pure Galilean transformations $\vec{v}$:

$$\vec{p} \to \vec{p}, \quad E \to E - \vec{p}\vec{v}, \quad \vec{q} \to \vec{q} - \vec{v}t, \quad t \to t \; .$$

Thus we define a group with 14 parameters which we call the Newton group. It contains the usual Galilean group and the Carol group defined by Levy-Leblond [4] as subgroups. It is also a contraction of the group that we have defined for the Lorentz particle and which we will call the Einstein group. The quantal solution of the corresponding imprimitivity systems can easily be defined. $\mathcal{L}$ is the same as for the quantal Lorentz particle, it is the lattice of orthogonal projectors of $L^2(\mathbb{R}^4, dvdt)$. $S(q)$ is induced by the following unitary transformations :

$$(U(\vec{\pi})\phi)(\vec{x},t) = \exp(i\hbar^{-1}\vec{\pi}\vec{x})\phi(\vec{x},t)$$

$$(U(w)\phi)(\vec{x},t) = \exp(-i\hbar^{-1}wt)\phi(\vec{x},t)$$

$$(U(a^\mu)\phi)(x^\mu) = \phi(x^\mu - a^\mu)$$

$$(U(R)\phi)(\vec{x},t) = \phi(R^{-1}\vec{x},t)$$

$$(U(\vec{v})\phi)(\vec{x},t) = \phi(\vec{x} + \vec{v}t,t) \; .$$

Finally, the observables are practically the same as for the quantal Lorentz particle, i.e. :

$$p^k = -i\hbar\partial_k \; , \quad E = i\hbar\partial_t \; , \quad q^k = x^k, \quad q^4 = t \; .$$

As is easy to see, such a model is not essentially different from the Lorentz one, just the group and its action are different.

b) The modified Lorentz particle : We can also impose a variable as superselection in our quantal Lorentz particle model. If we choose the coordinate t in some frame for such variable we obtain by integral decomposition a model which turns out to be almost the same as our previous quantal Galilean model. The lattice $\mathcal{L}$ is defined by a family of Hilbert space $\{H_t\}$, $t \in \mathbb{R}$ all identical to $L^2(\mathbb{R}^3, dv)$, and the observables are the same, but of course the energy E is not defined. This new quantal Lorentz particle model is not manifestly covariant. To emphasize the covariance we introduce a unit time-like vector $n^\mu(-g_{\mu\nu}n^\mu n^\nu = 1)$ pointing the direction of the time coordinate and then we can write t like $c^{-1}g_{\mu\nu}n^\mu q^\nu$. With the help of the vector n^μ we are able to define the superselection parameter in a way which is independent of the frame of coordinates. Obviously, n^{μ} breaks the space-time symmetry and we must justify this breaking of symmetry from the physical point of view. This will be clear after the discussion of the dynamics, but let us just remark now that in the Galilean model such symmetry breaking exists also. To decompose the space-time vectors q we must first define a unit time vector n and then we can write $q = \vec{q} + tn$. It is true that t does not depend on the decomposition, it is invariant under Galilean transformation, but this is not the case for $\vec{q}$ which is transformed into $\vec{q} + \vec{v}t$.

6. THE DYNAMICS

The usual way to describe the dynamics of a particle is to define the changes of the state as given by unitary transformations. Such unitary transfromations are supposed to form a one-parameter group representation. According to Stone's theorem such an evolution is completely defined by a self-adjoint operator K and the Schrödinger equation

$$i\partial_\tau \psi_\tau = K(\vec{p},E,\vec{q},t)\psi_\tau \tag{6.1}$$

where ψ represent the state of the system and τ the parameter. If the state is given in $L^2(\mathbb{R}^4,dvdt)$ we can choose :

$$K = \frac{1}{2M}(\vec{p} - \vec{A}(\vec{q},t))^2 + V(\vec{q},t) - E \tag{6.2}$$

in the Galilean case and

$$K = \frac{1}{2M}((\vec{p} - \vec{A}(\vec{q},t))^2 - \frac{1}{c^2}(E - V(\vec{q},t))^2) + \frac{1}{2} Mc^2 \tag{6.3}$$

in the relativistic case. Mathematically, this description is perfectly consistent, but it is difficult to interpret. Since the parameter τ is not a self-adjoint operator acting in $L^2(\mathbb{R}^4,dvdt)$ it is not an observable, it has no definite values and it must disappear at the end of any calculation. Thus, the equation (6.1) can only predict the trajectory and not the final state. In fact, during a real evolution, one always measures some physical variable and the final state corresponds to a given value of this variable. For this reason, we call such a variable which is continuously measured during the evolution, the control parameter [5]. To each possible control parameter corresponds a different physical situation and accordingly, a different type of evolution. From the classical point of view, these evolutions are all physically equivalent, since in classical theory the control of a variable does not in principle disturb the system. But in quantum theory the motions corresponding to the different evolutions cannot be the same; to each control variable these must correspond a different system with a different evolution and thus a different equation of motion.

As we have said, in the classical case all these motions are physically equivalent, nevertheless, it is useful, for comparison with the quantum case, to be able to write directly the corresponding canonical equations relative to a given control variable. To do this, let us first describe the motion in an intrinsic and geometrical way. In the phase space $\mathbb{R}^8 = (\vec{p}, E, \vec{q}, t)$ we consider the restriction Ω_Γ of the 2-form :

$$\Omega = d\vec{p} \wedge d\vec{q} - dE \wedge dt \tag{6.4}$$

on the manifold Γ defined by $K = 0$ where K is the function on $\mathbb{R}^8$ given by (6.2) or (6.3). The motion is given by the tangent vector field $\eta \in T\Gamma$ such that

$$i_\eta \Omega_\Gamma = 0 \, . \tag{6.5}$$

In the Galilean case if t is the control variable, we must write E as a function of the other variables

$$E = H(\vec{p}, \vec{q}, t) \equiv \frac{1}{2M}(\vec{p} - \vec{A}(\vec{q}, t))^2 + V(\vec{q}, t) \tag{6.6}$$

and we find for the canonical equations :

$$\frac{dp^i}{dt} = -\frac{\partial H}{\partial q^i}, \quad \frac{dE}{dt} = \frac{\partial H}{\partial t}, \quad \frac{dq^i}{dt} = \frac{\partial H}{\partial p^i} \, .$$

But if q^1 is the control variable, we must write p^1 as a function of the other variables :

$$p^1 = \pi^1(p^2, p^3, E, \vec{q}, t) \equiv \tag{6.7}$$

$$A^1(\vec{q}, t) \pm \left(2M(E - V(\vec{q}, t)) - (p^2 - A^2(\vec{q}, t))^2 - (p^3 - A^3(\vec{q}, t))^2\right)^{\frac{1}{2}}$$

and we find for the canonical equations :

$$\frac{dp^2}{dq^1} = \frac{\partial \pi^1}{\partial p^2}, \quad \frac{dp^3}{dq^1} = \frac{\partial \pi^1}{\partial q^3}, \quad \frac{dE}{dq^1} = -\frac{\partial \pi^1}{\partial t}, \quad \frac{dq^3}{dq^1} = -\frac{\partial \pi^1}{\partial p^3}, \quad \frac{dq^3}{dq^1} = -\frac{\partial \pi^1}{\partial p^3},$$

$$\frac{dt}{dq^1} = \frac{\partial \pi^1}{\partial E} \, .$$

In the relativistic case if t is the control variable we must do the same, i.e. write E as function of the other variables

$$E = H(\vec{p},\vec{q},t) \equiv c((\vec{p}-\vec{A}(\vec{q},t))^2 + M^2c^2)^{\frac{1}{2}} + V(\vec{q},t) \qquad (6.8)$$

and we find the same canonical equations.

The invariance of the 2-form (6.4) and of (6.5) on the action of the Newton group or of the Einstein group express the covariance of such dynamics. Another way to see this is, as in (6.1), to introduce a new parameter τ which is not an observable, i.e. in the classical case, which is not a function on the phase space Ω. With such a parameter the equation of motion (6.5) is equivalent to the canonical equations :

$$\frac{dp^i}{d\tau} = -\frac{\partial K}{\partial q^i},\ \frac{dE}{d\tau} = \frac{\partial K}{\partial t},\ \frac{dq^i}{d\tau} = \frac{\partial K}{\partial p^i},\ \frac{dt}{d\tau} = \frac{\partial K}{\partial E} \qquad (6.9)$$

where K is the superhamiltonian [6] given by (6.2) or (6.3) and where we have imposed initial condition $K(\vec{p},E,\vec{q},t) = 0$.

Having discussed the classical motion, we can consider the quantum motion. We will follow the correspondence principle. Let us consider the case where t, the time defined by the time-like vector $n^\mu = (0,0,0,c^{-1})$, is the control variable of the system. Since t takes continuously different actual values, t is a continuous superselection variable and the system is described by a family of Hilbert spaces all identical to $L^2(\mathbb{R}^3,dv)$. The equation of motion is given by the Schrödinger equation :

$$i\hbar\partial_t\psi_t = H_t(\vec{p},\vec{q})\psi_t \qquad (6.10)$$

where, according to (6.6) and (6.8), $H(\vec{p},\vec{q},t)$ is the family of operators

$$H_t(\vec{p},\vec{q}) = \frac{1}{2M}(\vec{p}-\vec{A}(\vec{q},t))^2 + V(\vec{q},t)$$

in the Galilean case and

$$H_t(\vec{p},\vec{q}) = c((\vec{p}-\vec{A}(\vec{q},t))^2 + M^2c^2)^{\frac{1}{2}} + V(\vec{q},t)$$

in the relativistic case.

The equation (6.10) is covariant but not manifestly covariant since the symmetry is broken by giving the time-vector n^μ explicitly. There is a relation between the equations of motion (6.1) and (6.10) corresponding in the classical case to the equivalence between the canonical equations given by (6.9) and those given by (6.6) or (6.8). It is easy to check that a solution of (6.10) considered as a function of $\vec{x}$ and t is a generalized eigenstate of some operator $K(\vec{p},E,\vec{q},t)$ for the eigenvalue 0.

If q^1 is the control parameter then it is also q^1 which is a superselection variable. The system is described by a family of Hilbert spaces and the corresponding "Schrödinger" equation is given by :

$$i\hbar\partial_{q^1}\psi_{q^1} = \pi^1{}_{q^1}(p^2,p^3,E,q^2,q^3,t)\psi_{q^1} \tag{6.11}$$

where the family of operators $\pi^1{}_{q^1}(p^2,p^3,E,q^2,q^3,t)$ is defined by (6.7) according to the correspondence principle (or by the analogous relativistic expression). Here also it is easy to check that a solution of (6.11) considered as a function of $\vec{x}$ and t is a generalized eigenstate of some operator $K(\vec{p},E,\vec{q},t)$ for the eigenvalue 0.

As an application of this new Schrödinger equation let us consider a free Galilean particle moving in one dimensional space from $q = 0$ to $q = q$. If q is the control parameter the "Schrödinger" equation is then

$$-i\hbar\partial_q\psi_q(t) = (2ME)^{\frac{1}{2}}\psi_q(t)$$

and the solution is given by

$$\psi_q(t) = e^{i\hbar^{-1}(2ME)^{\frac{1}{2}}q}\psi_o(t)$$

where is the initial state at $q = 0$. The expectation value of the time t measured at the position q can be calculated and we find

$$\langle t\rangle_q = \int dt\psi^*_q(t)t\psi_q(t) = \langle t\rangle_o + \langle\frac{M}{(2ME)^{\frac{1}{2}}}\rangle_o q$$

which is what we would expect from the physical point of view.

REFERENCES

Piron C. : (1976), Foundations of Quantum Physics, Benjamin Reading Mass. pp 64.

Mackey G.W. : (1968), Induced representations of Groups and Quantum Mechanics, Benjamin Reading Mass.

Giovannini N. and Piron C. : (1979), On the group-theoretical foundations of classical and quantal..., Helv. Phys. Acta, vol. 52.

Levy-Lebrond J.-M. : (1963), Journal of Math. Phys. 4, pp. 776.

Piron C. : (1978), A New Type of Schrödinger Equation, invited paper given at Tutzing (July 3th - 6th).
Misner C.W., Thorne K.S. and Wheeler J.A. : (1973), Gravitation, W.H. Freeman and Company, San Francisco, pp. 489.

ON CONFORMAL SPINORS AND SEMISPINORS

Paolo Budinich

International School for Advanced Studies
Strada Costiera, 11
Miramare-Grignano
34014 Trieste
Italy

1. INTRODUCTION

The main endeavour of elementary-particle theoretical physics is, since many years, to properly understand, formulate and solve the problem of internal symmetry which seems to be at the root of most physical phenomena at very small distances.

It was in atomic spectroscopy that internal symmetry made its first appearance in physics: and precisely when the Zeeman doubling of spectroscopic lines, due to the action of a magnetic field on the optical electrons, was discovered, and W. Pauli attributed the effect to an "internal quantum number" of the electron: the spin one, and postulated the Pauli-Schroedinger equation presenting an internal SU(2) symmetry for the electron and able to explain and predict the experiments. But that electron "internal symmetry" was properly understood only afterwards, with the Dirac equation, when it appeared as one of the consequences of relativistic covariance and of the fact that electrons, like all fermions, are quanta of fields transforming with the spinor representations of the Lorentz group.

The second major appearance of internal symmetry in physics was the discovery of "isotopic spin" acting on the neutron-proton doublet and ruling nuclear forces , again with a SU(2) internal

A. O. Barut (ed.), Quantum Theory, Groups, Fields and Particles, 251–269.

symmetry group. But this time, despite the discovery that similar SU(2) groups seem to govern also quite different and even more elementary forces, like the weak ones, and acting also on quite different spinor doublets like electron-neutrino, muon-neutrino, etc., no explanation and understanding, similar to that of electron spin, was found, and the complications of the newly discovered groups of symmetry: SU(4) flavour, SU(2) x U(1) electroweak, SU(3) colour, SU(5) ..... G.U.T., added little to the real understanding of the deep nature and origin of these misterious internal symmetries. In general they were added ad hoc, in a similar way to the one adopted by W. Pauli in postulating Pauli-Schroedinger equation, appropriate to explain the experiments, but of little help for the understanding of the deeper meaning of internal symmetry.

It is true that the mere fact of appearance of fermion doublets both in strong and weak interactions suggested to Heisenberg and to many other physicists, the idea that perhaps the conformal group, with the Poincare' one as a subgroup, whose spinors, by a general theorem, have eight components, may be at the origin of the fermion doublets and of SU(2) isospin internal symmetry; and even more so since it was known that Maxwell equations are conformally covariant, and it was recognized that high energy phenomena manifest so called "scaling": another consequence of the possible fundamental role of conformal symmetry, in natural laws, at small distances.

Should, at that time, had been shown that conformal covariance by itself imposes the doubling of Dirac equation, conformal group would have been accepted and studied as the origin of isotopic spin.

On the contrary, it was shown that massless Dirac equation, or better Weyl equations, by themselves, are conformally covariant and consequently the necessity of conformal group as the origin of fermion doublets on which conformal group acts seemed superfluous and the hope to understand geometrically isotopic spin, as a consequence of conformal covariance, was abandoned.

We wish here first to show that, while Dirac equation is indeed conformally covariant, its conformal covariance is not a "natural" but a "restricted" one and then to show that if a "natural" conformally covariant spinor field equation is adopted it forsees indeed, in Minkowski space-time $M^{3,1}$, the doubling of Dirac equation and the natural appearance of a U(2) internal symmetry.

2. THE CONFORMAL GROUP: TRANSFORMATIONS FROM $M^{4,2}$ TO $M^{3,1}$

It is known that the 15 parameter conformal group is isomorphic to the group O(4,2) acting in pseudoeuclidean space $M^{4,2}$ with coordinates and metric tensor:

$$\begin{matrix} & \eta_0, & \eta_1, & \eta_2, & \eta_3, & \eta_5, & \eta_6 \\ g_{ab} = & (1, & -1, & -1, & -1, & -1, & +1) \end{matrix} \tag{2.1}$$

the coordinates x_μ of Minkowski space (compactified) are generally obtained, following the early work of Dirac[1], through the introduction of 5 homogeneous coordinates:

$$x_\mu = \frac{\eta_\mu}{K} \qquad\qquad \alpha^2 = \frac{\eta^2}{K^2} \tag{2.2}$$

where $K = \eta_5 + \eta_6$

and for $\eta^2 = \alpha^2 = 0$ and $K \neq 0$. (2.3)

The O(4,2) transformations in $M^{4,2}$:

$$\eta'_a = \Lambda_a^{\ b} \eta_b \qquad\qquad a = 0, 1, 2, 3, 5, 6$$

may be expressed in terms of (2.2) coordinates:

$$\begin{aligned} x'_\mu &= \Lambda_\mu^{\ \nu} x_\nu + b_\mu \ ; \ \alpha' = \alpha && \text{(Poincare')} \\ x'_\mu &= \rho\, x_\mu \ ; \quad \alpha' = \rho\alpha && \text{(Dilatations)} \\ x'_\mu &= \omega^{-1}[x_\mu + c_\mu(x^2 - \alpha^2)]; \alpha' = \omega^{-1}\alpha && \text{(Special conformal)} \end{aligned} \tag{2.4}$$

with $\omega = 1 + 2c_\mu x^\mu + c^2 \ (x^2 - \alpha^2)$

where the 15 parameters $\Lambda_\mu^{\ \nu}$, b_μ, c_μ, ρ are expressed in terms of Λ_a^b. For $\alpha^2 = 0$ (which is conformally invariant) the transformations (2.4) become the standard conformal ones of space-time $M^{3,1}$.

Given an O(4,2) spinor or tensor field ϕ_β (η) in $M^{4,2}$ the transformed field in homogeneous coordinates x_μ, α^2 will be defined by:

$$\phi_\beta(x,\alpha^2) = (S^{-1})_\beta^\delta \ \phi_\delta \ (x, \alpha^2, K = 1) \tag{2.5}$$

where the non singular operator S is defined by:

$$S = e^{i\pi_\mu x^\mu} \tag{2.6}$$

with

$$\pi_\mu = s_{5\mu} + s_{6\mu} \tag{2.7}$$

and s_{ab} are a representation, in space of ϕ_β, of the intrinsec part of SO(4,2) generators I_{ab} (for spinors $\pi_\mu = \frac{-i}{2}(\Gamma_5 + \Gamma_6) \ \Gamma_\mu$). (2.5) ensures that in $M^{3,1}$ the generators of Poincare' translations are represented by:

$$P_\mu = i \frac{\partial}{\partial x_\mu}$$

for any field in any basis (including conformal semispinors).

3. CONFORMAL SPINORS

The Clifford algebra $C_{4,2}$ of $M^{4,2}$ may be constructed, as products, from the basic elements

$$\Gamma_0, \ \Gamma_1, \ \Gamma_2, \ \Gamma_3, \ \Gamma_5, \ \Gamma_6 \tag{3.1}$$

corresponding to the unit vectors of a Cartesian reference system in $M^{4,2}$, obeying

$$[\Gamma_a, \Gamma_b]_+ = 2 \, g_{ab} \tag{3.2}$$

Conformal spinors ξ, vectors of the space where $C_{4,2}$ is linearly represented, have $2^3 = 8$ components. The Lie algebra of SO(4,2) acting on them is:

$$s_{ab} = \frac{i}{4} [\Gamma_a, \Gamma_b]_- \tag{3.3}$$

There are two main basis of $C_{4,2}$ representations in spinor space.

a) The Cartan one where the SO(4,2) Casimir operator:

$$\Gamma_7 = i\,\Gamma_0\,\Gamma_1\,\Gamma_2\,\Gamma_3\,\Gamma_5\,\Gamma_6 \tag{3.4}$$

is represented by the Pauli matrix $1\otimes\sigma_3$ (all other Γ_a being block antidiagonal), and the spinors have the form

$$\xi^C = \begin{vmatrix} \phi_I \\ \phi_{II} \end{vmatrix} ;$$

consequently

$$\frac{1}{2}(1+\Gamma_7)\xi^C = \begin{vmatrix} \phi_I \\ o \end{vmatrix} ; \frac{1}{2}(1-\Gamma_7)\xi^C = \begin{vmatrix} o \\ \phi_{II} \end{vmatrix} ; \tag{3.5}$$

ϕ_I and ϕ_{II} are called semispinors and as eigenstates of the Casimir operator Γ_7 they are, each of them, irreducible basis for SO(4,2) inequivalent representations. One of their main properties is that for <u>any</u> reflection in $M^{4,2}$ (and consequently also in $M^{3,1}$) $\phi_I \leftrightarrow \phi_{II}$: they transform into each other and in fact Γ_7 anticommutes with all Γ_a.

b) The Dirac basis in which the generator $\Gamma_6\Gamma_5$ is represented by $1\otimes\sigma_3$ and

$$\xi^D = \begin{vmatrix} \psi_+ \\ \psi_- \end{vmatrix} \text{ with } \frac{1}{2}(1 \pm \Gamma_6\Gamma_5)\xi^D = \begin{cases} \begin{vmatrix} \psi_+ \\ o \end{vmatrix} \\ \begin{vmatrix} o \\ \psi_- \end{vmatrix} \end{cases} \tag{3.6}$$

Since $\Gamma_6\Gamma_5$ commutes with the Lorentz subalgebra $s_{\mu\nu}$ of s_{ab} in (3.3), $\psi_\pm$ are basis for irreducible representations of the Lorentz (disconnected) subgroup of the (disconnected) conformal group. In fact they are the well known Dirac spinors. For space-time reflections as known, they transform each of them into themselves.

4. "RESTRICTED" CONFORMAL COVARIANCE OF THE MASSLESS DIRAC EQUATION

Conformal covariance of massless Dirac equation:

$$i\gamma_\mu \frac{\partial}{\partial x_\mu}\psi(x) = o \tag{4.1}$$

has been proved by many authors[2]. Most of the procedures make use of the induced representation method, which starts by considering the representation of the "little group" built up by those transformations of the conformal group, which leave x = 0 invariant (Lorentz, dilatations, and special conformal transformations). However, on a Dirac spinor which is, as seen from (3.6), an eigenstate of $\Gamma_6\Gamma_5$ (generator of dilatations), special conformal transformations must be trivially represented since they do not commute with $\Gamma_6\Gamma_5$ and the generator K_μ of special conformal transformations must be represented by 0. Consequently, the operator $K_\mu(x)$ representing the generator of special conformal transformations on Dirac spinors must have the property:

$$K_\mu(0) = 0 \tag{4.2}$$

that is in x = 0 the representation of the conformal group is not faithful: only the Weyl subgroup may be faithfully represented. This "restricted" covariance may also be shown in a geometrically more expressive way; in fact it may be shown that equation (4.1) is equivalent (it may be transformed through non singular, invertible transformations) to the system of equations:

$$\begin{aligned} &\eta_a \Gamma^a \eta_b \frac{\partial}{\partial \eta_b} (\Gamma_5 + \Gamma_6)\, \xi(\eta) = 0 \\ &\eta_a \Gamma^a \xi(\eta) = 0 \end{aligned} \tag{4.3}$$

where the factor $(\Gamma_5 + \Gamma_6)$ (which selects one of the two Dirac spinors contained in ξ^D) shows that (4.1) is equivalent to an equation in $M^{4,2}$ which is covariant only for those transformations of O(4.2) which leave $\Gamma_5 + \Gamma_6$ invariant and these are precisely those for which, in $M^{3,1}$ (4.2) is satisfied.

The situation would be different for semispinors; in fact for them, being eigenstates of the Casimir operator Γ_7, no such difficulty arises and for them $K(0) \neq 0$ (e.g. $K_\mu(0) = \gamma_\mu (1 + \gamma_5)$) and one may have a faithful representation of the whole (connected) group SO(4,2) for all values of x acting on them.

This considered, one may conclude that the restricted nature of the massless Dirac equation covariance makes it not the ideal instrument for the study of the conformal properties of spinor field theories; and that an equation with <u>unrestricted</u> or "natural" conformal covariance must be used[1].

5. CONFORMAL SPINOR FIELD EQUATIONS "NATURALLY" COVARIANT

E. Cartan [3] first defined pure spinors, in complex Euclidean spaces C^{2n+1}, as equivalent to polarized isotopic n-planes of C^{2n+1} through a system of linear homogeneous equations of rank n + 1 in the spinor components and space coordinates.

For C^7 the equations are:

$$p_A \Gamma^A \xi = 0 \qquad A = 0, 1, 2, 3, 5, 6, 7 \tag{5.1}$$

with $p_A p^A = 0$.

The equation for C^6 may be obtained by putting $p_7 = 0$:

$$p_a \Gamma^a \xi = 0 \qquad a = 0, 1, 2, 3, 5, 6 \tag{5.2}$$

and, if of rank n, it defines two pure semispinors:

$$\frac{1}{2}(1 + \Gamma_7)\xi^C = \begin{vmatrix} \phi_I \\ o \end{vmatrix} \quad ; \quad \frac{1}{2}(1 - \Gamma_7)\xi^C = \begin{vmatrix} o \\ \phi_{II} \end{vmatrix}$$

(in the Cartan basis) as equivalent each to a polarized isotopic semi-3-planes of C^6 each one dual of the other.

Considering p_a as momentum coordinates, we may take (5.2) as the fundamental (free) equation of our spinor field theory in $M^{4,2}$:

$$\Gamma_a \frac{\partial}{\partial \eta_a} \xi(\eta) = 0 \tag{5.3}$$

which is manifestly and naturally O(4,2) covariant[2].

Equation (5.1) may be considered as corresponding to a particular case of the following obtained from (5.3) with the addition of O(4,2) covariant interaction terms:

$$\left[\Gamma_a \frac{\partial}{\partial \eta_a} + \Omega(\eta) \right] \xi(\eta) = 0 \tag{5.4}$$

with

$$\Omega(\eta) = g \ [B + iB_a \Gamma^a + \frac{i}{2} B_{[ab]}\Gamma^{[ab]} + \frac{1}{3!} B_{[abc]}\Gamma^{[abc]}] +$$

$$+ g' \ [B' + B_a' \ \Gamma^a + \frac{i}{2} B'_{[ab]}\Gamma^{[ab]}] \ \Gamma_7 . \tag{5.5}$$

We could then take in $M^{4,2}$ as the correspondent of (5.1):

$$[\Gamma_a \frac{\partial}{\partial \eta_a} + \phi' \ (\eta) \ \Gamma_7] \ \xi(\eta) = 0 \tag{5.6}$$

with $\phi'(\eta)$ O(4,2) pseudoscalar field.

We shall start from $(5.3)^3$. Through the transformations (2.2) it may be brought [7] to the form (for $\xi \ (\eta)$ homogeneous of degree - 2):

$$i\gamma_\mu \ \partial^\mu \ \Psi_+ \ (x,\alpha) + 2 \frac{\partial}{\partial \alpha^2} \ \Psi_- \ (x, \ \alpha) = 0$$

$$i\gamma_\mu \ \partial^\mu \ \Psi_- \ (x,\alpha) - 2 \ \alpha^2 \ \frac{\partial}{\partial \alpha^2} \ \Psi_+ \ (x, \ \alpha) = 0 \tag{5.7}$$

where the spinor in the Dirac basis $\xi^D(x,\alpha)$:

$$\frac{1}{2} (1 + \Gamma_6 \Gamma_5) \ \xi^D \ (x,\alpha) = \begin{vmatrix} \Psi_+ (x,\alpha) \\ o \end{vmatrix} \ ; \ \frac{1}{2} (1 - \Gamma_6 \Gamma_5) \ \xi^D \ (x,\alpha) = \begin{vmatrix} o \\ \Psi_- \ (x,\alpha) \end{vmatrix} \tag{5.8}$$

is obtained from $\xi^D(\eta)$ through (2, 5, 6, 7). Equation (5.7) is equivalent to (5.3) and consequently is O(4,2) naturally covariant as may be easily checked by applying (2.2) transformations. It is seen from (5.7) that Ψ_+ and Ψ_- differ in dimension by one unit; precisely their dimensions in terms of length are -5/2 and -3/2 respectively.

6. EQ. (5.3) IN $M^{3,1}$: ELEMENTARY SOLUTIONS

In order to interpret eq. (5.7) in $M^{3,1}$ space-time (obtained from (2.2) for $\alpha^2 = 0$) we must eliminate α^2 from the equations (alternatively go to the limit $\alpha^2 \to 0$). This may be done by separation of variables with the ansatz:

$$\Psi_{\pm}(x, \alpha) = f_{\pm}(\alpha)\, \chi_{\pm}(x) \tag{6.1}$$

for $f_{\pm}(\alpha)$ satisfying

$$2 \frac{\partial}{\partial \alpha^2} f_{-}(\alpha^2) = m_{-} f_{+}(\alpha^2);\; 2\alpha^2 \frac{\partial}{\partial \alpha^2} f_{+}(\alpha^2) = -m_{+} f_{-}(\alpha^2) \tag{6.2}$$

where $m_{\pm}$ are α (and x) independent, arbitrary, parameters; $f_{\pm}(\alpha)$ may now be eliminated from (5.7) which becomes:

$$(i\gamma_{\mu} \partial^{\mu} + \sigma_1 M)\, N(x) = 0 \tag{6.3}$$

where

$$M = \begin{vmatrix} m_+ & o \\ o & m_- \end{vmatrix} \quad \text{and } N(x) = \begin{vmatrix} \chi_+(x) \\ \chi_-(x) \end{vmatrix} \tag{6.4}$$

and is interpretable in $M^{3,1}$ since the α dependence has been eliminated.

In Cartan basis (6.3) becomes:

$$(i\gamma_{\mu} \partial^{\mu} + M)\, N^C = 0 \tag{6.3'}$$

where Dirac spinors $\chi_{\pm}$ are substituted by semispinors.

Solutions of (6.2) are given by:

$$\begin{aligned} f_- &= \alpha\, [c_- J_1(\sqrt{\alpha^2 m_-^2}) + d_- Y_1(\sqrt{\alpha^2 m_-^2})] \\ f_+ &= c_+ J_o(\sqrt{\alpha^2 m_+^2}) + d_+ Y_o(\sqrt{\alpha^2 m_+^2}) \end{aligned} \tag{6.5}$$

where J_i, Y_i are Bessel functions. It is seen from (6.1), (6.2) and (6.5) that f_+ and f_- have different dimensions and consequently the Dirac spinors $\chi_+(x)$, $\chi_-(x)$ have now the same dimension -5/2 in terms of length. Since $f_{\pm}$ are exact solutions of (6.2), equation (6.3) obtained from O(4,2) covariant (5.7) spontaneously breaks O(4,2) covariance, but also Weyl covariance since the parameters $m_{\pm}$ have the dimension of a mass but, being α and x independent, are not subject to dilatations. However, we may recover this covariance by substituting in (6.1), (6.2) and (6.5) $m_{\pm}$ with x dependent arbitrary scalar fields $\phi_{\pm}(x)$.

7. EQ. (5.3) IN $M^{3,1}$: SOLUTIONS WITH WEYL COVARIANCE

With the mentioned substitution the ansatz (6.1) becomes:

$$\Psi_{\pm} = f_{\pm}(\alpha, x)\, \chi_{\pm}(x) \tag{7.1}$$

with $f_{\pm}$ satisfying:

$$2 \frac{\partial}{\partial \alpha^2} f_-(\alpha, x) = \phi_-(x) f_+(\alpha, x) \; ; \; 2\alpha^2 \frac{\partial}{\partial \alpha^2} f_+(\alpha, x) = \tag{7.2}$$

$$= - \phi_+(x)\, f_-(\alpha, x)$$

and given by (6.5) where $\sqrt{\alpha^2 m_{\pm}^2}$ is substituted by $\sqrt{\alpha^2 \phi_{\pm}^2(x)}$ in the argument of the Bessel functions. $\Psi_{\pm}(x,\alpha)$ in (7.1) satisfy

$$\Box\, \Psi_{\pm}(x,\alpha) = \phi_{\pm}^2(x)\, \Psi_{\pm}(x,\alpha) \tag{7.3}$$

It is seen then that for dilatations:

$$\phi_{\pm}(x) \to \phi_{\pm}'(x') = \frac{1}{\rho} \phi_{\pm}(x) \tag{7.4}$$

and our equations may now be Weyl covariant. (6.3) is now substituted by:

$$[i\gamma_{\mu} D^{\mu} + \sigma_1 M(x)]\, N(x) = 0 \tag{7.5}$$

where:

$$M(x) = \begin{vmatrix} \phi_+(x) & o \\ o & \phi_-(x) \end{vmatrix}$$

and:

$$D^{\mu} = \lim_{\alpha^2 \to o} [\partial^{\mu} + \frac{1+\sigma_3}{2} \frac{\partial}{\partial x_{\mu}} \log f_+(\alpha,x) + \frac{1-\sigma_3}{2} \frac{\partial}{\partial x_{\mu}} \log f_-(\alpha,x)] \tag{7.6}$$

It is easily seen that due to the property of Bessel functions

$$\lim_{\alpha^2\to 0} \frac{\partial}{\partial x_\mu} \log f_+ (\alpha, x) = 0 \qquad (7.7)$$

and we are left with:

$$\lim_{\alpha^2\to 0} \frac{\partial}{\partial x_\mu} \log f_- (\alpha,x) = \pm \frac{\partial \log\phi_-(x)}{\partial x_\mu} = \pm i A_\mu^{(-)} \neq 0 \qquad (7.8)$$

where $\pm$ depends on the choice of the c_- and d_- constants in (6.5) and we are left with:

$$\gamma^\mu [\partial_\mu \pm \frac{i}{2} (1-\sigma_3) A_\mu^{(-)}] N = i[\phi_+(x)\sigma_- + \phi_-(x)\sigma_+]N \qquad (7.9)$$

$A_\mu^{(-)}$ could also be absorbed by a gauge field provided $\chi_-(x)$ interacts with one (ϕ_- should then be complex: in fact we need $\phi_+ = \phi_-^*$ for the hermiticity of (7.9)).

Since the equations (Bessel) fix only the values of the square of the arbitrary functions $\phi_\pm(x)$, these could also be represented by matrices (in the spirit of spinor-geometry [8]:

$$\phi_\pm(x) = \gamma_\mu W_\pm^\mu \quad ; \quad [= \gamma_\mu (1+\gamma_5) W_\pm^\mu] \qquad (7.10)$$

for:

$$[\phi_\pm(x)]^2 = W_\pm^\mu W_{\pm\mu} \neq 0 \quad [=0]$$

Considering that the pure spinors defined by (5.2) and (5.3) are semispinors which can only contain left and right handed Dirac spinors, we see then that (7.9) may represent the main feature of weak interactions (we shall see in § 10 that for $\chi_\pm(x)$ canonical a neutral-current-like term is also necessary).

The most characteristic feature of (7.9) is the asymmetry of the $A_\mu^{(-)}$ "interaction" with respect to the two Dirac spinors (it reminds the asymmetry in charge of most isospinor doublets); it is of geometrical origin and it may be connected with the asymmetry of dimension of the original spinors Ψ_+ and Ψ_- in (5.7) and perhaps to the topological structure of $M^{4,2}$ (similar terms appear in the Dirac equations of exotic fermions on a thorus [9]).

8. EQ. (5.6) IN $M^{3,1}$: SOLUTIONS WITH WEYL COVARIANCE

We shall now consider (5.6) which could be thought derived from the definition of pure conformal spinors in C^7 (and not semispinors) With the same methods as before, taking the Dirac basis in which

$$\Gamma_\mu = \gamma_\mu \otimes 1 \quad ; \quad \Gamma_5 = i\gamma_5 \otimes \sigma_1 \quad ; \quad \Gamma_6 = \tau_2 \gamma_5 \otimes \sigma_2 \tag{8.1}$$

and space reflection is represented by Γ_0, we arrive at

$$\gamma^\mu \left[\partial_\mu \pm \frac{i}{2}(1-\sigma_3)A_\mu^{(-)}\right] N = i\,\vec{\pi}\cdot\vec{\sigma}\,\gamma_5 N \tag{8.2}$$

where

$$\pi_{\frac{1}{2}}(x) = \frac{1}{\sqrt{2}}\left[\phi_+(x) \pm i\phi_-(x)\right]$$

$$\pi_3(x) = \lim_{\alpha\to o} \phi(x, \alpha)$$

which formally reminds su(2) symmetric pion nucleon interaction. Formally, since the π_i isotopic spin indices could only result if we express the $\pi_i(x)$ fields as bilinears in conformal spinors (and the same would apply to $W_\pm^\mu$ above; this would in fact fit in the spirit of spinor geometry [8]).

Up to now we have only studied the features of equations in $M^{3,1}$ obtained by partially solving exactly the "free" equations (5.3) and (5.6). The obtained equations then represent the geometrical frame in which physical laws in $M^{3,1}$ may be formulated if compatible with naturally conformal covariant spinor field equations (5.3) and (5.6).

We could now try to insert O(4,2) covariant interactions as in (5.4), (5.5) and reduce them with the above methods to $M^{3,1}$. For this we need to postulate the equations for the tensor fields in $M^{4,2}$ and correspondingly a Lagrangian formalism is appropriate. This has been already studied in ref. [10] and will be developed in a subsequent paper. We shall here briefly illustrate some features of the Lagrangian formalism of interest for the present lecture.

9. LAGRANGIAN FORMALISM

Let us introduce an action in $M^{4,2}$ from which (5.4) may be obtained as Euler-Lagrangian equations.

$$I = I_o + I_i = \int \bar{\xi}^D (\Gamma_a \overleftrightarrow{\partial}^a + \Omega)\, \xi^D d^6\eta \qquad (9.1)$$

with Ω given by (5.5) and

$$\bar{\xi}^D = i\, \xi^{D+}\, \Gamma_o \Gamma_6 \qquad (9.2)$$

(we may now verify that g and g' are real and the fields real for I real) (9.1) now implies (5.4) and

$$\bar{\xi}^D (\Gamma_a \overleftarrow{\partial}^a + \Omega) = 0 \qquad (9.3)$$

The action:

$$I_i = \int \bar{\xi}^D\, \Omega\, \xi^D d^6\eta \qquad (9.4)$$

which is, by definition, O(4,2) covariant may also be decomposed in O(3,1) invariant terms presenting an U(1,1) internal symmetry.

Let us now go to the Cartan basis through the unitary operator U^4:

$$\zeta^C = U\zeta^D \quad \text{and} \quad U = L\otimes 1 + R\otimes\sigma_1$$

$$\left.\begin{matrix} L \\ R \end{matrix}\right\} = \frac{1}{2}(1 \pm \gamma_5)$$

Since

$$\Gamma_a^C = U\, \Gamma_a^D\, U^{-1}$$

starting from the Dirac basis:

$$\Gamma_\mu^D = \gamma_\mu \otimes \sigma_3 \quad ; \quad \Gamma_5 = i \otimes \sigma_1 \quad ; \quad \Gamma_6 = 1 \otimes \sigma_2 \qquad (9.5)$$

where we get:

$$\bar{\xi}^D = i\xi^{D+}\Gamma_o^D \Gamma_6^D = \xi^{D+}\gamma_o \otimes \sigma_1 \qquad (9.6)$$

we get the Cartan one:

$$\Gamma_\mu^C = -i\gamma_\mu\gamma_5 \otimes \sigma_2; \; \Gamma_5^C = i \otimes \sigma_1; \; \Gamma_6^C = \gamma_5 \otimes \sigma_2 \tag{9.7}$$

where now

$$\bar{\xi}^C = i\,\xi^{C^+}\Gamma_0^C\,\Gamma_6^C = \xi^{C+}\gamma_o \otimes 1 \tag{9.8}$$

and:

$$I_i^C = \int \bar{\xi}^D U^{-1} U \Omega^D U^{-1} U \xi^D d^6\eta = \int \bar{\xi}^C \Omega^C \xi^C d^6\eta \tag{9.9}$$

one may now decompose the fields appearing in $\Omega^C(\eta)$ in O(3,1) covariance terms and one finds, due to the new Cartan basis in spinor space:

$$\Omega^C = g\,[A(\eta) + \vec{\sigma}\cdot\vec{B}(\eta)] + g'\,[A'(\eta) + \vec{\sigma}\cdot\vec{B}'(\eta)] \tag{9.10}$$

where

$$\begin{aligned} A(\eta) &= S(\mu) + i\gamma_\mu V^\mu(\mu) + \frac{i}{2}[\gamma_\mu, \gamma_\nu]_- T^{\mu\nu} \\ \vec{B}(\mu) &= \vec{S}(\mu) + i\gamma_\mu \vec{V}^\mu + \frac{i}{2}[\gamma_\mu, \gamma_\nu]_- \vec{T}^{\mu\nu} \end{aligned} \tag{9.11}$$

and A' B' have the same decomposition in terms of the corresponding pseudo tensors.

It is seen that the U(1,1) internal symmetry in the Dirac basis, that is for the Dirac doublets, is transformed in a U(2) one in the Cartan basis; that is for semispinor doublets[5].

We have seen that with the method of separation of variables the doublets of Dirac spinors with different dimensions contained on the conformal spinor have been transformed in doublets of Dirac spinors $\chi_\pm(x)$ in $M^{3,1}$ having both the same dimension: anomalous, however (-5/2) and not canonical (-3/2); we shall see that also this difficulty may be easily overcome in the frame of the Lagrangian formalism.

10. SEPARATION OF VARIABLES IN THE ACTION INTEGRALS

We shall now, for simplicity consider only the action I_o giving rise to the free equation (5.3). If we now substitute η_a variables with x_μ,α, K from (2.2); introduce (7.1) ansatz with $f_\pm$ satisfying

(7.2) by which O(4,2) covariance is broken, eliminate K integration through a δ function: $\delta(\eta.n - 1)$ where n is a unit vector in $M^{4,2}$ with components (00001,1), we arrive to an action integral of the form (for brevity we give only part of the cinetic terms):

$$I_o = \int d^4x \int_{-\infty}^{+\infty} d\alpha \, f_+^*(\alpha,x) f_-(\alpha x) \bar{\chi}_+(x) \overleftrightarrow{\not{D}} \chi_-(x) + \ldots \qquad (10.1)$$

where

$$\overrightarrow{\not{D}} = \gamma^\mu(\partial_\mu \pm iA_\mu^{(-)}); \overleftarrow{\not{D}} = \gamma^\mu \partial_\mu \qquad (10.2)$$

Assuming for simplicity $\phi_+(x) = \phi_-(x) = \phi(x)$ and introducing the dimensionless variable

$$y = \alpha \, \phi(x) \qquad (10.3)$$

we may separate the variables in the integrals in (10.1) and we obtain

$$I_o = \int_{-\infty}^{+\infty} f_+^*(y) f_-(y) \, dy \int \bar{\psi}_+(x) \overleftrightarrow{\not{A}} \psi_-(x) \, d^4x + \ldots \qquad (10.4)$$

where

$$\psi_\pm(x) = \phi^{-1}(x) \, \chi_\pm(x)$$

and

$$\overrightarrow{\not{A}} = \gamma^\mu(\overrightarrow{\partial}_\mu \pm iA_\mu^{(-)} + \frac{1}{\phi}\frac{\partial\phi}{\partial x\mu}); \quad \overleftarrow{\not{A}} = \gamma^\mu(\overleftarrow{\partial}_\mu + \frac{1}{\phi}\frac{\partial\phi}{\partial x\mu}) \qquad (10.5)$$

The fields $\psi_\pm(x)$ are now Dirac spinor fields having canonical dimensions -3/2 in terms of length. The integrals in y give rise to dimensionless constants and the action is only represented by the integrals in x_μ giving rise to equation

$$\gamma^\mu[\partial_\mu \pm \frac{i}{2}(1-\sigma_3)A_\mu^{(-)} + \frac{1}{\phi}\frac{\partial\phi}{\partial x\mu}] D(x) = i \, \sigma_1 \phi(x) D(x) \qquad (10.6)$$

and the conjugate one where now

$$D(x) = \begin{vmatrix} \psi_+(x) \\ \psi_-(x) \end{vmatrix} \qquad (10.7)$$

represents a doublet of canonical Dirac spinor fields. Again the reduction of the spinor field equation from $M^{4,2}$ to $M^{3,1}$ has been possible by partial solution of the original field equation and has brought to an equation on a doublet of canonical Dirac spinors; the spinors have been brought to canonical dimensions in $M^{3,1}$ through the introduction of the $\phi^{-1}\partial\phi/\partial x$ "interaction" which might be absorbed by a gauge field acting on both spinors of the doublet.

11. HIGHER DIMENSIONAL SPACES

If we adopt the hypothesis (spinor - geometry) that spinors are the fundamental elements for the description of both: space-time and geometry, and of physics in space-time, we know that $C_{4,2}$ or conformal spinors, are not enough to build up $M^{3,1}$ coordinates as well as the Poincare' algebra acting on it, and not less than $C_{5,3}$ spinors with $2^4 = 16$ components are necessary.

The procedure described before can be easily repeated, C^9 (complex) and $M^{5,3}$ substituting C^7 and $M^{4,2}$. We shall obtain in $M^{4,2}$ two O(4,2) covariant spinor field equations presenting formally an internal U(2) symmetry and then subsequently in $M^{3,1}$ a system of 4 Poincare' covariant spinor field equations with a U(4) internal symmetry. For C^9 spinors Θ purity conditions (equivalence with polarized, isotopic 4-planes) may be imposed which have the form of the following 10 equations:

$$\bar{\Theta}\,\Theta = 0; \quad \bar{\Theta}\,\Gamma_A\,\Theta = 0 \qquad A = 0, 1, 2, 3, 5, 6, 7, 8, 9 \quad (11.1)$$

As a consequence the U(4) symmetry will be reduced for vector pseudovector interactions from U(4) to U(2) x U(2).

Furthermore, the 4 Dirac spinors in $M^{3,1}$ obtained from Θ will have dimension $1^{-7/2}$ and one will need two interaction terms of the type $\phi_i^{-1}\partial\phi_i/\partial x_\mu$ to reduce them to canonical Dirac spinors (three of these will be necessary if one starts from C^{11} or $M^{6,4}$ in which case the equations of pureness (11.1) acting on octuplets of Dirac spinors include also tensors and are in number of 66).

12. CONCLUSION AND CONSEQUENCES

We have seen that naturally conformally covariant spinor field equations arising from the E. Cartan definition of pure spinors may be written either in the form (5.3) or (5.6) and may be partially solved, thus spontaneously breaking the original O(4,2) symmetry and becomming equations in $M^{3,1}$ space-time, presenting Poincare' times, formally, internal U(2) covariance acting on canonical Dirac or semispinor doublets. The situation is similar to the familiar one of Dirac equation which may be reduced, for slow motions, to the Pauli-Schroedinger one presenting Galilei xSU(2) spin (internal) covariance. The main difference between the two cases is that while in the latter case the original, relativistic, symmetry is broken by varying one of the parameters of the group (velocity) and both equations Dirac and Pauli-Schroedinger one, act in the same space-time; in the former case the symmetry is broken spontaneously; by inserting exact partial solutions in the original equations and then, by eliminating the dependence from the extra coordinate α, equations in $M^{3,1}$ space-time are obtained. But in both cases the internal symmetry results geometrically from the breaking of a higher covariance.

This picture may perhaps justify the hope that isotopic spin (or the first SU(2) flavour symmetry) may be understood geometrically as a consequence of the fundamental role of conformal covariance for the elementary physical laws. The hypothesis, if true, could have far reaching consequences for physics (specially in the case that higher dimensional spinors and corresponding groups are necessary for building up space-time geometry) on which we would like to express here some tentative, preliminary remarks:

I. Higher dimensional spinor field equations, by partial exact solutions, will be eventually brought to Poincare' times internal symmetry equations in space time $M^{3,1}$ but purity conditions may partially break the internal symmetry group and families of SU(2) internal symmetry systems will arise, each acting on a (canonical) Dirac spinor (or conformal semispinor) doublet of conformal origin.

II. Not only the experimentally observed "family" structure of spinor doublets would naturally arise, but also the hierarky since the dimensional differences of the spinors of the original multiplets will be naturally compensated, through the Lagrangian mechanism by "interactions" which eventually will be absorbed in gauge interactions; the higher dimensions of the higher spinors needing more compensating interactions.

III. The separation of internal from Poincare' symmetry should be maintained only for the approximate description of low energy phenomena in space-time (when scaling is not ruling) but for higher precision computations, even at low energy, one should develop a field theory fully covariant with respect to the higher groups and use (quantize) the spinor field equations (Lagrangians) prior to the insertion of exact partial solutions and to the corresponding spontaneous breaking of symmetries. Only afterwards the extra variables should be eliminated (possibly by integration) for the interpretation of the results in space-time $M^{3,1}$; in the same way as Dirac, and not Pauli-Schroedinger equation (Lagrangian) is at the basis of high-precision computations on slow electron motion phenomena like the Lamb-Shift one.

FOOTNOTES

[1] In a similar way, it may be proved that the spinor field equation on the plane $y_1\ y_2$: $\sum_{i=1}^{2} \sigma_i \frac{\partial}{\partial y_i} \phi\ (y) = 0$ is restricted Lorentz covariant and equivalent to the equations $x_\mu \gamma^\mu x_\nu \partial^\nu (\gamma_0+\gamma_3)\psi=0$; $x_\mu \gamma^\mu \psi=0$ in $M^{3,1}$. We think that these equations would not be the ideal instrument for the study of relativistic (even if massless) spinor field theories: in fact $\phi(y)$ is obtained from $\frac{1}{2}(1+\gamma_3\gamma_0)\psi$; and its chiral projection: $\frac{1}{4}(1+\gamma_5)\ (1+\gamma_0\gamma_5)\psi$ has only one non zero component.

[2] Equation (5.3) has been proposed first by Y. Muray [4] and later on studied by A.O. Barut [5] and also by R.L. Ingraham [6].

[3] Applying the projectors $\frac{1}{2}(1\pm\Gamma_7)$ it may be split in two independent equations for the semispinors.

[4] U may be defined from the property $U\Gamma_6\Gamma_5 U^{-1} = \Gamma_7$ in any basis.

[5] It may appear surprising that U(1,1) internal symmetry is transformed to a U(2) one through a unitary transformation U: but this one acts only on spinor space and spinor operators while the tensor ones appearing in $\Omega(\eta)$ remain unaltered. The actual compactification of U(1,1) may be understood by comparing (9.6) with (9.8).

REFERENCES

[1] Dirac P.A.M.: 1936, Ann. Math. 37, 429.

[2] Mack, G. and Salam, A.: 1969, Ann. of Physics, 53, 174; Todorov, I.: 1/81/EP ISAS preprint: to be consulted for further references.
[3] Cartan, E.: 1939, La theorie des spineurs, Paris.
[4] Muray, J.: 1958, Nucl. Phys., 6, 489.
[5] Barut, A.O. and Haugen, R.B.: 1972, Ann. of Phys. 71, 519; 1973, Nuovo Cimento 18A, 495, 511.
[6] Ingraham, R.L.: 1981, New Mexico State University preprint.
[7] Furlan, P.: 17/82/EP ISAS preprint.
[8] Budinich, P.: 14/82/EP ISAS preprint.
[9] Petry, H.R.: 1979, J. Math. Phys. 20(2), pp. 231-240.
[10] Budinich, P., Furlan, P. and Raczka, R.: 1979, Nuovo Cimento 52A, 191.

[2] [illegible] and [illegible] 19[illegible], [illegible] Series, [illegible]
[illegible] 1[illegible]/[illegible] 1[illegible] [illegible] for
[illegible]
[illegible] Paris.
[illegible] Phys. [illegible]
[illegible] 19[illegible] [illegible]
[illegible]
[6] [illegible]
[7] [illegible] preprint
[8] [illegible]
[9] [illegible]
[10] [illegible] Moscow [illegible]
[illegible]

CLASSICAL NONLINEAR DIRAC FIELD MODELS OF EXTENDED PARTICLES

ANTONIO F. RAÑADA

Departamento de Física Teórica, Universidad Complutense
Madrid-3

ABSTRACT

The present situation of the classical theory of nonlinear Dirac fields and, specially, of its particle-like solutions is reviewed. The emphasis is placed on a line of research which produced some models of extended particle which were applied to the nucleons and to particles with confined constituents.

I. INTRODUCTION

In this paper we shall review the efforts to build models of extended particles by means of nonlinear Dirac fields. First of all it is necessary to explain the sense in which the word classical is used. It is not related here to any kind of limit of any quantum theory when $\hbar \to 0$, but to a precise spacetime description of the waves and fields, which are ordinary functions or c-numbers. Those physicists to whom the idea of classical Dirac field seem contradictory may identify in this paper the terms classical and prequantum.

The main concept of the models which will be presented here is the finite energy solitary wave or particle-like solution (PLS from now on for short). It is a solution which propagates without changing its shape and which is very often called soliton by particle physicists. On the other hand, this term is used in mathematical physics in a more restricted sense to denote the solitary waves in (1+1) dimensions which keep their form after collisions and obey equations solvable by the IST method (1-3). We will adhere to this more restricted use, although it is yet unclear which is the adequate generalization of the concept of soliton in more than one space dimension. It would be most interesting to know if the waves which we will consider are

A. O. Barut (ed.), Quantum Theory, Groups, Fields and Particles, 271–291.

solitons in some sense. As it is impossible to give an answer to this question, except in some cases in which it is negative, we will instead speak just of solitary waves or PLS. But we can assert that the models to be discussed later offer good reasons for trying hard to extend soliton theory to more than one space dimension.

The plan of the paper is the following. In section II we present a bibliographical review of the use of classical nonlinear Dirac fields. In III we explain the Soler model of elementary fermion and in IV several realistic models which describe the main static properties of the nucleons. The stability properties of the Dirac solitary waves are discussed in V. Finally, in VI we show a model of extended particle with structure, whose PLS are bound states of confined constituents and which shows the characteristic pattern of three constituents and constituent--anticonstituent particles.

II. A BIBLIOGRAPHICAL REVIEW

Nonlinear phenomena have been one of the most popular topics during the last years. Nevertheless, it must be admitted that nonlinear classical fields have not received a general consideration. This is probably due to the mathematical difficulties which arise because of the nonrenormalizability of the Fermi and other nonlinear couplings. In this section we briefly review the most significant efforts to understand the basic properties of nonlinear spinors and to use them as a tool to attack the chalenge presented by extended particles.

Nonlinear selfcouplings of the spinor fields may arise as a consequence of the geometrical structure of the spacetime and, more precisely, because of the existence of a torsion. As soon as in 1938, Ivanenko (4-6) showed that a relativistic theory imposes in some cases a forth order selfcoupling. In 1950 Weyl (7) proved that, if the affine and the metric properties of the spacetime are taken as independent, the spinor field obeys either a linear equation in a space with torsion or a nonlinear one in a Riemannian space. As the selfaction is of spin-spin type, it allows the assignement of a dynamical role to the spin and offers a clue about the origine of the nonlinearities. This question was further clarified in some important papers by Utiyama (8), Kibble (9) and Sciama (10). In the simplest scheme the selfaction is of pseudovector type, but it can be shown that one can also get a scalar coupling (11). An excellent review of the problem may be found in reference (12).

Nonlinear quantum Dirac fields were used by Heisenberg (13-14) in his ambitious unified theory of elementary particles. They are presently the object of renewed interest since a widely known paper by Gross and Neveu (15, see also 16).

On the other hand, many authors considered in the last years the incorporation of extended classical solutions of nonlinear equations into standard quantum theories by quantization around them and by semiclassical approximation or the quantization of nonlinear theories (17-27, where references 17-20 and 25-27 are review articles). In this paper however we will only consider the classical properties of the nonlinear Dirac fields, whose study has followed a line quite unrelated to the above mentioned attempts which will not be discussed here.

The first efforts to obtain particle-like solutions of a classical field theory go back to 1903 and 1904, when Abraham (28) and Lorentz (29) tried to develop a purely electromagnetic model of the electron, the only particle known at the time. Although the idea is certainly interesting, it can even be considered as the basis of the actual conception of mass, the model was not quite successful, mainly because of the necessity of introducing an ad hoc cohesive force to compensate the electrostatic repulsion of the different parts of the particle (30). Later on in 1939, Rosen (31-32) tried the construction of classical relativistic theory of extended charged scalar particles in interaction with their own electromagnetic field. His main purpose was the search of a finite theory in which the characteristic divergences of the perturbation series were eliminated from the start. The attempt did not succeed because of two different reasons. First of all, no bound states could be found unless the sign of the Lagrangian density of the matter field was reversed and, in that case, they have negative energy. Moreover, to overcome the indeterminacy of the frequency Rosen looked for minima of the energy. Unfortunately they only appear at $\omega = 0$, uncoupling the fields and making the charge zero. In spite of these problems, the Rosen model has the value of being the first effort to go beyond the Abraham-Lorentz work and to build a consistent and free of divergences classical model of extended particles, in order to avoid the very difficult problems due to the appearance of infinities in the usual quantum theory.

Some years later, in two very interesting papers, Finkelstein et al. (33-34) considered the case of a spinor field with several types of fourth order selfcouplings. They quantized the model by imposing to the PLS a normalization condition and discussed the radial equation for different values of the angular momentum. They found that the discrete values of the rest energies are of the same order of magnitude, a result which they deemed discouraging. Later on, Wakano (35) calculated the classical rest energy of a linear spinor in interaction with its own electromagnetic field. By numerical analysis he found, as in the case of Rosen, a negative value, after assuming that the time component of the electromagnetic field dominates.

In 1970 Soler (36) proposed a model of elementary fermion as the PLS of a nonlinear Dirac field with scalar selfaction. The solitary waves form a family which depends continuously on the frequency and, to break this degeneracy, he used a minimum energy principle. Subsequently the electromagnetic interaction was introduced (37-39) in order to construct a model of extended charged fermion. Following an idea of Weyl (7), a connection of the nonlinearity with the geometrical properties of the spacetime was established (11). As this model does not seem only of academic interest but, quite on the contrary, it appears to be adequate to represent real particles, several models of the nucleon, to be described in section IV, were developed (40-43). In spite of their simplicity, the main properties of the nucleons such as mass, charge, spin, magnetic moments, electromagnetic radii and form factors are fairly well accounted for. A similar picture of the pion and the kaon was also studied (44). García and Usón (45) considered the corresponding PLS as elastic bodies, calculated the distribution of internal stresses and showed how the nonlinear terms provide with the cohesive forces needed by the Abraham-Lorentz model.

In the methodological development of this approach, the relation between classical nonlinear fields and quantum physics (46-48) and the problem of the linearization near a quantum mechanical state were investigated (49-51). The most recent result of this approach is the construction of a model of extended particle with confined constituents which will be presented in section VI (52-53).

In an interesting paper Mielke (54) offers a new approach to the problem of extended particles, relating the theory of the double solution with PLS, and gives a thorough review of the literature on the subject. In 1952, Rosen and Rosenstock (55) proved that, although the short range interaction of PLS is not of potential type, the long range one can be approximated by a Yukawa potential. This result is of interest since it shows a curious similarity between two very different schemes: the nonlinear interaction and the interchange of an intermediate object. G. Rosen (56) in a series of papers studied several properties of the evolution of scalar PLS, some of which can be also applied to spinors. Chu and Kaus (57) looked for a connection with quantum theory by studying the dependence on the angular momentum and the Regge trajectories. The main ideas of the Soler model were published again in 1977 by Rafelski (58) and in 1979 by Takahashi (59). Barut (60 and references therein) considered the classical aspects of the electromagnetic selfaction of a Dirac field as a first step towards its quantization. The nonlinear integrodifferential equations which arise have interesting properties as, for instance, the existence of magnetic levels and high energy narrow resonances which indicate that the electromagnetic forces may be stronger at small distances than what they are supposed to be. He further studied the consequences of this fact in the study of the forces between elementary particles.

In 1977 Werle (61) proposed a very interesting model with the nonpolynomial selfcoupling $(\bar{\psi}\psi)^{\sigma}$, $1/2 < \sigma < 1$. Its PLS, called droplets or bags by the author, have a sharp boundary, their support are contained inside a sphere and show appealing stability properties. Moreover, it turns out that the field equations have only confined, nondissipative solutions, a remarkable property which is due to the fractional character of the nonlinearity. This is because the selfcoupling can be considered as equivalent to an effective attractive potential which goes to infinity when the field vanishes, avoiding thus the dissipation of any solution (62-64).

No doubt, the PLS remind of the solitons. In fact, Skyrme (65) proposed in 1961 the use of those of the sine-Gordon equation as models of relativistic particles and the massive Thirring Model (66) can be solved by the IST method (67-68). But as there is not a theory of three dimensional solitons, several authors studied the stability properties which are certainly most relevant in any physical model. The first results, obtained by Hobart in 1963 (69) and Derrick in 1964 (70), are discouraging since they state that any real scalar PLS is unstable under deformations. The situation is actually more complex. Although most of the work relates to scalar fields (20), there is a number of interesting papers on spinors which will be considered in section V and which have been written by Soler (1975), Rybakov (1966), Bogolubsky (1979), Werle (1981), Alvarez and Carreras (1981), Alvarez (1982), Alvarez and Soler (1982) and Moore 1982) (61, 71-77).

To close this section, some authors have studied the mathematical properties of the classical Dirac fields, which seem to be somewhat more difficult to investigate than those of the scalars. There are a number of interesting results on the local and global existence and behaviour of solutions, conservation laws, bounds for the radii of the PLS, etc. (78-88).

III. THE SOLER MODEL OF ELEMENTARY FERMION

Let us consider a Dirac field with the most general fourth order selfcoupling. Finkelstein et al (34) proved that the lagrangian density has the form

$$L = \frac{i}{2}\{ \bar{\psi}\gamma^{\mu}\partial_{\mu}\psi - (\partial_{\mu}\bar{\psi})\gamma^{\mu}\psi\} - m\bar{\psi}\psi + \lambda\{ (\bar{\psi}\psi)^2 + b(\bar{\psi}\gamma^5\psi)^2 \} \qquad (1)$$

In other words, the nonlinearity can always be expressed as a sum of a scalar and a pseudoscalar terms. We will consider the case $b = 0$, $\lambda > 0$, which will be called the Soler model (36). It has the advantage that its S wave solutions can be factorized in spherical coordinates. It is not difficult to show that this model admits regular localized solutions which represent localizations of all the dynamical variables associated with the field. Moreover, and this establish a very important difference with the linear theory, they are simultaneously solitary waves with characteristic shape and square integrable, finite energy solutions. They will be called particles from now on. In the case of S waves with spin up they have the form

$$\psi = e^{i\Omega mt} \sqrt{\frac{m}{2\lambda}} \begin{pmatrix} H(\rho) \begin{pmatrix} 1 \\ 0 \end{pmatrix} \\ iF(\rho) \begin{pmatrix} \cos\theta \\ e^{i\phi} \sin\theta \end{pmatrix} \end{pmatrix} \qquad \rho = mr \tag{2}$$

where H, F are dimensionless radial functions. It is easy to show that if (2) is substituted into the field equations, the angular dependence is factorized and the following radial equations are obtained:

$$\begin{aligned} F' + \frac{2}{\rho} F + (1 - \Omega + F^2 - H^2)H &= 0 \\ H' + (1 + \Omega + F^2 - H^2)F &= 0 \end{aligned} \tag{3}$$

Being only interested in finite energy regular solutions, we assume $F(0) = 0$ to avoid singularities at $\rho = 0$. We have then a solution for each value of $H(0)$. As it is not possible to find analytical solutions, numerical analysis must be used. It turns out that there is a certain interval $0 < H(0) < H_M$ such that the corresponding solutions have the limit

$$\rho \to \infty \implies H \to \pm\sqrt{1 - \Omega} \quad , \quad F \to 0 \tag{4}$$

Between the cases $+\sqrt{1-\Omega}$ and $-\sqrt{1-\Omega}$ there are square integrable separatrices which tend to zero at infinity and which are characterized by the humber of nodes (36). If $H(0) > H_M$ all the solutions diverge at infinity.

The energy and spin of a solution can be obtained from the energy-momentum and angular momentum tensors and have the values:

$$E = \frac{2\pi}{\lambda m} \{\Omega \int_0^\infty (F^2 + H^2)\rho^2 d\rho + \frac{1}{2} \int_0^\infty (F^2 - H^2)^2 \rho^2 d\rho\} \tag{5}$$

$$S_3 = \frac{\pi}{\lambda m^2}\int_0^\infty (F^2 + H^2)\rho^2 d\rho \; ; \quad S_1 = S_2 = 0 \tag{6}$$

As the norm is

$$N = \int_{\mathbb{R}^3} \psi^+ \psi \, d^3\vec{r} = \frac{2\pi}{\lambda m^2}\int_0^\infty (F^2 + H^2)\rho^2 d\rho \tag{7}$$

one has

$$S = \frac{1}{2} N \tag{8}$$

As we see the norm plays a role similar to the action quantum $\hbar$. In spite of (8) the spin is $\frac{1}{2}$ from the kinematical point of view. For several values of n, n-1 being the number of nodes, a family of solutions which depends continuously on the frequency Ω was found. Consequently, at least for an interval of Ω, there seems to be a numerable sequence of solutions characterized by the number of nodes. The analogy with the well known linear Sturm-Liouville systems is very strong. The mathematical study of this analogy is a very interesting unsolved problem.

In order to determine a physical value of the frequency Soler used a minimum energy principle: the physical value of the frequency corresponds to a minimum of the function $E(\Omega)$. The numerical analysis suggests that there is a minimum for each value of n. Soler found that the ground state corresponds to the nodeless minimum with

$$\Omega = 0.936 \tag{9}$$

the energy and the norm being

$$E = \frac{23.60}{\lambda m} \quad ; \quad N = \frac{22.98}{\lambda m^2} \tag{10}$$

from which E = 1.026 m.N. If the norm is equal to one ($\hbar$ in our units) the relation between the energy and the mass parameter is E = 1.026 m. In other words, there is a mass renormalization of 2.6%.

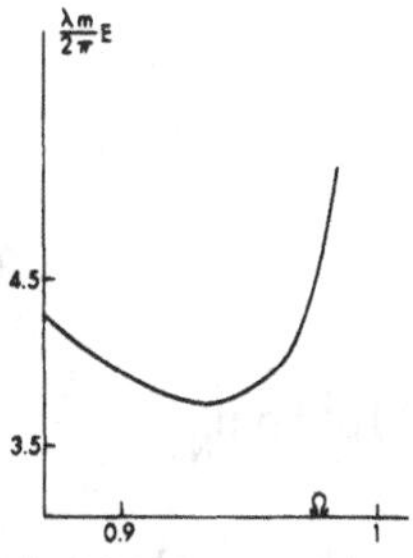

Figure 1. Shape of the curve $E(\Omega)$ for the nodeless solutions.

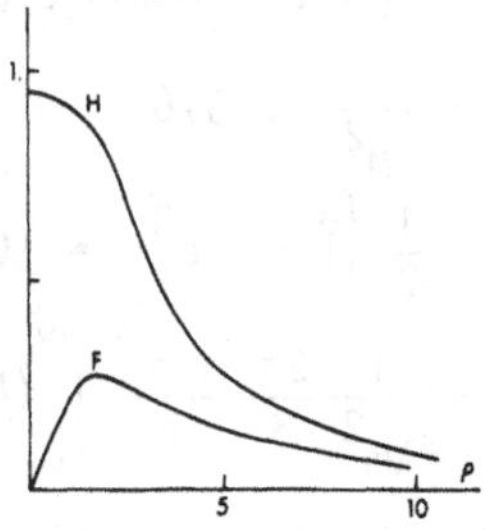

Figure 2. Radial functions F and H versus ρ for the minimum of $E(\Omega)$.

IV. SEVERAL MODELS OF THE NUCLEON

The previous considerations suggest the use of nonlinear Dirac equations to represent elementary particles. Several models of the nucleon have been proposed along this line. If the electromagnetic interaction is introduced the lagrangian density takes the form:

$$L = L_D + \lambda(\bar{\psi}\psi)^2 + L_{EM} \tag{11}$$

where L_D is the usual Dirac lagrangian density and

$$L_{EM} = -\frac{1}{4} F_{\alpha\beta}F^{\alpha\beta} - e\delta\bar{\psi}\gamma^{\mu}\psi\, A_{\mu} - \kappa \frac{e}{4M}\bar{\psi}\sigma^{\alpha\beta}\psi F_{\alpha\beta} \tag{12}$$

where $\delta = 0,1$ to describe neutral or charged particles and κ is the coupling constant of a Pauli term which is intended to describe the existence of an anomalous magnetic moment (40). As the nonlinear spinor term is stronger than the electromagnetic one we can treat the latter as a perturbation by developing everything in series of e^2. The zero order approximation gives then the Soler model. As its energy is $E = 1.026$ m we obtain the mass of the nucleon with a mass parameter $m \simeq 914$ Mev. Of course in that approximation there is no electromagnetic structure.

It turns out that, instead of e^2, it is more convenient to expand in series of $\varepsilon = \frac{e^2}{2\lambda M^2}$ which is the natural dimensionless parameter of the theory. The calculation uses standard techniques and the following results are obtained:

$$E = E_o + \delta E = E_o + \varepsilon E_1 \tag{13}$$

where

$$E_o = \frac{2\pi}{\lambda m} 3.7548 \quad ; \quad \Omega = 0.936 \tag{14}$$

$$\delta E = \frac{2\pi}{\lambda m} \varepsilon(11.922\,\delta^2 - 6.102\,\delta\kappa - 0.1075\,\kappa^2) \tag{15}$$

The norm and magnetic moment are

$$N = \frac{2\pi}{\lambda M^2} 3.6576$$

$$\vec{\mathcal{M}} = \frac{1}{2}\int \vec{r}\times\vec{j}\, d^3r = (0,0,\mathcal{M}) \tag{16}$$

$$\mathcal{M} = \frac{1}{\lambda m^2}\frac{2\pi}{3}\frac{M(p)}{m}\int_0^{\infty}(4\delta FH\rho + \kappa(F^2 + 3H^2))\rho^2 d\rho\, \mu_N$$

Where M(p) is the mass of the proton and μ_N is the nuclear magneton.

Three different kinds of solutions were studied:

a) The "Normal" solution which describes a charged particle without anomalous magnetic moment ($\delta = 1$, $\kappa = 0$)
b) The "Proton" solution ($\delta = 1$, $\kappa \neq 0$)
c) The "Neutron" solution ($\delta = 0$, $\kappa \neq 0$)

The value of ε or, equivalently, that of λm^2 is determined by the condition that the zero order norm be one (or $\hbar$ in nonnatural units). This gives $\varepsilon = .00199$. This relation, together with the values of the energy and magnetic moment determines λ, m and κ. In the case of the proton and neutron solutions the same value of m can be used so that the energy of the proton solution is M(p). A value for the neutron-proton mass difference is thus predicted. The details can be found in reference 40. The results concerning the proton and neutron solutions are the following:

Mass parameter $m = 913.7$ Mev. ; $\lambda m^2 = 22.981$
Proton solution: $E = 938.26$ Mev. ; $\delta E = .28$ Mev. $\kappa = 1.798$
Neutron solution $E = 937.81$ Mev. ; $\delta E = -\ .17$ Mev. $\kappa = -1.964$
Neutron-Proton mass difference: $E(n) - E(p) = -\ .45$ Mev

In table I we compare the electric and magnetic mean square radii predicted by this model with the experimental values (89): (1 unit = 1 fermi square)

Table I

	$\langle r^2 \rangle_{el}$	$\langle r^2 \rangle_{mg}$
Proton	$(.789)^2$	$(.908)^2$
Experimental value	$(.842 \pm .015)^2$	$(.843 \pm .028)^2$
Neutron	$-(.373)^2$	$(.919)^2$
Experimental value	$-(.340 \pm .004)^2$	$(.903 \pm .173)^2$

As we see the see the sizes of the distributions of charge and magnetization are very close to those of the nucleons. The form factors have consequently a very good behaviour at low momentum transfer as it turns out. The only bad result is the wrong value of the neutron-proton mass difference. However this problem remains as one of the puzzles of theoretical physics.

We must add that L. García (90) has calculated the electric and magnetic polarizabilities and found results close to the experimental data.

This model is too simple because it has no meson fields which are known to play an important role in the nucleon structure. To improve the model it may be convenient to introduce some pseudoscalar fields to represent the cloud of pions. Two such models were proposed. In one of them (42), which we will call model 2, there is a neutral pseudoscalar field ϕ coupled to ψ by a pseudoscalar coupling. In the other one (43), which we will call model 3, the same field ϕ interacts with ψ through a pseudovector coupling. More precisely, the lagrangian density includes, in addition to (1), the following terms:

In the model 2 $\quad \frac{1}{2}\left(\partial_\mu\phi\partial^\mu\phi - m'^2\phi^2\right) + G\bar{\psi}\gamma^5\psi\phi$ (17)

In the model 3 $\quad \frac{1}{2}\left(\partial_\mu\phi\partial^\mu\phi - m'^2\phi^2\right) + g\bar{\psi}\gamma^\mu\gamma^5\psi\partial_\mu\phi$ (18)

Where G and g are standard coupling constants, for which it is convenient to take the usual values in quantum field theory:

$$G^2/4\pi = 14.4 \qquad g^2/4\pi = f^2/m^2_{\pi^+} = G^2/16\pi M^2 \tag{19}$$

The most natural value for m' is 0.14 m; in such a way m' and m are in the same relation as the masses of the pion and the nucleon. It turns out, however, that the models show a very small sensitivity to the ratio m'/m.

The following form was assumed for the solutions: The angular dependence of ψ is taken as in model 1 (formula (2)) while the pseudoscalar field must be taken as

$$\phi = \alpha\Phi(\rho)\cos\theta \tag{20}$$

where $\alpha = G/\lambda M$ in model 2 and $\alpha = g/\lambda$ in model 3. This form of the fields does not separate the field equations but it corresponds to the lowest wave approximation in which ψ is in S wave while ϕ is in P wave. To obtain the radial equations for the functions H, F and Φ we can proceed in two different ways:

a) By means of a multipole expansion of the field equations.

b) By substitution of the above expressions in the Action Integral, integration over the angles and variation of the radial functions.

It is easy to show that the two procedures are completely equivalent. The calculations are explained in detail in references (42) and (43). The main results to be stressed are the following.

i.- The predictions of the first and the third models are very similar, although they use different values of λm^2. In other words, the effect of ϕ in pseudovector coupling and that of a variation of the selfcoupling constant are similar.

ii.- The second model predicts a value of 2.78 Mev for the neutron-proton mass differences with the right sign, and a mass renormalization of 8%. The radii however are too small by a factor 1.6, except in $\langle r^2\rangle_{el}$ of the neutron.

iii.- The radii of the models 1 and 3 are almost identical and agree fairly well with the experimental data.

A fourth model was studied in which the nonlinear coupling is of the type which Weyl showed to appear as a consequence of the existence of a torsion in the spacetime (7). The Lagrangian density has the form (41)

$$L = L_D + \lambda \bar{\psi} \gamma^{\mu} \gamma^5 \psi \bar{\psi} \gamma_{\mu} \gamma^5 \psi \tag{21}$$

It turns out that this can be written as (34):

$$L = L_D + \lambda \bar{\psi} \gamma^{\mu} \psi \bar{\psi} \gamma_{\mu} \psi \tag{22}$$

which is the Lagrangian density of the massive Thirring model in three space dimensions (66).

This fourth model seems to be very different from the first one with which it has in common that they have no meson fields. The numerical analysis shows only the existence of a branch of nodeless solutions with a curve $E(\Omega)$ with a very curious spiral shape. A careful search of solutions with nodes failed to find any one, presumably because they do not exist. The shape of the curve $E(\Omega)$ suggests that there is an infinite number of minima. If we take $m = 670$ Mev., $\lambda m^2 = 6.46$ the lowest minimum corresponds to a solitary wave with the mass and charge of the proton. The second and third minima have masses of 1,123 Mev and 1,184 Mev which are very close to those of the Λ(1,115 Mev) and the Σ(1,192 Mev). Although this is probably only a numerical coincidence it is encouraging to have a mass spectrum of the same order of magnitude as that of the baryons.

V. THE STABILITY OF THE DIRAC PARTICLE LIKE SOLUTIONS

It seems that the previous models may offer a useful tool to construct theories of extended particles. There is however a very important problem to consider: the stability of the solutions.

There are two different reasons to study this problem. First of all, it is not clear whether the solutions discussed previously can be reached with more than zero probability in a process of formation. This may seem strange since quantum mechanics does not attempt a description of the way in which a particle reaches its final stable shape in a scattering or formation reaction. On the other hand we can follow the evolution of a classical disturbance of the four dimensional continuum. We can therefore ask if the equations of motion push the disturbance towards the states previously identified as physical particles. Secondly, once the particle is formed it is submitted to perturbations due to other particles, radiation and so on. It could happen that these perturbations deform the solutions in such a way that the particles tend to disappear.

For these reasons particle like solutions must be stable if they are to be used to model extended particles even if these are of a decaying class. Because it must be stressed that theunstability of a solitary wave is of a different type as that of a particle as, say, the pion, which is due to the coupling to other fields which may be absent in the case of a PLS.

The simplest case to be studied is the $(\phi^*\phi)^2$ nonlinear Klein-Gordon equation (NLKGE), which is the closest counterpart to the $(\bar{\psi}\psi)^2$ case. It is known (69,70) that the static PLS of all the NLKGE are unstable against deformations. The $(\phi^*\phi)^2_4$ case was studied by Anderson and Derrick (91) who found that the spherically symmetric PLS which have the form

$$\phi = \theta(r)e^{-i\omega t} \tag{23}$$

are unstable against small deformations. They considered the problem from two different points of view. By a Poincaré linearization and by submitting the PLS to a random noise. They found that in a time of the order of 1/m the PLS disappears via one of the two modes which are called singular and disipative decays. In the first one the field shrinks and develops a singularity at its center. In the second one it tends point wise to zero but increases its radius beyond any limit in order to conserve the energy.

The stability properties of general NLKGE depend not only on the type of selfcoupling but also on the concept of stability since many different definitions can be used (92) and they are not equivalent. The two which are more frequently considered are the stability against deformations or perturbations and the energetic stability. Werle (61, see also 20) has sistematically studied these two kinds of stability for polynomial, logarithmic and fractional nonlinearities. It turns out that in some cases the PLS are stable, a specially interesting example being given by what he calls droplets (61).

It is known that the mathematical aspects of the nonlinear Dirac equations (NLDE) are in general more difficult to investigate than those of the scalar fields. Correspondingly perhaps, the stability properties seem to be more complex. Some authors claim that the PLS of the $(\bar{\psi}\psi)^2_2$ and the $(\bar{\psi}\psi)^2_4$ are unstable (62, 72,73) while others find a very stable behaviour (71,74-76). Let us consider their arguments.

Bogolubsky (73) requires, as a necessary condition for stability, the positivity of the second variation of the energy functional and shows that the family of charge preserving expansions of the PLS

$$\psi(\vec{r},t) \to \psi_a = a^{-n/2}\,\psi(\vec{r}/a,t) \tag{24}$$

where n is the dimension of the space and a is a scale parameter, induces a nonpositive definite $\delta^2 E$, as the energy

$$E(a) = \int T^{oo}(\psi_a\,,\,\partial^o\psi_a)d^3r \tag{25}$$

has not a minimum at $a=1$. He concludes that the PLS are not stable and states that computer experiments confirm this claim. Werle (62) considers the more restrictive condition of energetic stability, which requires that the energy functional has an absolute minimum (and not only a relative one) at a stable PLS. If this is the case, the function E(a) must also have a minimum at $a=1$. It is easy to prove that the $(\bar{\psi}\psi)_2^2$ and the $(\bar{\psi}\psi)_4^2$ equations are not stable according to this definition. He gives a detailed discussion of the polynomial, logarithmic and fractional nonlinearities and shows that, in some cases, the condition of energetic stability is verified. Moroever he proposes as a physical criterium to require, in addition to stability, the property of having only confined, nondissipative solutions for any permissible finite value of the energy, a condition which can not be verified in the case of a polynomial nonlinearity.

In the $(\bar{\psi}\psi)_4^2$ model the function E(a) has a maximum at $a=1$. and according to the previous authors, the model can not be stable. The physical interpretation is that, after any perturbation, the wave will either expand (if a increases) or contract (if a decreases) in close correspondence with the two kinds of decays observed by Anderson and Derrick in the scalar case.

On the other hand there are some numerical experiments which show that the PLS of the NLDE with scalar selfcoupling have a great degree of stability and contradict the previous arguments. Recently Alvarez and Carreras (74) studied the interaction dynamics for the PLS $(\bar{\psi}\psi)_2^2$ equation. Their main result can be stated by saying that the solitary waves, not only do not decay but they even keep their identity after the collisions as the solitons do. However, there is in this case some charge and energy transfer as well as some inelasticity. Moreover a very interesting phenomenon occurs in the collision of very slow solitary waves: the production of bound states as quasistationary configurations with a long lifetime (at least 520/m). The corresponding phenomenology is qualitatively similar to the Higgs one. Their conclusion is that, although it is a nonintegrable system, the Dirac equation with scalar selfcoupling in one space dimension has stable PLS.

In another paper, Alvarez and Soler (76) have studied the evolution from Cauchy data far from the PLS in the case of the $(\bar{\psi}\psi)_4^2$ equation. Lacking analytical techniques, they calculated numerically the evolution of the solutions which at the initial time have the form of a Gaussian. They report that, if the intensity of the field is above a certain value, it evolves towards a PLS giving up the extra energy and charge by radiation. In the other case the field is weak and the final state consists in a continuous radiation. This phenomenology is similar to that of the solitons and can only be understood if the Dirac PLS act as stable attractors in certain regions of the space of solutions, a most remarkable property.

They have also studied the evolution of contracted or expanded ψ_a with values of a close to 1. According to the above mentioned arguments the wave should decay either by contraction or by expansion. However they found that this is never the case in (1+1) dimensions nor in (1+3) dimensions, if the frequency is less than Ω_m, the minimum of the function $E(\Omega)$. Quite on the contrary, it goes to a PLS very close to the initial one with a slight change in the frequency and without preserving the initial shape. Curiously enough, the field radiates away negative energy if it is necessary to adjust itself to the final state, a process which is allowed by the nonpositivity of the energy density. This result clearly contradicts the claim by Bogolubsky.

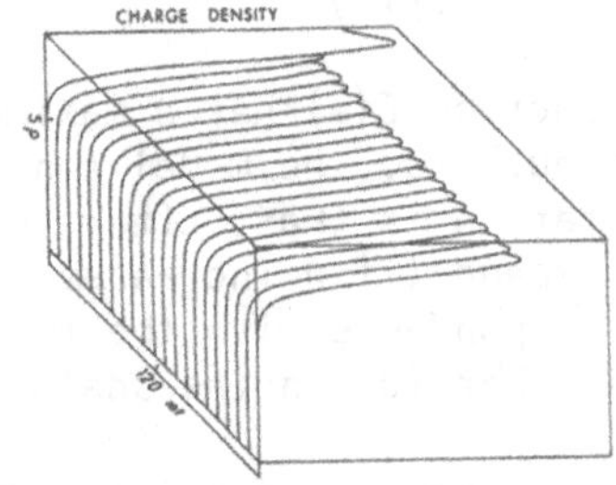

Figure 3. Charge density evolution from an expanded PLS of the form (24) with $a = 1.1$, $\Omega = 0.700$. The final state has more energy and less charge than the initial one and is another PLS with $\Omega = 0.7015 \pm 0.0005$

On the other hand, they report that PLS in the region $\Omega > \Omega_m$ are unstable, the final state being either another PLS with $\Omega < \Omega_m$ or $\psi = 0$. This result seems interesting since for any value of the charge Q there are two PLS with Ω greater and less than Ω_m respectively, one in each of the two branches of the curve $E(\Omega)$. PLS in the lower energy branch have $\Omega < \Omega_m$ and are stable while those in the high energy one have $\Omega > \Omega_m$ and are unstable. However, they found numerically that this unstability is very mild and does not resemble the explosive behaviour of other cases as that of the scalar PLS. This suggests that, if the variation of the field is written as

$$\delta\psi = (\eta e^{\Omega t} + \xi^* e^{\Omega^* t})e^{-i\omega t} \tag{26}$$

it does not correspond to an eigenvalue with $\mathrm{RE}\Omega \neq 0$, but to degenerate eigenvalues with $\mathrm{RE}\Omega = 0$. They have also found some mathematical arguments which indicate that this is indeed the case (75). The problem of the stability being a very important one it seems necessary to clarify this point and to reconcile the nonpositivity of $\delta^2 E$ with the results of the numerical experiments. In fact, there are some arguments which indicate that the condition $\delta^2 E > 0$, although suitable for scalars is not adeuqate in the case of spinors (70, see specially section 3b).

This condition arises from the generalization to field theory of a well known theorem of Classical Mechanics of discrete systems due to Lagrange which states that an equilibrium position corresponding to a minimum of the potential is stable and of the converse theorem at tributed to Lyapunov (93). However, they are only valid in the case of Lagrangian functions which are quadratic in the velocities for which the kinetic energy can not be negative. It is easy to understand why. Let us consider a system whose nonstandard Lagrangian is linear in the velocities (94)

$$L(q_k,\dot{q}_k) = \Sigma A_k(q)\dot{q}_k - V(q) \tag{27}$$

The Euler-Lagrange equations of motion are in this case

$$\begin{aligned} \Sigma\eta_{jk}\,\dot{q}_k &= \frac{\partial V}{\partial q_j}, \quad j = 1\ldots n \\ \eta_{jk} &= \frac{\partial A_k}{\partial q_j} - \frac{\partial A_j}{\partial q_k} \end{aligned} \tag{28}$$

If the antisymmetric matrix has an inverse η^{ik} (which occurs only if n is even) it is possible to determine the velocities as functions of the coordinates

$$\dot{q}_j = \Sigma\eta^{jk}\,\frac{\partial V}{\partial q_k} = \{q_k,V\}^* \tag{29}$$

where $\{\ \}^*$ is a generalized Poisson Bracket. As we see V acts as the generator of the time evolution, coincides with the Jacobi integral and is therefore a constant of the motion. There follows a consequence which may seem surprising: The maxima of the potential are stable equilibrium points or, in other words, the condition $\delta^2E>0$ does not hold in stable equilibrium points. It is very often stated in the textbooks that the fact that a field equation is first or second order is irrelevant since we may always put all of them in first order after the introduction of suitable suplementary functions. It is easy to show however that this can not be the case. For instance, the KG equation is usually written as a first order equation in two different ways, by defining a two component field $(\phi,\partial_o\phi)$ and by the Kemmer formalism. The first method leads to an equation which is not Lorentz invariant while the second one to a non hyperbolic system, any Cauchy surface being characteristic.

The relation between the condition $\delta^2E>0$ and the order of the field equation can be understood by the association of a dynamical system to the expansions of the PLS in the following way. Let us consider the $(\phi^*\phi)^2_2$ equation and the expansions

$$\phi_a = a^{-3/2}\,\phi(\vec{r}/a,t) \tag{30}$$

in which the scale parameter depends on the time $a = a(t)$. After substitution in the Lagrangian density and integration over the space variables we find (95)

$$L(a,\dot{a}) = \int L(\phi_a, \partial_\mu \phi_a)\, d^3\vec{r} = A\,\frac{\dot{a}^2}{a^2} - B - \frac{C}{a^2} + \frac{D}{a^3} = T - V \tag{31}$$

where B,C, and D are positive integrals. The function $L(a,\dot{a})$ can be considered as the Lagrangian of a dynamical system in the variable a. The point $a = 1$ is an equilibrium position corresponding to the fact that for this value there is a PLS. However, being a maximum of V, it is unstable and, after any perturbation, a will either increase (and the wave will expand in the dissipative decay mode) or decrease (and the wave will contract in the singular one), in correspondence with the unstability of the theory.

The spinor case $(\bar{\psi}\psi)^2_4$ is quite different. The corresponding Lagrangian es then

$$L(a,\dot{a}) = A\,\frac{\dot{a}}{a} - B - \frac{C}{a} + \frac{D}{a^3} = T - V \tag{32}$$

with A,B,C,D being integrals as before. This function, however does not define a dynamical system as it can be written as

$$L = \frac{d}{dt}\,(A \ln a) - V(a) \tag{33}$$

the Euler-Lagrange equations being $\frac{dV}{da} = 0$, whose only solution is $a = 1$, the maximum of the potential. We do not find therefore the tendency of a to increase or decrease as in the scalar case. It can be shown that the same thing happens in the case of more general distortions as

$$\psi \rightarrow (f_1 f_2 f_3)^{1/2}\,\psi(f_1 x, f_2 y, f_3 z, t) \tag{34}$$

where f_k are functions of several parameters, for instance $f_k = f_k(\dot{a},b)$.

The gives a clue about the solution to the discrepancy and indicates that the condition $\delta^2 E>0$ can not be used to study the stability of the Dirac PLS, as it was also stated by Derrick (70). Werle points out that even if this is the case when one has only Dirac fields, the introduction of other fields, as the electromagnetic one, might cause the system to go to other states with less energy so that the energetic stability criterium would become valid in the presence of photons or scalars. It is certainly interesting to consider this argument from which it follows that the cases of logarithmic or fractional nonlinearities should be carefully studied.

To conclude this section we could say that the stability of Dirac PLS is a complex and very relevant problem in which new interesting developments probably will arise.

VI. A MODEL OF EXTENDED PARTICLES WITH CONFINED CONSTITUENTS

The results of the preceding section suggest that the spinorial PLS may be useful to describe real extended particles. As all the objects of this kind existing in nature seem to be bound states of confined constituents, it is natural to look for a model in which all the PLS are bound states of several fields and in which there are no one field solitary waves, that is, in which only one field does not vanish. In such a model the basic fields would play a role similar to that of the quarks in the real world. The inexistence of one field PLS would parallel the impossibility of finding an isolated quark.

It turns out that such a model has recently been proposed (52,53). In it all the PLS are bound states of a system of six Dirac fields $\psi_k, \phi_k, k=1,2,3$, the ψ_k being charge conjugate to the ϕ_k. The Lagrangian density has the form

$$L = L_1 + L_2 + L_3 \tag{35}$$

where

$$L_1 = \sum^{3} \{ L_D(\psi_k) + L_{\bar{D}}(\phi_k) \} \tag{36a}$$

L_D being the usual Dirac Lagrangian density and $L_{\bar{D}}$ the samething except for a change in the sign of the derivative terms,

$$L_2 = \lambda/3(A^2 + B^2 + 4AB) \ , \ \lambda > 0 \tag{36b}$$

$$A = \Sigma\bar{\psi}_k\psi_k \quad ; \quad B = \Sigma\bar{\phi}_j\phi_j$$

and

$$L_3 = \lambda' \sum_{i<j} (\chi_{ij}\ \chi_{ij})^2$$

$$\chi_{ij} = \chi_i - \chi_j \quad ; \quad \chi_i = \psi_i + \gamma^5\phi_i \tag{36c}$$

L_1 gives the linear part of the field equations. Because of the mentioned change of sign, the conserved current is

$$j^\mu = \sum^{3} (\bar{\psi}_k\gamma^\mu\psi_k - \bar{\phi}_k\gamma^\mu\phi_k) \tag{37}$$

which justifies the interpretation of ψ and ϕ as mutually charge conjugate. L_2 provides the nonlinear attractive forces between the constituents. L_3 is an apparently complicated term which happens to vanish in the physically relevant PLS. Without going into the detaisl, which can be found in reference (52), the main properties of the model are the following:

i.- The fields of the constituents are confined because the set of the solutions do not contain one field solitary waves, that is, with only one nonvanishing field. Consequently, they can not manifest themselves as PLS of their own, although they appear as constituents of composite systems. This is due to the fact that each field acts as a source of the rest of them, so that they are never free.

ii.- All the PLS have zero triality, because the difference between the numbers of nonvanishing ψ's and ϕ's is always multiple of 3. In self explaining notation the PLS are of the types (3ψ), (3ϕ), (ψ,ϕ), $(2\psi,2\phi)$ and $(3\psi,3\phi)$. There are no PLS such as (ψ), (2ψ), $(3\psi,\phi)$ and so on. Associating the ψ's to quarks and the ϕ's to antiquarks, we find the states existing in nature and only those.

iii.- In the PLS the nonvanishing fields take the values $\psi_k = \psi_s$ $\phi_k = \gamma^2\gamma^5\psi_s^*$ where ψ_s is the solution of the Soler model. If $m = 390$ Mev and $\lambda m^2 \doteq 23$, the masses of the (3ψ) baryon solution and the $(n\psi, n\phi), n = 1,2,3$ meson solutions are 1,200 Mev and 800 Mev respectively. Introducing spin dependent forces in L_2, the meson solutions separate in a spin one of 800 Mev and a spin zero of 582 Mev.

iv.- Although the interaction between the constituents, is very strong the binding energy is very small or, in other words, they are very loosely bound.

As these properties parallel those of the hadrons and quarks, this model seem promising and worth of careful consideration.

REFERENCES

1. Scott, A., Chu, F., and McLaughlin, D.: 1973, Proc. IEEE 61, pp. 1443-1483
2. Calogero, F.: in Nonlinear Problems in Theoretical Physics, Lecture Notes in Physics 98, A.F. Rañada editor, Springer-Verlag 1979, pp. 29-34
3. Degasperis, A.: 1979, in reference 2, pp. 35-90
4. Ivanenko, D.: 1938, Soviet Physics 13, pp. 141-149
5. Ivanenko, D.: in Einstein Centenary Jubilee Volume, Italian Academy of Sciences, 1979.
6. Rodichev, V.: 1961, Sov. Phys. JETP 13, pp. 1029-1033
7. Weyl, H.: 1950, Phys. Rev. 77, pp. 699-701
8. Utiyama, R.: 1956, Phys. Rev. 101, pp. 1596-1607
9. Sciama, D.W.: Festschrift for Infeld, Pergamon Press, 1960, pp. 415-439
10. Kibble, T.W.B.: 1960, J. Math. Phys. 2, pp. 212-221
11. Rañada, A.F. and Soler, M.: 1972, J. Math. Phys. 13, pp.671-676
12. Hehl, F.W., von der Heyde, P., and Kerlick, G.D.: 1976, Rev. Mod. Phys. 48, pp. 393-416
13. Heisenberg, W.: 1953, Physica 19, pp. 897-908
14. Heisenberg, W.: 1957, Rev. Mod. Phys. 29, pp. 269-278

15. Gross, D.J., and Neveu, A.: 1974, Phys. Rev. D 10, pp. 3235-3253
16. Rembiesa, P.: 1981, Phys. Rev. D 24, pp. 1647-1661
17. Rajaraman, R.: 1975, Phys. Rep. 21, pp. 227-313
18. Jackiw, R.: 1975, Rev. Mod. Phys. 49, pp. 681-706
19. Neveu, A.: 1977, Rep. Prog. Phys. 40, pp. 709-728
20. Makhankov, V.G.: 1978, Phys. Rep. 35, pp. 1-128
21. Friedberg, R. and Lee, T.D.: 1977, Phys. Rev. D 15, pp. 1694-1711; 16, pp. 1096-1118; 1978, 18, pp. 2625-1631.
22. Dashen, R.F., Hasslacher, B., and Neveu, A.: 1974, Phys. Rev. D 10, pp. 4114-4142; 11, pp. 3424-3450; 1976, 12, pp. 2443-2458.
23. Gervais, J.L., Jevicki, A., and Sakita, B.: 1975, Phys. Rev. D 12, pp. 1038-1051
24. Coleman, S.: 1975, Phys. Rev. D 11, pp. 2088-2097.
25. Faddeev, L.D., and Korepin, V.E.: 1978, Phys. Rep. 42, pp.1-87
26. Gervais, J.L., and Neveu, A.: 1976, Phys. Rep. 23, pp. 234-374
27. de Vega, H.: Lectures at the Symposium on "Integrable Quantum Field Theories", Tvarmine, Finland, 1981
28. Abraham, M.: 1903, Ann. Phys. 10, p. 105; 1904, Phyz. Z. 5, p. 576
29. Lorentz, H.A.: 1904, Enzyclop. der Math. Wiss (Teubner, Leipzig) IV, p. 14
30. Rohrlich, F.: Classical Charged Particles, Addison-Wesley, Reading 1965
31. Rosen, N.: 1939, Phys. Rev. 55, pp. 94-101
32. Rosen, N., and Menius, A.: 1942, Phys. Rev. 62, pp. 436-437
33. Finkelstein, R., Lelevier, R., and Ruderman, M.: 1951, Phys. Rev. 83, pp. 326-332
34. Finkelstein, R., Fronsdal, C.F., and Kaus, P.: 1956, Phys. Rev. 103, pp. 1571-1579
35. Wakano, T.: 1966, Prog. Theor. Phys. (Kyoto) 35, pp. 1117-1132.
36. Soler, M.: 1970, Phys. Rev. D1, pp. 2766-2769
37. Soler, M.: 1973, Phys. Rev. D8, pp. 3424-3429
38. Rañada, A.F., and Soler, M.: 1973, Phys. Rev. D8, pp. 3430-3433
39. Rañada, A.F., Usón, J., and Vázquez, L.: 1980, Phys. Rev. D22, pp. 2422-2424
40. Rañada, A.F., Rañada, M.F., Soler, M., and Vázquez, L.: 1974, Phys. Rev. D10, pp. 517-525
41. Rañada, A.F.: 1978, J. Phys.A: Math. Gen. 11, pp. 341-346
42. Rañada, A.F., and Vázquez, L.: 1976, Prog. Theor. Phys.(Kyoto) 56, pp. 311-323
43. García, L. and Rañada, A.F.: 1980, Prog. Theor. Phys.(Kyoto) 64, pp. 671-693
44. Rañada, A.F. and Rañada, M.F.: 1977, J. Math. Phys. 18, pp. 2427-2431

45. García, L. and Usón, J.: 1980, Found. of Phys. 10, pp.137-149
46. Rañada, A.F.: 1977, Int. J. Theor. Phys. 16, pp. 795-812
47. Rañada, A.F. and Vázquez, L.: 1979, Phys. Rev. D19, pp. 493-495
48. Rañada, A.F. and Vázquez, L.: 1980, An. Fis. 76, pp. 139-141
49. Rañada, A.F. and Usón, J.: 1980, J. Math. Phys. 21, pp. 1205-1209
50. Rañada, A.F. and Usón, J.: 1981, J. Math. Phys. 22, pp. 2533-2538
51. Rañada, A.F., to appear in Old and New Questions in Physics, Essays in honour of W. Yourgrau, A. van der Merwe editor, Plenum Co.
52. Rañada, A.F. and Rañada, M.F., Preprint Universidad Complutense, 1982
53. Rañada, A.F., to appear in Dynamical Systems and Microphysics A. Avez, A. Blaquiere and A. Marzollo editors, Academic Press
54. Mielke, E.W.: 1981, to appear in Fortschr. Phys.
55. Rosen, N. and Rosenstock, H.B.: 1952, Phys. Rev. 85, pp. 257-259
56. Rosen, G.: 1965, J. Math. Phys.6, pp. 1269-1272; 1966, 7, p.2071; 1967, 8, pp. 573-581
57. Chu, S.Y. and Kaus, P.: 1977, Phys. Rev. D16, pp. 503-510
58. Rafelski, J.: 1977, Phys. Lett.B66, pp. 262-266
59. Takahashi, K.: 1979, J. Math. Phys. 20, pp. 1232-1238
60. Barut, A.O.: 1979, in reference 2, pp. 1-14
61. Werle, J.: 1977, Phys. Lett 71B, pp. 357-359; 71B, pp. 367-368
62. Werle, J.: 1981, Acta Physica Polonica B12, pp. 601-616
63. Simonov, Yu. A., and Tjon, J.A.: 1979, Phys. Lett. B85, pp. 380-384
64. Morris, T.F.: 1978, Phys. Lett.B76, pp. 337-339; 78, pp. 87-89
65. Skyrme, T.H.: 1961, Proc. Roy. Soc.260, p. 127; 262, p. 236
66. Thirring, W.E.: 1958, Ann. of Phys. 3, p. 91
67. Mikhailov, A.V.: 1976, JETP Letters 23, p. 320
68. Kaup, D.J. and Newell, A.C.: 1977, Lett. Nuovo Cimento 20, pp. 325-331
69. Hobart, R.H.: 1963, Proc. Phys. Soc. 82, pp. 201-203
70. Derrick, G.N.: 1964, J. Math. Phys. 5, pp. 1252
71. Soler, M., Memorias de la Academia de Ciencias, Madrid 1975
72. Rybakov, Yu.P., Messenger of the Moscow State University, ser III, n.1,72, in russian 1966
73. Bogolubsky, I.L.: 1979, Phys. Lett.A 73, pp. 87-90
74. Alvarez, A. and Carreras, B.: 1981, Phys. Lett.A 86, pp.327-332
75. Alvarez, A.: 1982, preprint Universidad Complutense, to be published.

76. Alvarez, A. and Soler, M.: 1982, Preprint Univ. Complutense, to be publised.
77. Moore, S.: 1982, Preprint Univ. de los Andes, Bogotá, to be published
78. Chadam, J.M.: 1972, J. Math. Phys. 13, pp. 597-604
79. Chadam, J.M.: 1973, J. Func. Analysis 13, pp. 173-184
80. Chadam, J.M. and Glassey, R.T.: 1974, Arch. Rat. Mech. Anal. 54, pp. 223-237
81. Chadam, M.M. and Glassey, R.T.: 1974, Proc. Amer. Math. Soc. 43, pp. 373-378
82. Glassey, R.T. and Strauss, W.: 1979, J. Math. Phys. 20, pp. 454-460
83. Vázquez, L.: 1977, J. Phys. A: Math. Gen. 10, pp. 1361-1368
84. Vázquez, L.: 1977, J. Math. Phys. 18, pp. 1343-1347
85. Vázquez, L.: 1977, Lett. N. Cimento 19, pp. 561-564
86. Galindo, A.: 1977, Lett. N. Cimento 20, pp. 210-212
87. Steeb, W.H.: 1980, J. Math. Phys. 21, pp. 1656-1658
88. Steeb, W.H., Erig, W. and Strauss, W.: 1981, J. Math. Phys. 22, pp. 970-977; 1982; 23, pp. 145-153
89. Meyer, U.: 1965, Berkhout CERN 65-24, vol.2
90. García, L.: 1982, private comunication, to be published.
91. Anderson, D.L.T. and Derrick, G.: 1971, J. Math. Phys. 11, pp. 1336-1346
92. Vázquez, L.: 1982, to appear in Il Nuovo Cimento B.
93. Meirovotch, Methods of Analytical Dynamics, McGraw-Hill 1970
94. Sudarshan, E.C.G. and Mukhunda, N., Classical Mechanics: A Modern Perspective, John Wiley, 1974
95. Laorga, R.: 1981, Graduate thesis, to be published

CONSTRAINT HAMILTONIAN DYNAMICS OF DIRECTLY INTERACTING RELATIVISTIC POINT PARTICLES*

I. T. Todorov
Institute of Nuclear Research and Nuclear Energy,
Bulgarian Academy of Sciences, Sofia 1113, Bulgaria

*Lecture presented at the Summer School in Mathematical Physics, Bogazici University, Bebek, Istanbul, Turkey, August 1979.

CONTENTS

A. O. Barut (ed.), Quantum Theory, Groups, Fields and Particles, 293–324.

INTRODUCTION

The study of relativistic single particle dynamics was started in the wake of the special theory of relativity [1,2]. However, the understanding of N particle interactions (especially for $N \geq 3$) in the framework of relativistic Hamiltonian mechanics of a system with a finite member of degrees of freedom is not yet complete*).

Without trying to follow all the intricate twists in the theoretical developments in this area let us mention a few landmarks.

Dirac's statement of the problem [4] seemed to have been answered negatively by the so called "no interaction theorem" [5,6]. The work of Wigner and Van Dam (see [7] and references therein) provided a consistent description of relativistic particle interaction at a price of giving up the Hamiltonian picture. The quasipotential approach [8-11] served as an useful framework for studying relativistic two-body problems arising in quantum field theory; however, it did not address the general problem of constructing a consistent N-particle Hamiltonian dynamics. The same can be said about the infinite-component wave equations technique [12,13]. S.N. Sokolow [16] was (to the best of our knowledge) the first to work out an off energy shell (not manifestly covariant) formulation of the (quantum) relativistic N-body problem which satisfies the cluster decomposition property for $N \geq 3$.

Dirac's formulation [17] of modern constraint Hamiltonian dynamics (for later development see [18,19] opened the way for a manifestly covariant approach to the relativistic N-particle problem**)[20-23]. Another interesting path in

---

*As pointed out by C. Piron the source of the difficulty--the ambiguity in relating the time parameters of different world lines--was already recognized by H.G. Wells in the last century [3].

**It is ironic to realize that neither of the authors of these papers seems to have been aware of his predecessors (F. Rohrlich [24] was the first to indicate the close relationships between two of the above references.) Of course, the content of the various papers is not quite identical; for instance, the Lagrangian approach of ref. [21] is much more restrictive than the Hamiltonian phase space formulation of ref. [22]. On the other hand, the first two articles of ref. [23] do not seem to contain anything new compared to the earlier work [22].

relativistic particle dynamics is being followed by C. Piron and his collaborators [25]. A lucid discussion of some of the recent work is presented in the review article by Leutwyler and Stern [27].

The present lectures are devoted to the constraing Hamiltonian approach of ref. [22] and its recent development.

In order to avoid any confusion on the basic premises and expectations, let us start by stating here our beliefs and understanding of the physical picture of relativistic particles' interaction and of the place of the Hamiltonian action-at-a-distance approach.

We believe that the true physical interaction between particles is always mediated by a field. The isolation, say, of a 2-body problem is only possible, physically, if the radiation effects can be neglected (which is indeed the case in electrodynamics, if we are interested in relativistic effects of order up to α^4 in the fine structure constant). Even then a retarded interaction (cp.[7] is presumably more realistic, as far as the details of particle trajectories at finite times are concerned. However, a (mathematically coherent) Hamiltonian formalism offers great computational simplifications, while still providing reliable asymptotic results (e.g., in evaluating the scattering matrix--both classical and quantum--as well as the quantum mechanical bound state energy levels). It turns out that there is a price to pay for these (technical) advantages. As it is demonstrated in ref.[28] particle world lines (say, in the 2-particle case) are "gauge dependent", in other words, they depend, in the presence of interaction, on the choice of the equal time surface. (It follows from the results of Wigner et al. [7] that this is not a consequence of the relativistic nature of interaction alone, but is rather enforced by our insistence in describing the dynamics in terms of a single-time Hamiltonian formalism.) We show, however, that the relative 2-particle motion and the (classical) scattering matrix are gauge independent.

The quantized form of the 2-particle constraint Hamiltonian dynamics is demonstrated to coincide with the local version [10, 11] of the relativistic quasipotential equation. This relation is used to determine the 2-particle "potential" for a given field theoretic interaction. The case of electromagnetic interaction is dealt with explicitly.

For the organization of material we refer to the rather detailed table of contents.

Most of the lectures review (with some minor changes of

emphasis) the work [22,28] of the last 4 years. Secs. 4A and 5B,C recall earlier work [11,40] on the quasipotential equation. Some new points can be found in Secs. 2B, 4B and 5A. A couple of open problems are discussed in Sec. 6.

1. HAMILTONIAN EQUATIONS OF MOTION AS A CONSEQUENCE OF RELATIVISTIC PHASE SPACE GEMOETRY

1A. Dual role of the mass-shell constraint for a single relativistic particle

The covariant form of Planck's variational principle

$$\delta A=0, \quad A=-m\int_{x_i}^{x_f} \sqrt{-dx^2}\ , \quad dx^2=d\underline{x}^2-(dx^0)^2 \tag{1.1}$$

is the most appropriate starting point for the dynamical description of a free relativistic particle of mass m (>0). (Geometrically it says that for the real particle motion the time-line path between the initial point (event) x_i and the final point x_f should have a maximal/Lorentzian/length.) The corresponding Lamgrangian density (for which $A=\int_{t_i}^{t_f} L dt$) is a first degree homogeneous function in the velocity:

$$L=-m\sqrt{-\dot{x}} \qquad (\dot{x}=\dot{\underline{x}}^2-(\dot{x}^0)^2). \tag{1.2}$$

Here $\dot{x}$ is the derivative of x with respect to an arbitrary "time-parameter on the world line. The homogeneity of L implies that the reparametrization invariant canonical momenta

$$p_\mu=\frac{\alpha x}{\alpha \dot{x}^\mu} = mu_\mu, \quad u=\frac{\dot{x}}{\sqrt{-\dot{x}^2}} \tag{1.3}$$

are not independent, but satisfy the mass-shell constraint

$$_{(m)} \equiv \tfrac{1}{2}(m^2+p^2)=0. \tag{1.4}$$

It is precisely the reparametrization (or "gauge") invariance of momenta (and of the equations of motion) that implies the singularity of the Lagrangian, which is expressed by the equation

$$\det\left(\frac{\partial^2 L}{\partial \dot{x}^\mu \partial \dot{x}_\nu}\right) = \det\ (-\dot{x}^2)^{-\frac{1}{2}}\left(\partial^\mu{}_\nu - \frac{\dot{x}^\mu \dot{x}_\nu}{\dot{x}^2}\right) =0. \tag{1.5}$$

Eq. (1.5), on the other hand, signals the existence of at least one constraint.

Furthermore, the canonical Hamiltonian vanishes:

$$H_{can} = p_\mu \dot{x}^\mu - L = \dot{x}^\mu \frac{\partial L}{\partial \dot{x}^\mu} - L = 0. \tag{1.6}$$

Thus, according to the Dirac theory, the total Hamiltonian is proportional to the constraint (1.4):

$$H_{tot}(=H_{can} + \lambda\ _{(m)}) = \lambda\ _{(m)}\ (\approx 0). \tag{1.7}$$

Here λ is a Langrange multiplier, which can be fixed by the choice of the time parameter. The large 8-dimensional phase space Γ of points (p,x) (for a spinless particle) has a natural Poincaré invariant Poisson bracket structure:

$$\{f,g\} = \frac{\partial f}{\partial p_\mu}\frac{\partial g}{\partial x^\mu} - \frac{\partial f}{\partial x^\mu}\frac{\partial g}{\partial p_\mu}\ . \tag{1.8}$$

The time evolution of an arbitrary dynamical variable is defined in terms of its Poisson brackets with the total Hamiltonian; in particular,

$$\dot{x} = \{H_{tot},\ x\} \approx \lambda p, \tag{1.9a}$$

$$\dot{p} = \{H_{tot},\ p\} \approx 0. \tag{1.9b}$$

[The <u>weak equality</u> sign $\approx$ is meant to indicate that the corresponding equation is only valid on the mass shell (1.4) (the constraint (1.4) is taken into account <u>after</u> evaluating the derivatives in the definition of the Poisson bracket).] The Lagrange multiplier can now be determined from (1.9a)(1.4):

$$\lambda = \frac{\sqrt{-\dot{x}^2}}{m}\ (>0). \tag{1.10}$$

Thus the constraint (1.4) not only reduces the number of variables in the large phase space Γ, but also determines the equations of motion for the on-shell dynamical variables.

The freedom in the choice of the time parameter t (or of the Lagrange multiplier) corresponds to the gauge freedom in the general theory of dynamical systems with first class constraints [18,19]. Although $\dot{x}$ depends on the choice of t (and hence, on λ)*, the normalized 4-velocity u (1.3) (and consequently the

---

*For instance, for $t=x^0$, we have $\dot{x}=(1,\underline{v})$ where $\underline{v} = \frac{1}{p^0}\ \underline{p} = (m^2+\underline{p}^2)^{-\frac{1}{2}}$, $\lambda = \frac{1}{m}\sqrt{1-\underline{v}^2}$ (this choice could be termed the "Newton-Wigner gauge" – cf. [29]. The covariant gauge $\zeta = -\frac{px}{m^2}$ (related

particle world line) are gauge independent.

We shall fix the direction of the time axis by requiring

$$\dot{x}^0>0; \qquad (1.11)$$

this implies energy positivity (because of (1.3)). Thus p belongs to the upper sheet Ω_m^+ of the mass hyperboloid (1.4)

$$\Omega_m^+ = \{p\varepsilon\ ^4;\ p^0=\sqrt{m^2+\underline{p}^2}\}\ . \qquad 1.12)$$

A dynamical variable that has (weakly) vanishing Poisson brackets with the constraints (in our case with (m) (1.4) is called an invariant observable (cf. [18]). It follows from (1.7) that each invariant observable is a fortiori a constant of the motion. It can be demonstrated that every single-particle observable is a function of the generators of the Poincaré group. This statement is also valid for a spinning particle. Indeed, the physical phase space of a relativistic particle (which is homeomorphic to $R^6 \times S^2$ in the case of a spinning particle) can be realized as an orbit [30] in the coadjoint representation of the Poincaré group [31].

We shall use the above phase-space geometric picture of the single particle dynamics as a guide in the subsequent study of relativistic particle interaction.

1B. Generalized N-particle mass shell

Our objective is to provide a phase space formulation of relativistic particle interaction, for which the laws of motion would follow from the geometry of the space -- much the same as in the 1-particle case. We shall deal, for the sake of simplicity, with spinless particles (although the inclusion of spin is not difficult--see [22] and Sec. 5C below).

A relativistic interacting particle does not belong (in general) to its mass shell--we know this even experimentally: the effective proton and neutron masses in a nucleus are smaller than the corresponding free particle masses. We shall define, following [22], the generalized mass shell $M(=M_N)$ of an N-particle system as a TN dimensional submanifold of the 8N dimensional "large phase space" Γ^N, singled out by N constraints of the type

---

to the proper time $m\zeta$) corresponds to the equation of motion $\dot{x}=p$ and Lagrange multiplier $\lambda=1$.

$$_k \equiv \tfrac{1}{2}(m_k^2 + p_k^2) + \phi_k \approx 0, \quad k-2,\ldots,N, \tag{1.13}$$

along with the inequalities

$$P^0 > |\underline{P}|, \quad Pp_k \; 0, < k=1,\ldots,N \quad (P=p_1+\ldots+p_n). \tag{1.14}$$

(Similar inequalities should also hold for each subsystem, provided the remaining particles are removed to spatial infinity.) The deviations ϕ_k from the free particle mass shell should satisfy the following conditions.

(i) Poincaré invariance: the functions ϕ_k only depend on the scalar products of the translationnally invariant vectors $x_{ij} = x_i - x_j$ and p_i. This implies the invariance of M with respect to the (diagonal) action of the Poincaré group with generators

$$P = p_1+\ldots+P_n \tag{1.15a}$$

$$L = p_1 \wedge x_1 + \ldots + p_n \wedge x_n \quad [P\wedge^x)\mu\nu = p_\mu x_\nu - p_\nu x_\mu]. \tag{1.15b}$$

(ii) Compatibility; the constraints (1.13) are first class:

$$\{ _i, _j \} = 0, \quad i,j = 1,\ldots,N. \tag{1.16}$$

(iii) The constraints can be solved with respect to particle energies: in every point of M

$$d \equiv \det\left(n \frac{\partial _i}{\partial _j}\right) > 0 \quad \text{for } n_0 (=-n^0) > |\underline{n}| . \tag{1.17}$$

(iv) Separability (or cluster decomposition property: if the 3-dimensional distances (in some reference frame) of the particles $i_1,\ldots,i_\nu (\nu<N)$ to all others $(j_1,\ldots,j_{n-\nu})$ tend to infinity, then all functions $\phi_1,\ldots,\phi_n$ have finite limits, such that ϕ_{j_1} do not depend on x_{i_k} and ϕ_{i_k} do not depend on x_{j_1} $(k=1,\ldots,\nu;$ $1=1,\ldots,N-\nu)$; moreover

$$\lim_{\substack{|x_{ij}| \to \infty \\ j=1,\ldots,i-1,i+1,\ldots,N}} \phi_i = 0 \quad (i-1,\ldots,N). \tag{1.18}$$

All ν-particle interactions $(\nu=2,\ldots,N-1)$, satisrying the above conditions, should be reproduced in this limit.

Note that the rising (or "confining") potentials of the type studied, e.g., in [27] can also be incorporated; only the separation of the kinetic term $\frac{1}{2}(m_k^2+p_k^2)$ becomes in such a case ambiguous.

(v) Restriction on admissible singularities: Eqs. (1.13) define a (smooth) submanifold of Γ^n without a boundary only if ϕ_k are smooth functions in the neighbourhood of M. From the point of view of physical applications, however, it is desirable to allow for some singular potentials. Singlar repulsive interactions (corresponding to $\phi_k \to \infty$) can be treated by excluding from M all points for which at least one of the ϕ_k is infinite. (A physical example of this type--the collision of elastic balls--is considered in ref. [28(b)].) We shall also allow weak singularities of the type $\phi_k \to -\infty$, including the Coulomb interaction.

1C. Two-particle interactions

Conditions (i)-(v) admit a wide class of 2-particle interactions. In order to display them we introduce appropriate kinematical variables.

We define the relative momentum p of a pair of particles as a linear combination of the particles' momenta p_1 and p_2 of the type

$$p=\mu_2 p_1-\mu_1 p_2, \quad \mu_1+\mu_2=1 \quad (\mu_1>0,\ \mu_2>0) \tag{1.19}$$

such that p is orthogonal to the total momentum $P=p_1+p_2$ on the surface

$$\Xi \; \tfrac{1}{2}(m_1^2+p_1^2-m_2^2-p_2^2)=0. \tag{1.20}$$

Assuming in addition that μ_k have zero Poisson brackets with the relative coordinate

$$x = x_1-x_2, \tag{1.21}$$

we find

$$\mu_1-\mu_2 = \frac{\mu_1^2-\mu_2^2}{\omega^2} \quad \text{or} \quad \mu_{1,2} = \frac{1}{2} \pm \frac{1}{2\omega^2}(m_1^2-m_2^2), \tag{1.22}$$

where

$$\omega^2= -P^2 \ (>0). \tag{1.23}$$

(The positivity of μ_k on the surface (1.20) is a consequence of the energy-positivity condition (1.14).)

The vectors p, x, and P are complemented to a canonical set by the centre of mass coordinate.

$$X_c = \frac{\partial}{\partial P}\,[(\mu_1 P+p)x + (\mu_2 P-p)x_2] = X + \frac{\mu_1\,\mu_2}{\omega^2}(xP)P, \qquad (1.24a)$$

where

$$X = \mu_1 x_1 + \mu_2 x_2, \qquad (1.24b)$$

the only non-vanishing Poisson brackets among these variables being

$$\{p_\nu\ x^\mu\} = \delta^\mu_\nu\ = \{P_\nu, X^\mu\}. \qquad (1.25)$$

The compatibility condition (1.16) assumes (for N=2) a more tractable form in the variables

$$_1 - {}_2 = \ +D(\approx 0), \quad =pP, \quad D=\phi_1-\phi_2, \qquad (1.26)$$

$$H=\mu_2\ {}_1+\mu_1\ {}_2=\tfrac{1}{2}[p^2-b^2(\omega)]+\phi, \quad \phi=\mu_2\phi_1+\mu_1\phi_1, \qquad (1.27)$$

where

$$b^2(\omega)= \frac{1}{4\omega^2}\,[\omega^2-2(m_1^2+m_2^2)\omega^2+(m_1^2-m_2^2)^2] \qquad (1.28)$$

is the value of the relative momentum square on the free 2-particle mass shell. Inserting $_1=H+\mu_1(\ +D)$, $_2=H-\mu_2(\ +D)$ in Eq. (1.16), we find

$$\{\ _1,\ _2\}=\{\ +D,H\}=P\frac{\partial\phi}{\partial x} - P\frac{\partial D}{\partial x} +\{D,\phi\}\approx 0. \qquad (1.29)$$

The simple solution

$$D=0=P\,\frac{\partial\phi}{\partial x} \qquad (1.30)$$

of this equation involves enough freedom to include the quasi-potential equations studied in [9-11] (in the spinless case). The last equation (1.30) is satisfied if ϕ depends only on the scalar products of the momenta and the proejction x_1 of the relative coordinate on the 3-plane orthogonal to P:

$$x_1=x\ =x-\frac{xP}{P^2}\,P \quad (\ ^\mu{}_\nu=\delta^\mu{}_\nu+\hat{P}^\mu\hat{P}_\nu, \hat{P}^\mu=\frac{P^\mu}{\omega}, \hat{P}^2=-1) \qquad (1.31a)$$

$$x_1^2=x^2+(x\hat{P})^2\equiv r^2\geq 0. \qquad (1.31b)$$

In order to define a single time Hamiltonian we have to

choose an equal time surface. We shall restrict our attention to equal-time planes of the type

$$nx=0 \ , \text{ where } n_0(=-n^0)>|\underline{n}|. \quad \{\ _k,n\}=0. \tag{1.32}$$

The most general linear combination of the constraints, whose Boisson bracket with nx vanishes (weakly) is

$$\lambda H^{(n)}=\lambda[H-\frac{1}{nP}(np+n\frac{\partial\phi}{\partial p})\ ](\approx 0). \tag{1.33}$$

Assuming that ϕ is independent of pP, or, in other words, that

$$P\frac{\partial\phi}{\partial p}=0 \tag{1.34}$$

(cf.[22]), we conclude that the Hamiltonian preserving the Yukawa-Markov condition

$$-\hat{P}x\approx 0 \tag{1.35}$$

is proportional to H (1.27).

2. RELATIVISTIC ADDITION OF INTERACTIONS

In the case of three or more particles the construction of functions $_k$ satisfying conditions (i)-(v) is a much more complicated problem. The main difficulty appears in trying to combine compatibility and cluster decomposition. (If we drop condition (ii), the remaining conditions are fulfillfed by a simple addition of interactions--see, e.g., [32].) Mutze [33] has argued (in a different formulation of the relativistic many body problem) that only N-particle forces can appear among N interacting particles (if one of the particles is taken to infinity, then the remaining N-1 particles become free). Thus, we are faced with a non-trivial problem: is there any non-trivial solution for $N\geq 3$ of the separability requirement (iv) (along with conditions (i)-(iii))? We shall demonstrate (at least formally), applying to our approach Sokolov's idea about a relativistic addition of itneractions (see ref. [16]), that assumptions (i)-(v) are indeed compatible also for $N\geq 3$. Regrettably, the subsequent argument is not very constructive, since it presumes the knowledge of the Moeller wave operators.

2A. The Classical Wave Operators

We shall make use of notation and techniques of ref. [34].

To each smooth function f(p,q) in phase space we make correspond the vector field (first order differential operator)

$$L_f = \frac{\partial f}{\partial p}\frac{\partial}{\partial q} - \frac{\partial f}{\partial q}\frac{\partial}{\partial p}\ , \qquad (2.1)$$

called the Liouville operator (generated by f). The operators (2.1) give rise to a representation of the Lie algebra of Poisson brackets:

$$[L_f, L_g] = L_{\{f,g\}}. \qquad (2.2)$$

Let H_0 and H be a pair of functions ("Hamiltonians"), for which the strong limits

$$W_{\pm} = W_{\pm}(H, H\ ) = s\ \lim_{t\to\pm\infty} e^{tL_H} e^{-tL_{H_0}} \qquad (2.3)$$

exist with respect to the L^1-norm

$$\|f\| = \ _{\Gamma} |f(p,q)| \ \frac{dpdq}{2\pi}\ . \qquad (2.4)$$

The operators $W_{\pm}$ are the classical wave (or Moeller) operators. (We note that the existence of such operators is established for a wider class of quantum mechanical Hamiltonians than in the classical theory--see, e.g., the discussion of existence of global solutions of the classical equations of motion in [35].)

The operators $W_{\pm}$ have the intertwining property

$$e^{tL_H} W_{\pm} = W_{\pm} e^{tL_{H_0}}\ , \text{ so that } L_H W_{\pm} = W_{\pm} L_{H_0} \qquad (2.5)$$

(It is an immediate consequence of their definition).

If, on the other hand,   is another function on phase space such that $\{H,\ \} = 0 = \{H_0,\ \}$, then we observe that

$$W_{\pm}(H, H_0) = W_{\pm}(\lambda H + \ , \lambda H_0 + \ ). \qquad (2.6)$$

(It is only this special case of the Birman-Kato invariance principle that will be used in the sequel.)

If the Hamiltonians H and H_0 do not lead to a bounded motion (or to a bound state in the quantum mechanical context), then

$$W_{\pm} H_0 = H. \qquad (2.7)$$

This is true, in particular, if H_0 is a free Hamiltonian while

$H=H_0+\phi$, where ϕ is a repulsive potential. In general, if $H-H_0=\phi$, we have

$$W_{\pm}H_0=\lim_{t\to\pm\infty} e^{tL_H}(H-\phi)=H-\lim_{t\to\pm\infty}\phi(t).$$

In the 2-particle problem of Sec. 1C, if $\phi=\phi(r,w)$ where $r=\sqrt{x_1^2}$ - see (1.31), then $\phi(t)=\phi(r(t),\omega)$ (where $r(t)$ is a solution of the equations of motion). The absence of bounded motion means in this case $r(t)\to\infty$ for $t\to\pm\infty$, so that $\phi(t)\to 0$ in view of (1.18).

2B. Three Particle Constraints Satisfying Compatibility and Cluster, in Terms of Classical Wave Operators

For a given pair of 2-particle Hamiltonians

$$H_0=\tfrac{1}{2}(p^2-b^2(\omega)),\quad H=H_0+\phi \tag{2.8}$$

(with no bound-states) and known asymptotic form $\ ^0_k=\frac{1}{2}(m_k^2+p_k^2)$, $k=1,2$ of the functions $\ _k$ (for $r\to\infty$) we can reconstruct the 2-particle constraints for any value of r by applying Eq. (2.7) (and using the fact that $W_\pm$ are linear operators). Indeed, since

$$L_H\ =\{H,\ \}=0=L_{H_0}\ , \tag{2.9}$$

for (defined by (1.20)), we have $W_\pm$ = and hence, according to (2.7) and (2.6),

$$W_\pm\tfrac{1}{2}(m_1^2+p_1^2)=W_\pm(H_0+\mu_1\ )=H+\mu_1\ =\tfrac{1}{2}(m_1^2+p_1^2)+\phi\equiv\ _1, \tag{2.10a}$$

$$W_\pm\tfrac{1}{2}(m_2^2+p_2^2)=W_\pm(H_0-M_2\ )=H-\mu_2\ =\tfrac{1}{2}(m_2^2+p_2^2)+\phi\equiv\ _2. \tag{2.10b}$$

The compatibility condition (1.29) is then a consequence of the trivial relation $\{\ ^0_1,\ ^0_2\}=0$, because

$$W_\pm\{f,g\}=\{W_\pm f,W_\pm g\}. \tag{2.11}$$

We shall demonstrate that this way of deriving the constraints is also applicable to the N-particle case (for $N\geq 3$). In order to keep notation simple, we shall restrict our discussion to 3-particle interaction (generated by 2-body forces).

Let (i,j,k) be any permutation of $(1,2,3)$. We assume that for each pair (i,j) (when k is taken "behind the moon") we are given 2-particle constraints of the above type with interaction functions

$$\phi_{ij}=\phi_{ij}(r_{ij},w_{ij}),\quad\text{where } r_{ij}^2=x_{ij}^2+(x_{ij}\hat{P}_{ij})^2,$$

$$x_{ij}=x_i-x_j,\quad P_{ij}=p_i+p_j=w_{ij}\hat{P}_{ij},\quad w^2_{ij}=-P^2_{ij}(>0) \tag{2.12}$$

which forbid bounded motion. Let further $\lambda_1,\lambda_2,\lambda_3$ be arbitrary positive functions of the dynamcial variables. Then the pair of Hamiltonians

$$H_0^{(\lambda)}=\sum_{k=1}^{3}\frac{\lambda k}{2}(m_k+p_k),\quad H^{(\lambda)}=H_0^{(\lambda)}+(\lambda_1+\lambda_2)\phi_{12}+(\lambda_2+\lambda_3)$$

$$\phi_{23}+(\lambda_1+\lambda_3)\phi_{13} \tag{2.13}$$

agrees with the cluster property in the following sense. If one of the particles, say K, is taken to spatial infinity, so that $\phi_{ik}\to 0$ and $\phi_{jk}\to 0$, then the limit

$$W_{\pm}^{(k)}=\lim_{x_k\to\infty} W_{\pm}(H^{(\lambda)},H_0^{\lambda})$$

satisfies

$$W_{\pm}{}^{(k)}\tfrac{1}{2}(m^2_\ell+p^2_\ell)=\tfrac{1}{2}(m^2_\ell+p^2_\ell)+\phi_{ij},\ \ell=i,j;\ w_{\pm}{}^{(k)}\ \frac{m^2_k+p^2_k}{2}=$$

$$\tfrac{1}{2}(m^2_k+p^2_k)(\approx 0). \tag{2.14}$$

(In deriving (2.14) we have used the invariance property (2.6).) We shall restrict the choice of the multipliers λ_k by demanding that Eq. (2.13) reproduces the usual addition of interactions in the non-relativistic limit (for small velocities and small ratios ϕ_{ij}/m^2_ℓ). This gives

$$\lambda k\to\frac{\lambda}{m_k},\quad \text{for } p^2_k+(\hat{P}p_k)^2<<m^2_k,\ k=1,2,3. \tag{2.15}$$

This condition is met, for example, for λ satisfying

$$\{H_0{}^{(\lambda)}\,x_{ij}P\}=0,\ \text{i.e. for } \lambda_k=\frac{2}{|\hat{P}p_k|}. \tag{2.16}$$

It is now easily demonstrated (at least on a formal level) that the constraints

$$_k\equiv W_{\pm}(H^{(\lambda)},H_0^{(\lambda)})\tfrac{1}{2}(m^2_k+p^2_k)\approx 0,\ k=1,2,3 \tag{2.17}$$

satisfy all requirements (i)-(v) (and reproduce in the appropriate limit the non-relativistic addition of interactions).

In practice, the construction of the wave operators is,

clearly, not an easy task. Assuming, however, that the 2-particle interactions involve a small coupling constant, say $e_i e_j \sim \alpha$, one can write $_k$ as

$$_k = \tfrac{1}{2}(m_k^2 + p_k^2) + \phi_{ik} + \phi_{jk} + \phi_k \; . \tag{2.18}$$

where the correction ϕ_k to the naive addition of interaction terms ϕ_{ek} can be found*) at least on lowest order in α.

3. REPARAMETRIZATION INVARIANT CHARACTERISTICS OF THE 2-PARTICLE MOTION

3A. Dependence of Interacting Particles' World Lines on the Choice of Equal Time Surface

For the sake of simplicity, we shall restrict our attention in this section to the study of a system of two spinless particles with constraints of the type

$$H = (p - \quad (w)) + \quad (r,w) \quad 0, \quad r = [x \; + (xP) \quad] \tag{3.1}$$

$$= pP \approx 0 \quad p = \mu_2 p_1 - \mu_1 p_2 \quad (\mu_1 + \mu_2 = 1, \; \mu_1 - \mu_2 = \frac{m_1^2 \; m_2^2}{w^2}). \tag{3.2}$$

According to (1.33) the Hamiltonian $H^{(n)}$ preserving the gauge condition nx=0 (1.32) is, in this case, proportional to

$$H^{(n)} = H - \frac{nP}{nP} \quad . \tag{3.3}$$

If the interaction is non-trivial, so that $\frac{\partial\phi}{\partial r} \neq 0$, then the normalized 4-velocities $u_k = (-\dot{x}_k^2)^{-\frac{1}{2}} \dot{x}_k$, where

$$\dot{x}_1 = \{H^{(n)}, x_1\} = (\mu_2 - \frac{np}{nP}) p_1 + \frac{\partial\phi}{\partial P} \; , \tag{3.4a}$$

$$\dot{x}_2 = \{H^{(n)}, x_2\} = (\mu_1 + \frac{np}{nP}) p_2 + \frac{\partial\phi}{\partial P} \; , \tag{3.4b}$$

$$\frac{\partial\phi}{\partial P} = - \frac{\partial\phi}{\partial W} \hat{P} + \frac{xP}{w^2 r} \frac{\partial\phi}{\partial r} x_1 \tag{3.5}$$

depend on the direction of n. It follows that in the presence of

---

*Work on these lines is in progress (in collaboration with S. Bidikov. A perturbative approach to the relativistic addition of direct interactions has also been developed in an alternative formulation of the N-particle dynamics in ref. [36].

a (smooth) non-vanishing interaction ϕ the particle world lines depend on the choice of the equal-time plane.

For someone accustomed to regard world lines as something fundamental in classical relativistic dynamics such a "gauge dependence" of particle trajectories may appear as a disaster for the entire set-up. It is therefore important to realize how general is this property. The problem was studied thoroughly in ref. [28]; we shall summarize here the results.

The question is: what is the most general set of constraints such that the particle world lines are independent of n and how do the particle trajectories look like for such constraints? The first result says that if particle world lines are gauge invariant then the constraints can be chosen in the form

$$_1 = \,_1(\tfrac{1}{2}x^2,\ xp_1,\ \tfrac{1}{2}p_1{}^2)\approx 0, \qquad _2 = \,_2(\tfrac{1}{2}x^2,\ xp_2,\ \tfrac{1}{2}p_2{}^2)\approx 0. \tag{3.6}$$

(The statement is true locally--in the neighbourhood of every point of a dense open subset of the generalized mass shell--see Propositions 1 and 2 of Appendix A.3 to ref. [28(b)].)

Using the Poisson bracket relations

$$\{\tfrac{1}{2}p_1^2, xp_2\}=\{xp_1, \tfrac{1}{2}p_2^2\}=p_1p_1, \quad \{\tfrac{1}{2}p_1^2, \tfrac{1}{2}x^2\}=xp_1, \quad \{\tfrac{1}{2}x^2, \tfrac{1}{2}p_2^2\}=xp_2,$$

$$\{xp_1, \tfrac{1}{2}x^2\}=\{\tfrac{1}{2}x^2, xp_2\}=x^2, \quad \{xp_1, xp_2\}=xP \tag{3.7}$$

we can write the compatibility condition (1.16) in the form

$$\{\ _1,\ _2\}=\left(\frac{\partial\ _1}{\partial\frac{1}{2}p_1^2}\ \frac{\partial\ _2}{\partial p_2x}+\frac{\partial\ _1}{\partial p_1x}\ \frac{\partial\ _2}{\partial\frac{1}{2}p_2^2}\right)P_1P_2+\frac{\partial\ _1}{\partial\frac{1}{2}p_1^2}\ \frac{\partial\ _2}{\partial\frac{1}{2}x^2}\ x\ P_1+$$

$$\frac{\partial\ _1}{\partial\frac{1}{2}x^2}\ \frac{\partial\ _2}{\partial\frac{1}{2}p_2^2}\ x\ p_2+\frac{\partial\ _1}{\partial xp_1}\ \frac{\partial\ _2}{\partial xp_2}\ xP+x^2\left(\frac{\partial\ _1}{\partial xp_1}\ \frac{\partial\ _2}{\partial\frac{1}{2}x^2}+\frac{\partial\ _1}{\partial\frac{1}{2}x^2}\ \frac{\partial\ _2}{\partial xp_2}\right)\approx 0. \tag{3.8}$$

The subsequent argument (in the proof of Lemma 2 of Sec. 2B) of ref. (28(b)] is based on the fact that the scalar product p_1p_2 only appears as a factor in the first term in the above equation, so that its coefficient should vanish. We note that this argument is only correct for space-time dimensions $D \geq 3$; it fails in the two dimensional case, in which p_1p_2 can be expressed as a function of the remaining five scalar products $\frac{1}{2}p_k, xp_k$, and $\frac{1}{2}x^2$. Moreover, the results of ref. [37] indicate the existence of a non-trivial gauge invariant dynamics in the 2-dimensional case.

The conclusion in the general case, $D \geq 3$, (including the physical case of interest, $D=4$), is that only locally straight world lines are independent of the choice of the equal time surface. This means that only the free motion and the elastic collision of finite size balls, for which $\phi=0$ at all points where the particles have well defined velocities, are "gauge independent". As stated in the introduction this result just shows the limitations of the Hamiltonian single time approach to the relativistic particle dynamics--we do not wish to assert that classical relativistic world lines are actually unobservable.

We do not regard, however, the frame dependence of relativistic trajectories as too high a price for preserving the Hamiltonian formalism. As it will be demonstrated in the next subsection, the truly relevant physical characteristics of the particle motion are gauge invariant. We only wish to stress at this point that frame dependence of world lines should not be confused with lack of Lorentz invariance. Taking $n=-\hat{P}$ (that is, adopting the Yukawa-Markov gauge condition (1.35) for which $H^{(n)}=H$ (3.1) we get a manifestly invariant formulation of the relativistic 2-body problem. The need for a wider-gauge-invariance becomes apparent when one proceeds to the N-particle problem for $N\geq 3$, since there is no distinguished frame in that case, equally suited to the description of the system as a whole and of its various subsystems.

3B. Gauge invariance of the 2-particle relative motion and of the scattering matrix

It follows from (3.4) that the relative velocity

$$\dot{x}(=\dot{x}_1-\dot{x}_2)=p-\frac{np}{nP}P \tag{3.9}$$

is always n-dependent. However, its projection on the 3-plane orthogonal to P is frame independent:

$$\dot{x}_\perp(=x+(\hat{P}\dot{x})\hat{P})=p. \tag{3.10}$$

The same is true for the (orthogonal) relative acceleration:

$$\ddot{x}_\perp=\dot{p}=-\frac{\partial\phi}{\partial r}\frac{x_\perp}{r} \quad (r=|x_\perp|=\sqrt{x_\perp^2}). \tag{3.11}$$

Thus the relative motion is gauge invariant.

The gauge invariance of the scattering matrix (if it exists) is a simple consequence of the invariance of the straight line parts of particle trajectories.

A 2-particle scattering system (as opposed to a "planetary" bounded system) is characterized by the existence of the limits

$$\lim_{t\to\pm\infty} p(t) = p_{in}^{out} \equiv p^{as} \neq 0, \quad as = \{{}^{out}_{in} , \tag{3.12a}$$

$$\lim_{t\to\pm\infty} [x(t) - p^{as} t] = a^{as} \tag{3.12b}$$

($1a^{as}$) is called impact parameters). Since the sum $p_1(t)+p_2(t)=P$ is time independent, the limits

$$\lim_{t\to\pm\infty} p_1(t) = \mu_1 P + p^{as} \equiv p_1^{as}, \quad \lim_{t\to\pm\infty} p_2(t) = \mu_2 P - p^{as} \equiv p_2^{as} \tag{3.13}$$

exist and

$$p_1^{out} + p_2^{out} = P = p_1^{in} + p_2^{in} . \tag{3.14}$$

Eqs. (3.12) (and the nonvanishing of p^{as}) imply that

$$|x(t)| \equiv r(t) \to \infty \quad \text{for } t \to \pm\infty \tag{3.15}$$

Hence, the interaction ϕ vanishes at infinite times (because of (1.18)), so that

$$m_k^2 + (p_k^{as})^2 = 0, \quad k=1,2 \tag{3.16}$$

Although the limits (3.12) are originally defined for fixed n (and given Hamiltonian H that governs the time evolution) they are actually n-independent because of the gauge invariance of the relative motion.

The classical scattering operator is defined in terms of the Moeller wave operators (2.3) as

$$S = W_+^*(H,H_0) W_-(H,H\ ) \tag{3.17a}$$

where

$$W_\pm^*(H_1 H_0) = \text{s-}\lim_{t\to\infty} e^{tL_H} e^{-tL_H} \tag{3.17b}$$

hence it is gauge invariant provided the wave operators are. The gauge invariance of $W_\pm$, on the other hand, is guaranteed by (2.6). (It is thus related to the Birman-Kato invariance principle.)

In more elementary terms, we could say that the classical scattering matrix is gauge invariant, since it transforms the

4-vectors $u_k^{in}(=\frac{1}{m_k}p_k^{in})$ and a_k^{in} into u_k^{out} and a_k^{out}, all of which are frame-independent.

4. EXAMPLE OF RELATIVISTIC TWO-PARTICLE DYNAMICS: SPINLESS CHARGED PARTICLES

4A. Relativistic reduced mass and minimal Coulomb interaction

We could already appreciate the relative simplicity of 2-particle interactions, which enabled us to satisfy all requirements (i)-(v) of Sec. 1B, and led to the program of constructing N particle interactions out of 2-body forces. The 2-particle case yields a remarkable simplification also in the nonrelativistic picture. Only in this case can one reduce the relative motion to that of an effective particle in an external field. We shall demonstrate that the concept of an effective particle can be extended to a relativistic 2-particle system.

The key classical notion of reduced mass m is defined by the relation

$$mM = m_1 m_2 \tag{4.1}$$

where M is the total (non-relativistic) mass: $M = m_1+m_2$. We estend it to the relativistic case by retaining (4.1), but with M replaced by the relativistic total mass w of the system. In otehr words, we define (following ref. [10]) the relativistic reduced mass by

$$m_w = \frac{m_1 m_2}{w} . \tag{4.2}$$

The 3-momentum of the effective particle in the centre of mass frame (or in the rest frame of the "source" in the language of single particle picture) is identified with the relative momentum $p=(o,\underline{p})$. The on-shell value of $\underline{p}^2$ is $f^2(w)$ (1.28). The energy E of the effective particle is then defined by

$$E = \sqrt{m_w^2 + f^2(w)} = \frac{1}{2w}(w^2-m_1^2-m_2^2). \tag{4.3}$$

The free Hamiltonian H_0(2.8) (that preserves the Yukawa-Markov gauge condition (1.35)) can be written as the mass-shell condition for the effective particle:

$$H_0=\tfrac{1}{2}(m_w^2 +\underline{p}^2-E^2) \quad (H_0\approx 0 \text{ for } \phi=0/\text{e.g., for } r\to\infty/). \tag{4.4}$$

(We stick from now on to the centre of mass frame.)

Consider next a system of two charged particles (of charges e_1 and e_2) whose interaction is described in the non-relativistic limit by the Coulomb potential

$$V_0(r) = \frac{e_1 e_2}{4\pi r} \quad (r=\sqrt{x^2}). \tag{4.5}$$

Combining (4.4) and (4.5) one can introduce a minimal (instantaneous) relativistic Coulomb interaction by the Hamiltonian constraint

$$H_{Coul} = \tfrac{1}{2}[m_w^2+\underline{p}^2-(E-V_0)^2]\approx 0. \tag{4.6}$$

(The second constraint ≈ 0 (3.2) is kept unchanged.)

Such a definition of relativistic Coulomb interaction may appear somewhat arbitrary. (Indeed, one does encounter alternative approaches in the literature--see, e.g., [38].) Happily, we have a reliable physical test: comparison with field theory. Its systematic application in the quantum domain is embodied in the quasipotential approach [8-11]. As indicated in Sec. 5C below, it leads to an extra term in H_{Coul}, coming from the vector potential. For the time being we shall regard the constraint (4.6) as defining a simplified model for the relativistic Coulomb interaction, which will be used to illustrate the efficiency of the approach on a concrete problem.

4B. Scattering on a central potential. Relativistic Rutherford formula

For a fixed w the Hamiltonian H (3.1) has the same structure as the non-relativistic Hamiltonian. This observation allows to extend known non-relativistic results (including exact solutions) to the relativistic case.

We shall consider as an illustration some general properties of 2-particle scattering (for central forces) and the exact solution for the "relativistic Coulomb potential" (4.6)(4.5).

Angular momentum conservation (implied by the rotation invariance of the Hamiltonian constraint (3.1) makes it convenient to choose one of the coordinate axes (say, the third axis) along $\underline{\ell}$:

$$\underline{\ell}\equiv\underline{p}\wedge\underline{r}=(0,0,\ell). \tag{4.7}$$

Since

$$\ell^2=p^2r^2-(\underline{p}\,\underline{r})^2 \quad (p^2=\underline{p}^2, r^2=\underline{r}^2) \tag{4.8}$$

and

$$\dot{r} = \{H,r\} = \frac{1}{r}\underline{p}\,\underline{r}, \tag{4.9}$$

we find (using 3.1):

$$p^2 = f^2 - 2\phi(r,w) = \dot{r}^2 + \frac{\ell^2}{r^2}\,. \tag{4.10}$$

We shall now make more precise the restriction ((v) of Sec. 1B) on a possible attractive singularity at r=0 by assuming

$$\frac{\ell^2}{r^2} + 2\phi(r,w) \geq -m_1 m_2 \quad \text{for } r\to 0. \tag{4.11}$$

(This assumption ensures that Eq. (4.10) has a real solution $w^2 (\geq m_1^2+m_2^2)$ for arbitrarily small r and $\dot{r}$.) Note that for

$$\phi(r,w) = EV_0(r) - \tfrac{1}{2}V_0^2(r),\quad V_0(r) = \frac{e_1e_2}{4\pi r} \equiv \pm\frac{\alpha}{r},\quad \alpha = \frac{|e_1e_2|}{4\pi}, \tag{4.12}$$

condition (4.11) is only satisfied for $\ell \geq \alpha$.

The scattering problme is distinguished by the boundary condition $r(\zeta)\to\infty$ for $\zeta\to\pm\infty$. The distance $r(\zeta)$ should reach a (non-negative) minimum r_0 for some finite ζ_0 which will be chosen as the origin of the time scale. Since $\dot{r}$ vanishes at $\zeta=\zeta_0(=0)$ the value r_0 can be determined as the lartest root of the equation

$$\ell_0^2(w) - 2\phi(r_0,w) - \frac{\ell^2}{r_0{}^2} = 0; \tag{4.13}$$

according to (4.10) $\dot{r}(\zeta)$ assumes the form

$$\dot{r}(\zeta) = [\ell_0^2(w) - 2\phi(r,w) - \frac{\ell^2}{r_0(\zeta)}]^{\frac{1}{2}}\ \text{sign}\zeta\ /\text{for } r(0)=r_0, \dot{r}(0)=0/. \tag{4.14}$$

It is not difficult to write down the equation for the $\underline{r}$-space trajectory in polar coordinates

$$\underline{r} = r(\cos\ \ ,\ \sin\ \ ,\ 0) \tag{4.15a}$$

in which

$$\underline{p} = \dot{r}(\cos\ \ - r\ '\text{sine}\ \ ,\ \text{sine}\ \ + r\ '\cos\ \ ,0),\qquad ' = \frac{d\ell}{dr} \tag{4.15b}$$

Inserting in (4.10) we find

$$(r\ ')^2 (= \frac{\ell^2}{(r\dot{r})^2}) = \frac{\ell^2}{(f^2-2\phi)r^2 1\ell^2} \,. \tag{4.16}$$

This equation is easily integrated for

$$2\phi = 2EV_0(r) = \pm \frac{w^2-m_1^2-m_2^2}{w} \frac{\alpha}{r} \,. \tag{4.17}$$

The corresponding Hamiltonian H (which is obtained from H_{Coul} (4.6) by neglecting the $-\frac{1}{2}V_0^2$-term) admits the O(4)-symmetry of the non-relativistic Coulomb problem. The Runge-Lenz vector

$$\underline{A} = \mp \frac{1}{\alpha E} \underline{p} \wedge \underline{L} + \frac{\underline{r}}{r} \quad (\mp \frac{1}{\alpha} = - \frac{4\pi}{e_1 e_2}) \tag{4.18}$$

is conserved,

$$\{H, \underline{A}\} = 0 \tag{4.19}$$

and hence determines a time independent direction in the scattering plane, which will be identified with the first axis:

$$\underline{A} = (\varepsilon, 0, 0), \quad \varepsilon^2 = 1 + \frac{f^2 \ell^2}{\alpha^2 E^2} \,. \tag{4.20}$$

In terms of the eccentricity ε the $\underline{r}$-space orbit can be written in the form

$$(\frac{1}{r} \underline{r}\,\underline{A} =)\varepsilon \cos = 1 \pm \frac{\ell^2}{\alpha}, \quad \varepsilon \sin \ = \pm \frac{\ell}{\alpha} \dot{r}, \ r_0 = \frac{\alpha}{2}(\varepsilon \pm 1), \tag{4.21}$$

where $\dot{r}$ is given by (4.14) (the sign $\pm$ coincides with the relative sign of the charges - cf. (4.12).

The trajectory corresponding to the Coulomb Hamiltonian (4.6) is obtained from here by the substitution $\ell^2 \to \ell^2 - \alpha^2 (\geq 0)$.

We are now prepared to determine the scattering characteristics. Noting that $r(\zeta) \to \infty$ for $\zeta \to \pm\infty$, we derive from (4.14), (4.16) and (4.21) the following asymptotic relations

$$r \to \pm f, \quad r\ ' \to 0 \quad \text{for } \zeta \to \pm\infty, \tag{4.22}$$

$$\cos \ \to \frac{1}{\varepsilon} \quad \text{for } \zeta \to \pm\infty \qquad \sin \ \begin{cases} \pm \frac{\ell f}{\alpha E \varepsilon} & \text{for } \zeta \to \infty \\ \mp \frac{\ell f}{\alpha E \varepsilon} & \text{for } \zeta \to -\infty. \end{cases} \tag{4.23}$$

With these relations in hand we can readily evaluate p^{as} from (4.15b):

$$\underline{p}^{in} = -f(\frac{1}{\varepsilon}, \pm\frac{\sqrt{\varepsilon^2-1}}{\varepsilon}, 0), \quad \underline{p}^{out} = f(\frac{1}{\varepsilon}, \mp\frac{\sqrt{\varepsilon^2-1}}{\varepsilon}, 0). \tag{4.24}$$

The <u>scattering angle</u> Θ between $\underline{p}^{in}$ and $\underline{p}^{out}$ is given by

$$\cos\Theta = \frac{1}{f^2}\underline{p}^{in}\underline{p}^{out} = 1 - \frac{2}{\varepsilon^2} \quad (0 \le \Theta \le \pi). \tag{4.25}$$

Using (4.20) and (4.25) we can express the angular momentum ℓ and hence the impact parameter $a = \frac{\ell}{f}$ in terms of w and Θ (which are commonly regarded as independent variables for the scattering problem). The (differential) elastic scattering cross section is defined in terms of $a(w,\Theta)$ in a standard fashion (see, e.g., [39]):

$$\frac{d\sigma}{d\Omega} = -\frac{a(w,\Theta)}{\sin\Theta}\frac{\partial a(w,\Theta)}{\partial\Theta}. \tag{4.26}$$

For the simplified Coulomb problem the impact parameter is expressed as

$$(w,\Theta) = \frac{\ell(w,\Theta)}{f\omega} = \frac{\alpha E}{f^2(\omega)} \operatorname{ctg}\frac{\Theta}{2}; \tag{4.27}$$

this leads to the following rleativistic Rutherford formula for the Coulomb differential cross section:

$$\frac{d\sigma}{d\Omega} = \frac{\alpha^2 E^2}{4f^4 \sin^4\frac{\Theta}{2}}. \tag{4.28}$$

5. LOCAL QUASIPOTENTIAL EQUATION FOR QUANTIZED RELATIVE TWO-PARTICLE MOTION

5A. A rigged Hilbert space description of 2-particle motion

The canonical phase space formulation of point particle dynamics is specially adapted for quantization. We shall restrict our present discussion to a brief quantum description of the relative 2-particle motion (regarding the total momentum P as an outside parameter, similar to the energy eigenvalue E in the stationary non-relativistic Schrodinger equation).

Consider the (locally convex) topological vector space S_P of infinitely smooth fast decreasing in $x = x + (x\hat{P})\hat{P}$ functions $\Psi(x) = \Psi(x,P)$ that satisfy the differential equation

$$P \frac{\partial}{\partial x} \Psi(x,P)=0 \quad (P^0>|\underline{P}|). \tag{5.1}$$

Let n be an arbitrary time-like vector normalized by

$$|n\hat{P}|=1 \quad \text{where } \hat{P}= \frac{1}{w}P \quad (\hat{P}^2= -1) \tag{5.2}$$

Then the scalar product

$$(\Phi,\Psi)(=(\Phi,\Psi)_{\hat{P}}) = \overline{\Phi}(x)\Psi(x)\delta(nx)d^4x \tag{5.3}$$

does not depend on n. Indeed, every solution has the form

$$\Psi(x_1P) = \Psi(p,P)\delta(p\hat{P})e^{-ipx} \frac{d^4p}{(x\pi)^3}, \tag{5.4}$$

which gives, upon inserting in (5.3), the manifestly n-independent expression

$$(\Phi,\Psi)= \overline{\quad(p,P)}\ \Psi(p,P)\delta(p\hat{P}) \frac{d^4p}{(x\pi)^3}. \tag{5.5}$$

The bound state wave functions (corresponding to some discrete "eigenvalues" of $\omega=\sqrt{-P^2}$) are vector in the Hilbert space completion H_P of S_P. By contrast, the scattering states (corresponding to the "continuous spectrum" $w\geq m_1+m_2$) are not normalizable; they belong to a wider space S_P^* H_P of generalized states.

In order To Define S_P^*, we ahve to specify a (locally convex) topology in S_P (stronger than the one defined by the scalar product) with respect to which S_P is a complete topological space. A standard choice is provided by the Frechet space topology, given by the denumerable set of seminorms

$$P_k(\Psi)=(\Psi,(\lambda x\Pi x - \frac{\partial}{\partial x}\Pi\frac{\partial}{\partial x})^k\Psi)^{\frac{1}{2}} \quad (\lambda>0 \text{ fixed}),\ k=0,1,2,\ldots, \tag{5.6}$$

where $\Pi^\mu_\nu = \delta^\mu_\nu + \hat{P}^\mu\hat{P}_\nu$ $(\Pi^2=\Pi)$. We then define S_P^* as the space of continuous (in the above topology) linear functionals on S_P.

We identify the physical relative coordinate with the 4-vector X (1.31) orthogonal to P. Every polynomially bounded (smooth) function f(x) (in particular, every component x^μ of x) defines a <u>multiplier</u> in S_P: if (x,P) S_P, then f(x)Ψ(x,P). Indeed, it is easily verified that $P\frac{\partial}{\partial x}f(x)=0$; consequently, if $\Psi_{(x)}$ satisfies (5.1), so does f(x)Ψ(x).

The quantized relative momentum p can be defined either as the differential operator

$$p=i\,\frac{\partial}{\partial x} \tag{5.7a}$$

or as its orthogonal (to P) projection

$$p=\Pi p=i\left(\frac{\partial}{\partial x}+\hat{P}\left(\hat{P}\,\frac{\partial}{\partial x}\right)\right) \tag{5.7b}$$

the two definitions being equivalent because of (5.1). In both cases

$$[p_\mu,p_\nu]=0(=[x^\mu,x^\nu]\quad [p_\nu,x^\mu]=i\Pi^\mu{}_\nu(=i\{p_\nu,x^\mu). \tag{5.8}$$

If we insist on having a canonical pair of coordinate and momentum, we have to give up manifest covariance and work in the rest frame of P. The centre-of-mass-frame three vectors $\underline{r}$ and $\underline{p}$ (related to x and p by $x=(o,\underline{r})$ and $p=(o,\underline{p})$ are then readily verified to satisfy canonical commutation relations.

5B. Quasipotential equation as quantum Hamiltonian constraint

Given an interaction (an energy dependent central potential) $\phi(r,w)$ $(r=|\underline{r}|=\sqrt{x^2})$ we shall postulate that the wave function Ψ satisfies the stationary type Schrodinger equation

$$H\Psi=[\tfrac{1}{2}(\underline{p}^2-f^2(\omega))+\phi(r,w)]\Psi(x,P)=0. \tag{5.9}$$

This is the quantum mechanical expression of the Hamiltonian constraint $H\approx 0$. Eq. (5.9) is a (local) partial differential equation. In the centre of mass frame it assumes a more familiar form

$$[-\tfrac{1}{2}(\Delta+f^2(w)+\phi(r,\omega)]\Psi(r,w)=0 \tag{5.10}$$

which coincides with the local quasipotential equation of refs. [10,11].

The identification of the quantized Hamiltonian constraint with a relativistic quasipotential equation involves an important benefit: it provides a systematic method for evaluating the interaction function $\phi(r,w)$. We shall briefly recall the general procedure for determining ϕ from the Feyman expansion of the corresponding scattering amplitude (see [8-11]). The momentum space quasipotential

$$V=V_W(\underline{k})=\frac{1}{w}\quad\phi(r,w)e^{i\underline{k}r}d^3r \tag{5.11}$$

is related to the 2-particle scattering amplitude $T=T_w(\underline{p},\underline{q})$ by a Lippmann-Schwinger type equation

$$T+V+V*GT\equiv T_w(p,q)+V_w(p-q)+$$
$$+SV_W(\underline{p}-\underline{k})G_w(\underline{k})T_w(\underline{k},\underline{q})\frac{d^3k}{(2\pi)^3} = 0. \qquad (5.12)$$

where

$$G_w(\underline{k})= \frac{1}{2W}\frac{1}{\underline{k}^2-f^2(w)-i0} \qquad (5.13)$$

In quantum field theory (with a fixed Lagrangian) we have a well defined (Feynman diagram) perturbation series $T=T_1+T_2+\ldots$ for the scattering amplitude. We derive a corresponding expansion $V-V_1+V_2+\ldots$ for the quasipotential as the interative solution of (5.12):

$$V_1= -T_1, \quad V_2= -T_2+T_1* G\ T_1, \text{ etc.} \qquad (5.14)$$

This procedure allows to determine the quasipotential with any desired degree of precision, at least for small coupling (this is the case of quantum electrodynamics/QED/; the same should be true in the validity domain of asymptotic freedom for quantum chromodynamics). Inserting even the very first (Born) term V_1 in the "Schrodinger" equation (5.10) already allows to obtain non-perturbative results such as the bound state energy levels.

5C. Example: Relativistic energy spectrum of hydrogen like systems

Consider as an example the electromagnetic interaction of an electron (or a μ^-) with a spinless (positively charged) nucleus: According to [11,40] the Born QED diagram along with the minimal coupling trick of Sec. 4A leads to the following centre of mass expression for the "Hamiltonian" H:

$$H= -\tfrac{1}{2}(\Delta+f^2(w))-E\ \frac{\alpha}{r}-\tfrac{1}{2}\frac{\alpha^2}{r^2}+\pi\ \frac{\alpha}{w}\delta(\underline{r})+\frac{\alpha^2}{8w^2r^4}+$$
$$+\pi\ \frac{\alpha}{w}\frac{E_2}{E_1+m_1}\delta(\underline{r})+\frac{\alpha}{2wr^3}\ (1+\frac{E_2}{E_1+m_1})\underline{\ell\sigma}. \qquad (5.15)$$

Here the index 1 refers to the spin $\tfrac{1}{2}$ particle,

$$E_{1,2} = \frac{1}{2W}\ [w^2\underline{+}(m_1^2-m_2^2)] \qquad (5.16)$$

are the energy values for particles 1 and 2, $\underline{\sigma} = (\sigma_1,\sigma_2,\sigma_3)$ are Pauli matrices, $\underline{\ell} = -i\underline{r}\wedge\underline{\nabla}$ is the orbital angular momentum of the system. The nonsymmetric terms with respect to the substitution $1\rightleftarrows 2$ (in the second line of Eq. (5.15)) reflect the asymmetric assignment of spin to the particles (they would disappear in the zero spin case).

It is remarkable that the corresponding energy eigenvalues [evaluated from the Schrodinger equation H =0 with H given by (5.15)],

$$W_{nj} = M - \frac{m\alpha^2}{2n^2} + \frac{m\alpha^4}{8n^4}\left(3 - \frac{m}{M}\right) - \frac{m\alpha^4}{Mn^3}\left(\frac{m_1 - m}{2\ell+1} + \frac{m_2 + m}{2j+1}\right)$$

$$/M = m_1 + m_2, \quad m = \frac{m_1 m_2}{M}/,$$

where $j(=\ell+\frac{1}{2})$ is the total angular momentum, give the right answer up to (including) α^4. (That was demonstrated in ref.[40] by computing the contribution of higher order Feynman diagrams to the energy levels.)

It would be interesting to have a non-perturbative evaluation of magnetic effects generated by the vector potential (they should come from the last four terms in the right hand side of (5.15), starting with $\frac{\pi\alpha}{w}\delta(\underline{r})$). They are expected to give rise to resonance states of the atomic system at short distances (cf.[41]). In order to treat Eq. (5.10) in a non-perturbative way, it would first be necessary to regularize the δ-function terms in (5.15) by replacing $\delta(\underline{r})$ with, say,

$$\delta_\varepsilon(\underline{r}) = \frac{1}{\pi^2}\frac{\varepsilon}{(r^2+\varepsilon^2)^2}\ ; \tag{5.18}$$

if we take $\varepsilon \sim e^{-\frac{1}{\alpha}}$, then δ_ε would be undistinguishable from $\delta(\underline{r})$ to any finite order of perturbation theory. After such a regularization it becomes clear that the dominant term in the right hand side of (5.15) at short distances is the repulsive potential $\frac{\alpha^2}{8w^2r^2}$. This guarantees that H can be defined rigorously as a bounded below operator so that a non-perturbative study of Eq. (5.10) really makes sense.

6. DISCUSSION. OPEN PROBLEMS.

The aim of these notes was twofold.

First, to demonstrate the possibility of a consistent Hamiltonian formulation of relativistic point particle dynamics. We have discsused some peculiarities of the resulting picture (notably, the dependence of interacting particles' world lines on the choice of an equal time plane, $nx_{12}=0$, discussed in Sec. 3, as well as the n-independence of the scattering matrix).

Second, to display the realistic features and practical usefulness of the present framework on both classical and quantum level (Secs. 4 and 5).

The work in neither direction is completed. We shall cite two unresolved problems which we regard as deserving attention (along with the problem of a non-perturbative evaluation of magnetic effects in the quantum theory of interacting charged particles with spin, mentioned at the end of Sec. 5C.).

The first question (already referred to in Sec. 2B.) has to do with the actual construction of 3-particle constraints for given 2-particle forces. This question has both theoretical and practical significance. Only after its solution we would be entitlted to speak about a consistent theory of directly interacting relativistic point particles. On the other hand, even an approximation to the resulting 3-particle forces (accurate, in the case of electromagnetic interactions, up to order α^2) would be useful for the study of mesomolecular processes, to cite one example (see, e.g., [42]).

The second problem is more practical in nature and should not involve serious difficulties.

The study of charmonium and other (heavy) quark-antiquark systems has recently gained some theoretical respectability. Within a single phenomenological parameter the quark-antiquark potential is evaluated from the Born diagram of quantum chromodynamics with proper account of the momentum dependence of the effective charge (derived by a standard renormalization group argument)--see [43]. It should be instructive to study in a systematic fashion the relativistic corrections to existing calculations (using the quasipotential framework of Sec. 5).

*

I am grateful to Dr. S.N. Sokolov and to the participants of the quantum field theory seminar of the Steklov Mathematical Institute in Moscow for enlightening discussions. I would like to thank Professor A.O. Barut and Professor E. Inönü for their hospitality in Istanbul, as well as to the participants of the Summer School in Mathematical Physics at the Bozaziçi University, for useful remarks.

REFERENCES

1. H. Poincaré, Comptes Rendues Acad. Sci. 140, 1504 (1905); Rendiconti del Circolo Matematico di Palermo 21, 129 (1906).

2. M. Planck, Verhandl. Deutsch. Phys. Ges. 8, 136 (1906).

3. H.G. Wells, The Time Machine (1895); this reference has attracted the attention of S. Weinberg, Gravitation and Cosmology (J. Wiley, N.Y., 1972)/see Chapter 2/.

4. P.A.M. Dirac, Rev. Mod. Phys. 21, 392 (1949).

5. D.G. Currie, T.F. Jordan, E.C.G. Sudarshan, Rev. Mod. Phys. 35, 350, 1032 (1963); for later development and further references see T.F. Jordan, Phys. Rev. D15, 3575 (1977).

6. H. Leutwyler, Nuovo Cim. 37, 556 (1965).

7. E.P. Wigner, Relativistic interactions of classical particles, in: Coral Gables Conference on Fundamental Interactions at High Energy, ed. T. Gudehus et al. (Gordon and Breach, N.Y. 1969), pp. 344-355, and references therein.

8. A.A. Logunov, A.N. Tavkhelidze, Nuovo Cim. 29, 380 (1963); A.A. Logunov, A.N. Tavkhelidze, I.T. Todorov, O.A. Khrustalev, Nuovo Cim. 30, 134 (1963).

9. V.A. Matveev, R.M. Muradyan, A.N. Tavkhelidze, Relativistic-covariant wave equations for N particles in quantum field theory. JINR-preprint P2-3900, Dubna (1968); P.N. Bogolubov, Teor. Mat. Fiz. 5, 244 (1970) [English transl.: Theor. Math. Phys. 5, 1121 (1970)].

10. I.T. Todorov, Phys. Rev. D3, 2351 (1971); see also "Quasi-potential approach to the two-body problem in quantum field theory". In: Properties of Fundamental Interactions, Vol. 9C, ed. A. Zichichi (Editrice Compositori, Bologna, 1973), pp. 951-979.

11. V.A. Rizov, I.T. Todorov, Elem. Chast. i Atom. Yad. 6, 669 (1975) English transl.: Sov. J. Part. Nucl. 6, 269 (1975).

12. Y. Nambu, Phys. Rev. 160, 1171 (1967).

13. A.O. Barut, D. Corrigan, H. Kleinert, Phys. Rev. Letters 20, 167 (1968); Phys. Rev. 167, 166 (1968); A.O. Barut, Phys. Rev. Letters 20, 893 (1968); A.O. Barut, A. Baiquni, Phys.

Rev. 184, 1342 (1969).

14. C. Fronsdal, L.E. Lundberg, Phys. Rev. D1, 3247 (1970); C. Fronsdal, Phys. Rev. D4, 1686 (1971).

15. C. Itzykson, V.G. Kadyshevsky, I.T. Todorov, Phys. Rev. D1, 2823 (1970); I.T. Todorov, "On the three dimensional formulation of the relativistic two-body problem". In: Lecture Notes in Theoretical Physics XII. Univ. of Colorado, 1969, ed. W. Brittin and K. Mahanthappa (Gordon and Breach, N.Y., 1972), pp. 183-212.

16. S.N. Sokolov, Teor. Mat. Fiz. 36, 193 (1978); S.N. Sokolov, A.N. Shatky, Teor. Mat. Fiz. 37, 291 (1978).

17. P.A.M. Dirac, Canad. J. Math. 2, 129 (1950); Proc. Roy. Soc. A246, 326 (1958); Lectures on Quantum Mechanics (Belfer Graduate School of Science, Yeshiva Univ., N.Y., 1964).

18. L.D. Faddeev, Teor. Mat. Fiz. 1, 3 (1969).

19. A.J. Hanson, T. Regge, C. Teitelboim, Constraint Hamiltonian Systems (Academic Nazionale dei Lincei, Roma, 1976).

20. Ph. Droz-Vincent, Lett. Nuovo Cim. 1, 839 (1969); Physica Scripta 2, 129 (1970); Rep. Math. Phys. 8, 79 (1975); Ann. Inst. H. Poincaré 18A, 57 (1975), L. Bel, J. Martin, Ann. Inst. H. Poincaré 22A, 173 (1975).

21. T. Takabayasi, Prog. Theor. Phys. 54, 563 (1975); ibid. 57, 331 (1977); ibid. 60, 595 (1978); ibid. 61, 1235 (1979); K. Kamimura, Prog. Theor. Phys. 58, 1947 (1978); S. Kojima, Prog. Theor. Phys. 61, 960 (1979); M. Kalb, P. Van Alstine, Relativistic two-body systems from invariant singular actions: a Hamiltonian formulation, Yale Univ. preprint, New Haven (1976).

22. I.T. Todorov, Dynamics of relativistic point particles as a problem with constraints, Commun. JINR, E2-10125 (1976).

23. A. Komar, Phys. Rev. D18, 1881, 1887, 3617 (1978).

24. F. Rohrlich, Ann. Phys. (N.Y.) 117, 292 (1979).

25. L.P. Horwitz, C. Piron, Helv. Phys. Acta 46, 316 (1973); C. Piron, F. Reuse, Helv. Phys. Acta 51, 146 (1978); F. Reuse, Helv. Phys. Acta 51, 157 (1978); A relativistic two body model for the hydrogen atom, Université de Genève

preprint (1979).

26. M. Pauri, G.M. Prosperi, J. Math. Phys. 16, 1503 (1975); ibid. 17, 1468 (1976).

27. H. Leutwyler, J. Stern, Ann. Phys. (N.Y.) 112, 94 (1978).

28. V.V. Molotkov, I.T. Todorov, (a) On the notion of world lines and scattering matrix for directly interacting relativistic point particles, Commun. JINR, E2-12270, Dubna (1979); (b) Frame dependence of world lines for directly interacting classical relativistic particles, Internal Report IC/79/59, Trieste (1979).

29. T.D. Newton, E.P. Wigner, Rev. Mod. Phys. 21, 400 (1949).

30. A.A. Kirillov, Elements of Representation Theory ("Nauka", M., 1972); see, in particular, Sec. 15.

31. A.G. Reyman, in: Differential Geometry, Lie Groups and Mechanics, ed. L.D. Faddeev, Zap. Nauch. Sem. LOMI 37 (Nauka, Leningrad, 1973), pp. 47-52.

32. H.W. Crater, Phys. Rev. D18, 2872 (1978).

33. U. Mutze, J. Math. Phys. 19, 231 (1978).

34. S.N. Sokolov, Classical analogues of the Moeller operators, of the Pearson example, and of the Birman-Kato invariance principle, preprint TH 2614-CERN (1978).

35. M. Reed, B. Simon, Methods of Modern Mathematical Physics, II: Fourier Analusis, Self-Adjointness (N.Y., Academic Press, 1975).

36. L.L. Foldy, R.A. Krajcik, Phys. Rev. D12, 1700 (1975).

37. G.P. Georjadze, A.K. Pogrebkov, M.K. Polivanov, Teor. Mat. Fiz. (Moscow) 40, 221 (1979).

38. L. Bel, A. Salas, J.M. Sanchez, Phys. Rev. D7, 1099 (1973); L. Bel, J. Martin, Phys. Rev. D8, 4347 (1973).

39. H. Goldstein, Classical Mechanics (Addison-Wesley, Reading, Mass., 1972) (see Eq. (3-62) of Sec. 3.7).

40. V.A. Rizov, I.T. Todorov, B.L. Aneva, Nucl. Phys. B98, 447-471 (1975).

41. A.O. Barut, G.L. Bornzin, J. Math. Phys. 12, 841 (1971);

A.O. Barut, Magnetic resonances between massive and massless spin ½ particles with magnetic moments, J. Math. Phys. 21, 568 (1980).

42. S.S. Gershtein, L.I. Ponomarev, Mesomolecular processes induced by μ^- and π^- -mesons. In: Muon Physics, Ed. Y.W. Hughes and C.S. Wu (Academic Press, New York, 1975), pp. 141-233; L.I. Ponamarev, Mesic atomic molecular processes in the hydrogen isotope mixtures, Proceedings of the Sixth International Conference on Atomic Physics, August 17-22 (Plenum Press, Riga, 1978), pp. 182-206; D.D. Bakalov, S.I. Vinisky, Spin effects in the three body problem with electromagnetic interaction, JINR preprint P4-12736, Dubna (1979).

43. J.L. Richardson, Phys. Letters, 82B, 272 (1979); J. Rafalski, K.D. Viollier, Test of QCD with heavy quark bound states, preprint TH 2673, CERN (1979).

THEORY OF DYNAMICAL SYSTEMS BOUND BY MAGNETIC FORCES

A. O. Barut

Department of Physics, The University of Colorado,
Boulder, Colorado 80309, U.S.A.

I. INTRODUCTION

The basic dynamical model of atomic and molecular phenomena since the work of Rutherford and Bohr seventy years ago is a system of stable nuclei and electrons interacting essentially by the long-ranged electric Coulomb forces. The spin interactions cause only small perturbations to atomic states. There is, however, an important difference with the analogous classical Kepler problem, namely the exchange interactions between indistinguishable electrons due to Pauli principle.

That there is another structure of matter formed from stable electrons and protons and neutrinos and bound by short-ranged magnetic forces has been recognized only very recently. These new states of matter are semi-stable, occur at much shorter distances than the atomic levels, and at much higher masses. The hypothesis has been advanced that these states should be identified with the nuclear and hadronic matter. Many qualitative and quantitative aspects of this hypothesis have been studied in detail[1]. In this work we shall formulate the mathematical problem of magnetic dynamical systems at various levels, in classical, semi-classical, quantum and relativistic quantum field theory. We shall state some results on exactly soluble models, and discuss some of the new open problems. It seems that magnetism rather than being a small perturbation to electricity may play an equally fundamental role in understanding the structure of matter. It is therefore worthwhile to extend all the mathematical results on the dynamical systems with Coulomb forces to systems with magnetic forces including exchange effects.

A. O. Barut (ed.), Quantum Theory, Groups, Fields and Particles, 325–332.

II. SOME SIMPLE MODELS

The simplest classical model is the motion of a charged particle in the field of a point magnetic dipole, the so-called Störmer problem[2] (model of cosmic ray particles in the earth's magnetic field). This is a mathematically interesting dynamical system in its own right showing non-integrable, "chaotic" behavior[3]. For us one of the most important features of this model is the existence of resonances, i.e., trapped particles. The classical relativistic charge-dipole Hamiltonian shows that the resonances occur at distances of the order of $r \sim \frac{g_E}{\ell_z} \frac{\alpha}{m}$ with a total positive resonance energy $E \sim \frac{\ell_z}{g_E} \frac{m}{\alpha}$, where m is the mass of the cosmic ray particle, α the fine structure constant, ℓ_z the z-component (in the direction of the dipole) of the angular momentum of the particle, and g_E the magnetic g-factor of the earth, both in units of $\hbar$.

The quantum-mechanical version of this model also shows resonances. For an elementary particle dipole, now, g_E is of the order 2, ℓ_z of the order of 1 or 2. Hence for an electron bound to a fixed point-dipole, $r \sim \frac{\alpha}{m} = 2.8$ fermi, $E \sim \frac{m}{\alpha} \sim 70$ MeV. These give us the order of magnitudes of the magnetic phenomena, namely precisely in the nuclear size and in the nuclear or hadronic mass range.

Many other non-relativistic and relativistic models have been studied along these lines[4].

In the region where magnetic effects are large and dominant, Coulomb effects are relatively small and vice versa, so that we obtain two seemingly distinct, separated and qualitatively different phenomena, a long-range electric region and a short-ranged magnetic region.

III. MAGNETIC INTERACTIONS IN QUANTUM ELECTRODYNAMICS

We distinguish between two types of magnetic forces; those coming from the minimal coupling and those coming from the anomalous Pauli-Oppenheimer coupling. The Lagrangian density for a number of basic fermion fields Ψ_j, j=1,2,..., with charges e_i, interacting with the electromagnetic field A_μ is

$$L = \sum_j \{-\bar{\Psi}_j(\gamma^\mu \partial_\mu + m_j)\Psi_j + i\, e_j \bar{\Psi}_j \gamma^\mu \Psi_j A_\mu + a_j \bar{\Psi}_j \sigma^{\mu\nu} \Psi_j F_{\mu\nu}\} - \tfrac{1}{4} F_{\mu\nu} F^{\mu\nu} \quad (1)$$

where a_j are the anomalous magnetic moments of the fermions.

If we express A_μ, using field equations, in terms of the electromagnetic currents, the total interaction action can be written as

$$\int L_{int} d^4x = \sum_{j,k} \int d^4x\, d^4y \Big\{ -i\, e_j e_k \bar{\Psi}_j(x)\gamma^\mu \Psi_j(x) D(x-y) \bar{\Psi}_k(y)\gamma_\mu \Psi_k(y)$$

$$+ 4e_j a_k \bar{\Psi}_j(x)\gamma^\mu \Psi_j(x)\partial^\lambda D(x-y)\bar{\Psi}_k(y)\sigma_{\lambda\mu}\Psi_k(y)$$

$$- 4i\, a_j a_k \bar{\Psi}_j(x)\sigma^{\mu\nu}\Psi_j(x)\partial_\mu \partial^\lambda D(x-y)\bar{\Psi}_k \sigma_{\lambda\nu}\Psi_k(y), \qquad (2)$$

where D(x-y) is the Green's function.

For stationary problems we assume that the current distributions will be time-independent. It is then possible to pass from field theory to a many-body Hamiltonian theory. The Hamiltonian density is

$$H = \sum_j \Psi_j^+ (-i\vec{\alpha}\cdot\nabla + \beta m_j)\Psi_j - L_{int} \qquad (3)$$

and a total many-body energy operator $\hat{H}$ for interparticle interactions can be defined by the relation

$$\int H d^3x = \int d\vec{x}\, d\vec{y}\, \Psi_j^+(x)\Psi_k^+(y)\hat{H}\Psi_j(x)\Psi_k(y) \qquad (4)$$

with the normalization condition

$$\int d^3x \Psi_j^+(x)\Psi_j(x) = 1. \qquad (5)$$

As a result of this procedure we obtain the following Hamiltonian valid for stationary states[5]:

$$\hat{H} = \sum_j (i\vec{\alpha}_j\cdot\nabla + \beta_j m_j) + \sum_{j\neq k} \frac{e_j e_k}{4\pi} \frac{1-\vec{\alpha}_j\cdot\vec{\alpha}_k}{|\vec{x}-\vec{y}|}$$

$$- \sum_{j\neq k} \frac{e_j e_k}{\pi} \vec{\alpha}_j \cdot (\beta\vec{\sigma})_k \wedge \frac{(\vec{x}-\vec{y})}{|\vec{x}-\vec{y}|^3}$$

$$- i \sum_{j\neq k} \frac{e_j e_k}{\pi} \mathbb{1}_j (\beta\vec{\sigma})_k \cdot \frac{(\vec{x}-\vec{y})}{|\vec{x}-\vec{y}|^3}$$

$$- \sum_{j\neq k} \frac{a_j a_k}{\pi} \left[\frac{3(\beta\vec{\sigma})_j\cdot(\vec{x}-\vec{y})(\beta\vec{\sigma})_k\cdot(\vec{x}-\vec{y})}{|\vec{x}-\vec{y}|^5} - \frac{(\beta\vec{\sigma})_j\cdot(\beta\vec{\sigma})_k}{|\vec{x}-\vec{y}|^3} \right.$$

$$+\frac{8\pi}{3}(\beta\vec{\sigma})_j\cdot(\beta\vec{\sigma})_k\delta(\vec{x}-\vec{y})\Bigg]$$

$$+\sum_{j\neq k}\frac{a_j a_k}{\pi}\left[\frac{3(\beta\vec{\alpha})_j\cdot(\vec{x}-\vec{y})(\beta\vec{\alpha})_k\cdot(\vec{x}-\vec{y})}{|\vec{x}-\vec{y}|^5}-\frac{(\beta\vec{\alpha})_j\cdot(\beta\vec{\alpha})_k}{|\vec{x}-\vec{y}|^3}\right.$$

$$\left.-\frac{4\pi}{3}(\beta\vec{\alpha})_j\cdot(\beta\vec{\alpha})_k\delta(\vec{x}-\vec{y})\right]. \qquad (6)$$

We recognize in this Hamiltonian the sum of free-Dirac Hamiltonians, the Breit interaction (charge-charge potential), and the new charge-dipole ($e_j a_k$-terms) and dipole-dipole (the $a_j a_k$-terms) potentials due to the anomalous magnetic moments. In the third term we see the vector potential of the magnetic dipole k, namely

$$\vec{A}_k = a_k \frac{\vec{\sigma}_k \wedge \vec{r}}{r^3} \qquad (7)$$

which has been used in models mentioned in Section II, and finally in the 5th-term the famous tensor potential (spin-spin interaction). But we also notice that there are also potentials terms with spin matrices $\vec{\sigma}_k$ replaced by the Dirac matrices $\sigma_k^{\rightarrow}$. These latter terms vanish in the non-relativistic limit, that is why they had not made their appearance in physics so far. However, they could be very important in relativistic problems.

The "relativistic potential" in (6) is so defined that the scattering amplitude calculated from it in the Born approximation coincides with the Born approximation of the field theory. This can be seen from Eq. (2), when the fields Ψ_j are approximated by plane-wave solutions. The solution of the field equations can be written formally by iteration starting from a plane-wave solution and this expansion corresponds to sum of Feynman diagrams. By reducing the problem to a many-body Hamiltonian dynamics and finding localized nonperturbative solutions of this Hamiltonian we are in fact avoiding this perturbation series and the infinities associated with it. A solution of the Hamiltonian also provides a solution to the coupled field equations. Of course what is not included in this approach so far is the self-energy terms which after renormalization will give rise to Lamb shift type mass splitting.

Another effect of the self-energy is to give to the electron an anomalous magnetic moment which is calculated in quantum electrodynamics. This anomalous magnetic moment in turn interacts with the other charged particles. The resultant coupling is of the Pauli-type, namely the third term in our basic Langrangian (1). Thus the coefficients a_j of this term can be interpreted as coming partly from anomalous magnetic moment, and partly from a possible intrinsic anomalous magnetic moment. In the former case, these coefficients may be even functions of space-time. A complete answer can be obtained if the renormalization of the self-energy

$$\int d^4x d^4y \quad \bar{\Psi}(x)\gamma^\mu \Psi(x) \; D(xy)\Psi(y)\gamma_\mu \Psi(y)$$

can be carried out nonperturbatively.

In the Hamiltonian (6) the Breit potential, $e_j e_k \frac{1-\vec{\alpha}_j \cdot \vec{\alpha}_k}{r}$, by itself, when reduced to the non-relativistic form, also contains <u>induced</u> spin-orbit and spin-spin terms, but with a coefficient $(e_j e_k)$ and not $(e_j a_k)$ or $(a_j a_k)$. These are the terms used in the fine or hypofine structure of positronium or atoms.

A crucial question is therefore the size of the Pauli-coupling term for the stable particles, the electron, the neutrino and the proton. For the neutrino we shall assume a small anomalous magnetic moment ($a \sim 10^{-10}$ Bohr magnetons). For the electron in addition to the radiative anomalous magnetic magnetic a small intrinsic term is within experimental limits. The proton has of course of large anomalous magnetic moment for which however we do not have yet a good theory.

IV. ON THE SOLUTIONS OF DYNAMICAL EQUATIONS

The next problem is to solve the dynamical equations with the Hamiltonian (6). Even for the 2-body problems, this is a set of 16 x 16 first order coupled differential equations. It is remarkable that even the Breit interaction has not been treated exactly. In fact, the reason the existence of magnetic resonances has not been recognized, say, 50 years ago, is that the spin-orbit and spin-spin interactions have been, up to the present time, treated as a perturbation to a solvable spin-independent dynamics, for example, the Coulomb potential.

It is possible to separate completely the radial and angular parts of the Hamiltonian (6) for a 2-body problem. The radial problem then reduces to the matrix dynamical system

$$\frac{du}{dr} = M(r)u, \tag{8}$$

where u is an N-component radial wave function, and M(r) an

N x N - matrix. This system is now being studied for the general magnetic interactions. Here we shall indicate some simple exactly soluble special cases of (8). These include a particle in the field of a fixed dipole and a relativistic Dirac particle with anomalous magnetic moment in the field of a fixed point charge. These and similar models finally reduce to a radial wave equation.

$$\left[\frac{d^2}{dr} + \lambda^2 - V(\lambda,\ell,j;r)\right] u(r)=0, \quad r \geq 0$$

$$u(0)=0. \qquad \lambda^2=E^2-m^2, \tag{9}$$

where the radial potential is in general energy dependent and of the form

$$V(r) = \frac{\nu_1}{r} + \frac{\nu_2}{r^2} + \frac{\nu_3}{r^3} + \frac{\nu_4}{r^4} \quad . \tag{10}$$

The term $\frac{\nu_4}{r^4}$ is always positive and corresponds to a repulsive core at very short distances. The interesting case is when the quantum numbers are such that the $\frac{\nu_3}{r^3}$ -term is negative (attractive). This leads to a deep potential well followed by the repulsive centrifugal term $\frac{\nu_2}{r^2}$. In this case narrow resonances occur, i.e. the particles are captured in the deep potential well, and eventually decay via barrier penetration.

The eigenvalue problem (9) can be solved via recursion relations. If one separates the behavior of the wave function at the singular points r=0 and r = ∞ by setting

$$u(r) = F(r)e^{-1/r}\, r^{\nu}\, e^{i\lambda r}$$

then one can make a power series expansion in the form

$$F(r) = \sum_{n=0}^{\infty} a_n r^n$$

and obtain recursion relations for the coefficients a_n. For special values of the energy values λ, this power series has finitely many terms, corresponding to resonances.

For under some conditions on the coefficients ν_3 and ν_4 the recursion formulas become particularly simple, and we have the following

Theorem[6]: the Sturm-Liouville equation (9) for potentials given in Eq. (10), with ν_4 normalized to 1 by scaling, such that

$$\nu_3 = -2(M+1), \quad M=0,1,2,..., \tag{11}$$

has (M+1) discrete, in general complex, eigenvalues λ given by the vanishing of the following determinant of order (M+1):

$$\begin{vmatrix} D & 2 & 0 & 0 & 0 & 0 & \cdot \\ -2i\lambda M & D-2M & 4 & 0 & 0 & 0 & \cdot \\ 0 & -2i\lambda(M-1) & D+2(1-2M) & 6 & 0 & 0 & \cdot \\ 0 & 0 & -2i\lambda(M-2) & D+3(2-2M) & 8 & 0 & \cdot \\ \cdot & \cdot & \cdot & \cdot & \cdot & \cdot & \cdot \\ \cdot & \cdot & \cdot & \cdot & \cdot & \cdot & \cdot \\ \cdot & \cdot & \cdot & \cdot & -2i\lambda & D+M(-M-1) & \cdot \end{vmatrix} = 0, \tag{12}$$

where

$$D = M^2 + M + 2i\lambda - \nu_2 .$$

Example: $M = 1$. Then $D = 2+2i\lambda-\nu_2$ and eq. (12) gives a pair of complex-conjugate poles at

$$\lambda = \tfrac{1}{2}\left[-i(\nu_2-2) \pm \sqrt{2(\nu_2-2)}\right]$$

which coincide in the λ-plane for $\nu_2=0$ at $\lambda=0$. For $\nu_2<2$, we have bound states, for $\nu_2<2$ resonances. Physically ν_2-controlls the height of the potential barrier.

CONCLUSIONS

We have briefly indicated the physical background and mathematical problems arising in the study of dynamical systems under magnetic interactions.

The unexpected new structures formed from charged particles with magnetic moments were discovered when the short ranged magnetic forces have been treated nonperturbatively. They reveal extremely rich physical phenomena at short distances (or high energies) which depend in an essential way on the spin quantum numbers of the particles and which have many of the characteristics of nuclear and hadronic states and processes. In this short-distance region, magnetic forces are no longer small perturbations. There is also a rich set of new mathematical problems. Yet the basic starting point is very conservative and simple, namely electrodynamics with magnetic moment interactions. No new hypothetical particles or interactions are introduced. The group-theoretical problems in building up the hadronic and leptonic multiplets have been discussed elsewhere[1,7].

REFERENCES

1. For recent reviews see A.O. Barut, in "Group Theory--1980," AIP Conference Proceedings, Vol. 71, pp. 73-108 (1981), ed. T. Seligman; in "Quantum electrodynamics in Strong Fields" (Plenum Press 1982), ed. W. Greiner; and Surveys in High Energy Physics, Vol. 1, pp. 113-130 (1980).

2. C. Störmer, Z.f. Astrophys. 1, p. 237 (1930). J. McConnell and E. Schrödinger, Proc. Roy. Irish Acad. 49A, p. 259 (1944).

3. A.J. Dragt and J.M. Finn, J. Geophys. Res. 81, p. 2327 (1976). M. Brown, SIAM Review, 23, p. 61 (1971).

4. A.O. Barut and J. Kraus, J. Math. Phys. 17, p. 506 (1976). A.O. Barut, J. Math. Phys. 21, p. 568 (1980). A.O. Barut and G. Strobel, KINAM, Revista de Physica 4, 151 (1982).

5. A.O. Barut and B-W. Xu, Physica Scripta 26, 126 (1982).

6. A.O. Barut, M. Berrondo and G. Garcia-Calderoñ, J. Math. Phys. 21, p. 1851 (1980).

7. S.A. Basri and A.O. Barut, Intern. J. Theor. Phys. (in press).

Index

MATHEMATICAL PHYSICS STUDIES

Volume 1

Local Jet Bundle Formulation of Bäcklund Transformations

by F. A. E. Pirani, D. C. Robinson, and W. F. Shadwick

1979, viii + 132 pp. ISBN 90-277-1036-8

Volume 2

Non-Relativistic Quantum Dynamics

by W. O. Amrein

1981, viii + 237 pp. ISBN 90-277-1324-3

Volume 3

Differential Geometry and Mathematical Physics

edited by M. Cahen, M. De Wilde, L. Lemaire, and L. Vanhecke

1983, vii + 188 pp. ISBN 90-277-1508-4